Lecture Notes in Computer Science   10332

Commenced Publication in 1973
Founding and Former Series Editors:
Gerhard Goos, Juris Hartmanis, and Jan van Leeuwen

More information about this series at http://www.springer.com/series/7410

Shlomi Dolev · Sachin Lodha (Eds.)

Cyber Security Cryptography and Machine Learning

First International Conference, CSCML 2017
Beer-Sheva, Israel, June 29–30, 2017
Proceedings

 Springer

Editors
Shlomi Dolev
Ben-Gurion University of the Negev
Beer-Sheva
Israel

Sachin Lodha
Tata Consultancy Services (India)
Pune, Maharashtra
India

ISSN 0302-9743 ISSN 1611-3349 (electronic)
Lecture Notes in Computer Science
ISBN 978-3-319-60079-6 ISBN 978-3-319-60080-2 (eBook)
DOI 10.1007/978-3-319-60080-2

Library of Congress Control Number: 2017943048

LNCS Sublibrary: SL4 – Security and Cryptology

Printed on acid-free paper

This Springer imprint is published by Springer Nature
The registered company is Springer International Publishing AG
The registered company address is: Gewerbestrasse 11, 6330 Cham, Switzerland

Preface

CSCML, the International Symposium on Cyber Security Cryptography and Machine Learning, is an international forum for researchers, entrepreneurs, and practitioners in the theory, design, analysis, implementation, or application of cyber security, cryptography, and machine-learning systems and networks, and, in particular, of conceptually innovative topics in this area. Information technology has become crucial to our everyday life in indispensable infrastructures of our society and therefore is also a target of attacks by malicious parties. Cyber security is one of the most important fields of research today because of these phenomena. The two, sometimes competing, fields of research, cryptography and machine learning, are the most important building blocks of cyber security, as cryptography hides information by avoiding the possibility to extract any useful information pattern while machine learning searches for meaningful information patterns. The subjects covered by the symposium include cyber security design; secure software development methodologies; formal methods, semantics, and verification of secure systems; fault tolerance, reliability, availability of distributed secure systems; game-theoretic approaches to secure computing; automatic recovery self-stabilizing, and self-organizing systems; communication, authentication, and identification security; cyber security for mobile and Internet of Things; cyber security of corporations; security and privacy for cloud, edge, and fog computing; cryptography; cryptographic implementation analysis and construction; secure multi-party computation; privacy-enhancing technologies and anonymity; post-quantum cryptography and security; machine learning and big data; anomaly detection and malware identification; business intelligence and security; digital forensics, digital rights management; trust management and reputation systems; and information retrieval, risk analysis, DoS.

The first edition of CSCML took place during June 29–30, 2017, in Beer-Sheva, Israel.

This volume contains 17 contributions selected by the Program Committee and four brief announcements. All submitted papers were read and evaluated by Program Committee members, assisted by external reviewers. We are grateful for the EasyChair system in assisting the reviewing process.

The support of Ben-Gurion University of the Negev (BGU), in particular the BGU Lynne and William Frankel Center for Computer Science, the BGU Cyber Security Research Center, and BGN, also the support of IBM, DELLEMC, JVP, Deutsche Telekom Capital Partners, Glilot, Magma, Pitango, and BaseCamp, is also gratefully acknowledged.

April 2017

Shlomi Dolev
Sachin Lodha

Organization

CSCML, the International Symposium on Cyber Security Cryptography and Machine Learning, is an international forum for researchers, entrepreneurs, and practitioners in the theory, design, analysis, implementation, or application of cyber security, cryptography, and machine-learning systems and networks, and, in particular, of conceptually innovative topics in this field.

Founding Steering Committee

Orna Berry	DELLEMC, Israel
Shlomi Dolev (Chair)	Ben-Gurion University, Israel
Yuval Elovici	Ben-Gurion University, Israel
Ehud Gudes	Ben-Gurion University, Israel
Jonathan Katz	University of Maryland, USA
Rafail Ostrovsky	UCLA, USA
Jeffrey D. Ullman	Stanford University, USA
Kalyan Veeramachaneni	MIT, USA
Yaron Wolfsthal	IBM, Israel
Moti Yung	Columbia University and Snapchat, USA

Organizing Committee

Program Chairs

Shlomi Dolev	Ben-Gurion University of the Negev, Israel
Sachin Lodha	Tata Consultancy Services, India

Organizing Chair

Timi Budai	Ben-Gurion University of the Negev, Israel

Program Committee

Ran Achituv	Magma Ventures, Israel
Yehuda Afek	Tel-Aviv University, Israel
Adi Akavia	Tel-Aviv Yaffo Academic College, Israel
Amir Averbuch	Tel-Aviv University, Israel
Roberto Baldoni	Università di Roma "La Sapienza", Italy
Michael Ben-Or	Hebrew University, Israel
Anat Bremler-Barr	IDC Herzliya, Israel
Yves-Alexandre de Montjoye	Imperial College London, UK
Itai Dinur	Ben-Gurion University, Israel
Shlomi Dolev (Co-chair)	Ben-Gurion University, Israel

Additional Reviewers

Vijayanand Banahatti	Tata Consultancy Services, India
Silvia Bonomi	Università di Roma "La Sapienza", Italy
Antonella Del Pozzo	Università di Roma "La Sapienza", Italy
Manish Shukla	Tata Consultancy Services, India
Ajeet Kumar Singh	Tata Consultancy Services, India

Sponsors

Ben-Gurion University of the Negev

CBG
Cyber@Ben-Gurion
University of the Negev

BGN Ltd

DELL EMC

BaseCamp
Innovation Center

IBM

JVP

Contents

Efficient, Reusable Fuzzy Extractors from LWE

Daniel Apon[2](✉), Chongwon Cho[1], Karim Eldefrawy[1], and Jonathan Katz[2]

[1] Information and Systems Science Laboratory, HRL Laboratories,
Los Angeles, USA
ccho@hrl.com
[2] University of Maryland, College Park, USA
{dapon,jkatz}@cs.umd.edu

Abstract. A fuzzy extractor (FE) enables reproducible generation of high-quality randomness from noisy inputs having sufficient min-entropy. FEs have been proposed for deriving cryptographic keys from biometric data. FEs rely in their operation on a public "helper string" that is guaranteed not to leak too much information about the original input. Unfortunately, this guarantee may not hold when *multiple* independent helper strings are generated from correlated inputs; *reusable* FEs are needed in that case. Although the notion of reusable FEs was introduced in 2004, it has received little attention since then.

In this paper, we first analyze an FE proposed by Fuller et al. (Asiacrypt 2013) based on the learning-with-errors (LWE) assumption, and show that it is *not* reusable. This is interesting as the first natural example of a non-reusable FE. We then show how to adapt their construction to obtain reusable FEs. Of independent interest, we show a generic technique for strengthening the notion of reusability achieved by an FE in the random-oracle model.

1 Introduction

Consider using biometric data as a source for generating cryptographic keys. For example, assume Alice wants to use her biometric data (e.g., fingerprint) w to generate a cryptographic key that she will then use to encrypt her data before storing it on a public server. In a naive approach, Alice could use w itself as the key to encrypt the data. There are two problems with this approach: first, when Alice re-scans her biometric data at a later point in time, it is likely she will recover a value w' that is close, but not equal, to the initial value w. Alice will be unable to recover her original data with such a noisy key if she uses a

This research is based upon work supported in part by the Office of the Director of National Intelligence (ODNI), Intelligence Advanced Research Projects Activity (IARPA). The views and conclusions contained herein are those of the authors and should not be interpreted as necessarily representing the official policies, either expressed or implied, of ODNI, IARPA, or the U.S. Government. The U.S. Government is authorized to reproduce and distribute reprints for governmental purposes notwithstanding any copyright annotation therein.

K. Eldefrawy—Currently at SRI International: karim.eldefrawy@sri.com.

© Springer International Publishing AG 2017
S. Dolev and S. Lodha (Eds.): CSCML 2017, LNCS 10332, pp. 1–18, 2017.
DOI: 10.1007/978-3-319-60080-2_1

standard encryption scheme. Second, w is not a uniform string, and thus it is unclear what security is obtained when using w as a key.

Fuzzy extractors. *Fuzzy Extractors (FEs)* provide a solution to the above challenges. An FE, first formally introduced by Dodis et al. [6], consists of a pair of algorithms (Gen, Rec) that work as follows: the generation algorithm Gen takes as input a value (e.g., biometric data) w, and outputs a pair of values (pub, r), where the first of these is called the "helper string." The recovery algorithm Rec takes as input pub along with a value w', and outputs r if w' is "sufficiently close" to the original value w. The security guarantee, roughly speaking, is that r is uniform—or at least computationally indistinguishable from uniform—for an adversary who is given pub, as long as the original input (i.e., w) comes from a distribution with sufficiently high min-entropy.

FEs can address the scenario described earlier. Alice can run Gen on the initial scan of her biometric data to compute (pub, r) ← Gen(w); she will use r to encrypt her data, and send the resulting ciphertext along with pub to the server. When she wishes to recover her data at some later point in time, she will obtain a fresh scan w' of her biometric data and the server will send to Alice the ciphertext and pub; Alice will compute $r = $ Rec(pub, w') and use r to decrypt the ciphertext. This ensures security, even from the server, since the key used for encryption (i.e., r) is uniform even conditioned on pub.

Reusability. One might hope that a FE would remain "secure" even if used on multiple, related input strings $w_1, \ldots$. Concretely, consider a setting in which a user relies on different scans $w_1, \ldots, w_\ell$ of their biometric data when interacting with ℓ servers such that each server (independently) computes (pub$_i$, r_i) ← Gen(w_i) as above. (We stress that even though the same underlying biometric feature is used every time, the $\{w_i\}$ represent independent scans of that feature; thus, the elements in $\{w_i\}$ will be close to each other but will not necessarily be identical.) Each of the public values pub$_i$ may become known to an adversary, and the original definition of FE does not provide any guarantees in this case. Boyen [4] was the first to highlight this issue, and he showed (contrived) constructions of FEs that are secure when used once, but that completely leak the underlying values $w_1, \ldots, w_\ell$ if used multiple times. On the positive side, Boyen defined a notion of reusability for FEs (called *outsider security*) and showed that the code-based construction of Dodis et al. [6] is reusable when a linear code is used. (We discuss definitions of reusability in further detail in Sect. 2.2.)

Somewhat surprisingly, there was no subsequent progress on reusable FEs until the recent work of Canetti et al. [5]. The primary advantage of their scheme is that it achieves reusability under very weak assumptions on the different scans $w_1, \ldots, w_\ell$. (In contrast, Boyen assumed that $w_i = w \oplus \delta_i$ for a small shift δ_i known to the adversary.) The scheme of Canetti et al. can also be used for sources of lower entropy rate than prior work, if the distribution of the $\{w_i\}$ satisfies a certain assumption. For completeness, however, we note that their scheme also has several disadvantages relative to the reusable scheme analyzed by Boyen: it tolerates a lower error rate, has computational—rather than information-theoretic—security, and relies on the random-oracle model.[1]

1.1 Our Contributions

In this work, we propose a new, indistinguishability-based definition for FEs, which can be viewed as adopting aspects of the definitions of reusability given by Boyen [4] and Canetti et al. [5]. Informally, our definition says that if $(\mathsf{pub}_i, r_i) \leftarrow \mathsf{Gen}(w_i)$ are computed as above, then an adversary cannot distinguish r_1 from a uniform string even given pub_1 and $\{(\mathsf{pub}_i, r_i)\}_{i>1}$.

We then show that the recent computational FE proposed by Fuller et al. [8] (the FMR-FE) is *not* reusable in a very strong sense: from the public information pub_1 and pub_2 of two instances of the scheme, an attacker can learn the original input data w_1 and w_2.[2] Fuller et al. do not claim reusability in their paper, but our result is nevertheless interesting as it gives the first *natural* example of an FE that is not reusable. On the other hand, we observe that their construction can achieve a weaker form of reusability if a common random string is available that can be used by the generation algorithm.

We then show several constructions of reusable FEs. First, we show a generic approach that can be used to convert any FE that achieves "weak" reusability to one that achieves our stronger notion of reusability, in the random-oracle model.[3] This approach can, in particular, be applied to the variant of the FMR-FE scheme described above (that assumes a common random string), which leads to an efficient construction achieving our strong notion of reusability based on the decisional *learning-with-errors* (LWE) assumption in the random-oracle model.

Finally, we show a construction of a (strongly) reusable FE that does not rely on the random-oracle model. It also relies on the LWE assumption, though with a super-polynomial modulus. Although we view this construction as being mainly of theoretical interest, we remark that it is more efficient than the scheme proposed by Canetti et al. [5], and achieves better parameters than the reusable scheme analyzed by Boyen [4].

1.2 Paper Organization

In Sect. 2 we review existing definitions of reusable FEs and introduce our new definition of reusability. We analyze the reusability of the FE proposed by Fuller et al. [8] in Sect. 3. We show how to modify their construction so that it achieves

---

[1] Technically, Canetti et al. rely on the assumption that "digital lockers" exist. All known constructions of digital lockers without random oracles require non-standard assumptions; in practice, digital lockers would most likely be instantiated with a hash function modeled as a random oracle.

[2] Huth et al. [10, Theorem 5] claim that the construction of Fuller et al. is reusable, but their proof is incorrect.

[3] Alamélou et al. [2] show a transformation with a similar goal, but it only applies to FEs for the set-difference metric on sets over exponential-size universes.

a weak notion of reusability, and then show a generic transformation that can, in particular, be applied to that result to obtain strong reusability. In Sect. 5, we present an LWE-based reusable FE (without random oracles). Finally, in Sect. 6 we provide a comparison of the estimated performance of our constructions compared to known reusable FEs.

2   Definitions

We let $H_\infty(\cdot)$ denote min-entropy, and let SD denote statistical distance.

2.1   Fuzzy Extractors

Let $\mathcal{M}$ be a metric space with distance metric d. We begin by reviewing the notion of fuzzy extractors (FEs).

Definition 1 (Fuzzy Extractor). *Let* $\Pi = (\mathsf{Gen}, \mathsf{Rec})$ *be such that* Gen *takes as input* $w \in \mathcal{M}$ *and outputs* (pub, r) *with* $r \in \{0,1\}^\ell$, *and where* Rec *takes as input* pub *and* $w' \in \mathcal{M}$ *and outputs a string* $r' \in \{0,1\}^\ell$ *or an error symbol* $\perp$. *We say that* Π *is an* $(\mathcal{M}, \ell, t, \epsilon)$-*fuzzy extractor for class of distributions* $\mathcal{W}$ *if:*

Correctness: *For any* $w, w' \in \mathcal{M}$ *with* $d(w, w') \leq t$, *if* $\mathsf{Gen}(w)$ *outputs* pub, r, *then* $\mathsf{Rec}(\mathsf{pub}, w') = r$.
Security: *For any adversary* $\mathcal{A}$ *and distribution* $W \in \mathcal{W}$, *the probability that* $\mathcal{A}$ *succeeds in the following experiment is at most* $1/2 + \epsilon$:
    1. w is sampled from W, and then $(\mathsf{pub}, r_0) \leftarrow \mathsf{Gen}(w)$ is computed.
    2. Uniform $r_1 \in \{0,1\}^\ell$ and $b \in \{0,1\}$ are chosen, and $\mathcal{A}$ is given (pub, r_b).
    3. $\mathcal{A}$ outputs b', and succeeds if $b = b'$.

The above definition is information-theoretic. For a *computational* $(\mathcal{M}, \ell, t, \epsilon)$-*fuzzy extractor* we require security to hold only for computationally bounded adversaries.

Other models. In this work we will also consider two other models in which FEs can be defined. First, in the *random-oracle model* we assume that Gen and Rec (as well as the adversary) have access to a uniform function H chosen at the outset of the experiment. In the *common random-string model* we assume that a uniform string is chosen by a trusted party, and made available to Gen and Rec (and the adversary). All our definitions can easily be adapted to either of these models.

2.2   Reusability of Fuzzy Extractors

Definition 1 provides a basic notion of security for FEs. As discussed in the Introduction, however, it does not ensure security if Gen is computed multiple times on the same (or related) inputs. Security in that setting is called *reusability*. Several definitions of reusability have been proposed in prior works [4,5]. We begin by reviewing prior definitions, and then suggest our own. In all cases, we describe an information-theoretic version of the definition, but a computational version can be obtained in the natural way.

Let $\Pi = (\mathsf{Gen}, \mathsf{Rec})$ be an $(\mathcal{M}, \ell, t, \epsilon)$-fuzzy extractor for class of distributions $\mathcal{W}$. The original definition suggested by Boyen [4, Definition 6] (adapted to the FE case rather than a fuzzy sketch[4]) considers a set Δ of permutations on $\mathcal{M}$, and requires that the success probability of any attacker in the following experiment should be small:

1. $\mathcal{A}$ specifies a distribution $W \in \mathcal{W}$.
2. w^* is sampled from W.
3. $\mathcal{A}$ may adaptively make queries of the following form:
   - $\mathcal{A}$ outputs a perturbation $\delta \in \Delta$.
   - In response, $\mathsf{Gen}(\delta(w))$ is run to obtain pub, r, and $\mathcal{A}$ is given pub.
4. $\mathcal{A}$ outputs w', and *succeeds* if $w' = w^*$.

Informally, then, the attacker is given the public output pub generated by several independent executions of Gen on a series of inputs related (in an adversarially chosen way) to an original value w^*; the definition then guarantees that the attacker cannot learn w^*.

Canetti et al. [5, Definition 2] consider a stronger definition, which requires that the success probability of any attacker should be close to $1/2$ in the following experiment:

1. $\mathcal{A}$ specifies a collection of (correlated) random variables $(W^*, W_2, \ldots, W_\rho)$ where each $W_i \in \mathcal{W}$.
2. Values $w^*, w_2, \ldots, w_\rho$ are sampled from $(W^*, W_2, \ldots, W_\rho)$.
3. Compute $(\mathsf{pub}^*, r^*) \leftarrow \mathsf{Gen}(w^*)$.
4. For each $2 \le i \le \rho$, compute $(\mathsf{pub}_i, r_i) \leftarrow \mathsf{Gen}(w_i)$. Give to $\mathcal{A}$ the values pub^* and $\{(\mathsf{pub}_i, r_i)\}_{i=2}^\rho$.
5. Choose $b \leftarrow \{0, 1\}$. If $b = 0$, give r^* to $\mathcal{A}$; otherwise, choose $u \leftarrow \{0, 1\}^\ell$ and give u to $\mathcal{A}$.
6. $\mathcal{A}$ outputs a bit b', and *succeeds* if $b' = b$.

This definition is stronger than Boyen's definition in several respects. First, it allows the attacker to request Gen to be run on a sequence of inputs that are correlated in an arbitrary way with the original value w^*; in fact, there is not even any requirement that w^* be "close" to w_i in any sense. Second, the definition gives the attacker the extracted strings $\{r_i\}$ and not just the public values $\{\mathsf{pub}_i\}$. Finally, it is required that the attacker cannot distinguish r^* from a uniform string, rather than requiring that the attacker cannot determine w^*.

New definitions. The benefit of the definition of Canetti et al. is that it refers to indistinguishability of the extracted string r^* from a uniform string (rather than inability of guessing w^* as in Boyen's definition). However, it seems too strong since it allows for arbitrarily correlated[5] random variables $W^*, W_2, \ldots, W_\rho$,

---

[4] A fuzzy sketch [6] is a precursor to a fuzzy extractor, but we do not rely on this notion directly in our work.

[5] Though whether this is realistic depends on whether errors in the biometric readings are dependent or independent of the underlying biometric.

rather than adopting a model of perturbations similar to the one considered by Boyen. We combine aspects of the previous definitions to obtain our definitions of *weak* and *strong* reusability, both of which focus on indistinguishability of the extracted string and which limit the possible perturbations being considered. The definition of weak reusability gives the adversary access to pub alone, whereas in strong reusability the adversary is also given the extracted string r.

For the next two definitions, we specialize to having $\mathcal{M}$ be $\{0,1\}^m$ under the Hamming distance metric; for $x \in \{0,1\}^m$ we let $d(x) = d(x,\mathbf{0})$ denote the Hamming weight of x.

Definition 2 (Weakly Reusable FE). *Let $\Pi = (\mathsf{Gen}, \mathsf{Rec})$ be an $(\mathcal{M}, \ell, t, \epsilon)$-fuzzy extractor for class of distributions $\mathcal{W}$. We say that Π is ϵ-weakly reusable if any adversary A succeeds with probability at most $1/2 + \epsilon$ in the following:*

1. *A specifies a distribution $W \in \mathcal{W}$.*
2. *A value w^* is sampled from W, and $\mathsf{Gen}(w^*)$ is run to obtain pub^*, r^*. The value pub^* is given to A.*
3. *A may adaptively make queries of the following form:*
   (a) A outputs a shift $\delta \in \mathcal{M}$ with $d(\delta) \leq t$.
   (b) $\mathsf{Gen}(w + \delta)$ is run to obtain pub and r, and A is given pub.
4. *Choose $b \leftarrow \{0,1\}$. If $b = 0$, give r^* to A; otherwise, choose $u \leftarrow \{0,1\}^\ell$ and give u to A.*
5. *A outputs a bit b', and succeeds if $b' = b$.*

Definition 3 (Reusable FE). *Let $\Pi = (\mathsf{Gen}, \mathsf{Rec})$ be an $(\mathcal{M}, \ell, t, \epsilon)$-fuzzy extractor for class of distributions $\mathcal{W}$. We say that Π is ϵ-reusable if any adversary A succeeds with probability at most $1/2 + \epsilon$ in the following:*

1. *A specifies a distribution $W \in \mathcal{W}$.*
2. *A value w^* is sampled from W, and $\mathsf{Gen}(w^*)$ is run to obtain pub^*, r^*. The value pub^* is given to A.*
3. *A may adaptively make queries of the following form:*
   (a) A outputs a shift $\delta \in \mathcal{M}$ with $d(\delta) \leq t$.
   (b) $\mathsf{Gen}(w + \delta)$ is run to obtain pub and r, which are then both given to A.
4. *Choose $b \leftarrow \{0,1\}$. If $b = 0$, give r^* to A; otherwise, choose $u \leftarrow \{0,1\}^\ell$ and give u to A.*
5. *A outputs a bit b', and succeeds if $b' = b$.*

2.3   The Learning-With-Errors Assumption

The learning-with-errors (LWE) assumption was introduced by Regev [13], who showed that solving it *on the average* is as hard as quantumly solving several standard lattice problems *in the worst case*. We rely on the decisional version of the assumption:

Definition 4 (Decisional LWE). *For an integer $q \geq 2$, and a distribution χ over $\mathbb{Z}_q$, the learning-with-errors problem $\mathsf{LWE}_{n,m,q,\chi}$ is to distinguish between the following distributions:*

$$\{\mathbf{A}, \boldsymbol{b} = \mathbf{A}\boldsymbol{s} + \boldsymbol{e}\} \text{ and } \{\mathbf{A}, \boldsymbol{u}\}$$

where $\mathbf{A} \xleftarrow{\$} \mathbb{Z}_q^{m \times n}$, $\boldsymbol{s} \xleftarrow{\$} \mathbb{Z}_q^n$, $\boldsymbol{u} \xleftarrow{\$} \mathbb{Z}_q^m$, and $\boldsymbol{e} \xleftarrow{\$} \chi^m$.

Typically, the error distribution χ under which LWE is considered is the *discrete Gaussian distribution* $\mathcal{D}_{\mathbb{Z},\alpha}$ (where α is the width of the samples), but this is not the only possibility. For example, we can restrict the modulus q to be a polynomially-small prime integer and use the reduction of Döttling and Müller-Quade [7, Corollary 1]. Alternately, we can simply forego any reduction to a "classical" lattice problem and simply take the above as an assumption based on currently known algorithms for solving the LWE problem.

Akavia et al. [1] showed that LWE has many simultaneously hardcore bits. More formally:

Lemma 1 ([1], Lemma 2). *Assume $\mathsf{LWE}_{n-k,m,q,\chi}$ is hard. Then the following pairs of distributions are hard to distinguish:*

$$\{\mathbf{A}, \boldsymbol{b} = \mathbf{A}\boldsymbol{s} + \boldsymbol{e}, \boldsymbol{s}_{1,\dots,k}\} \text{ and } \{\mathbf{A}, \boldsymbol{b} = \mathbf{A}\boldsymbol{s} + \boldsymbol{e}, \boldsymbol{u}_{1,\dots,k}\}$$

where $\mathbf{A} \xleftarrow{\$} \mathbb{Z}_q^{m \times n}$, $\boldsymbol{s}, \boldsymbol{u} \xleftarrow{\$} \mathbb{Z}_q^n$, and $\boldsymbol{e} \xleftarrow{\$} \chi^m$.

The above results primarily make use of LWE in a regime with a polynomial modulus. We will also (implicitly) make use of the following result of Goldwasser et al. on the "inherent robustness" of LWE with a *superpolynomially-large* modulus and Discrete Gaussian error distribution:

Lemma 2 ([9], Theorem 5). *Let $k \geq \log q$, and let $\mathcal{H}$ be the class of all (possibly randomized) functions $h : \{0,1\}^n \to \{0,1\}^*$ that are 2^{-k} hard to invert (for polynomial-time algorithms); i.e. given $h(\boldsymbol{s})$, no PPT algorithm can find $\boldsymbol{s}$ with probability better than 2^{-k}. Assume that $\mathsf{LWE}_{\ell,m,q,\mathcal{D}_{\mathbb{Z},\beta}}$, where $\ell = \frac{k-\omega(\log(n))}{\log(q)}$, is hard.*

Then for any superpolynomial $q = q(n)$, any $m = \mathsf{poly}(n)$, any $\alpha, \beta \in (0,q)$ such that $\beta/\alpha = \mathsf{negl}(n)$, the following pairs of distributions are hard to distinguish:

$$\{\mathbf{A}, \boldsymbol{b} = \mathbf{A}\boldsymbol{s} + \boldsymbol{e}, h(\boldsymbol{s})\} \text{ and } \{\mathbf{A}, \boldsymbol{u}, h(\boldsymbol{s})\}$$

where $\mathbf{A} \xleftarrow{\$} \mathbb{Z}_q^{m \times n}$, $\boldsymbol{s} \xleftarrow{\$} \mathbb{Z}_q^n$, $\boldsymbol{u} \xleftarrow{\$} \mathbb{Z}_q^m$, and $\boldsymbol{e} \xleftarrow{\$} \mathcal{D}_{\mathbb{Z},\alpha}^m$.

3 Reusability Analysis of Prior Work

In this section, we begin by reviewing the details of the FE due to Fuller et al. [8]; then we explore and discuss various security vulnerabilities that arise when considering reusability (i.e., multiple, simultaneous enrollments).

3.1   Review of the Construction

We first recall the computational FE construction proposed by Fuller et al. [8]; we refer to that construction as FMR-FE. The security of FMR-EF depends on the distribution W over the source. They proved the security relying on the LWE assumption when **(i)** W is the uniform distribution over $\mathbb{Z}_q^n$, and when **(ii)** W is a particular, structured, non-uniform distribution – namely: a *symbol-fixing* source as originally defined by [11]. Their construction follows the "code-offset" paradigm due to [6, Sect. 5], instantiated with a random linear code (i.e., as given by LWE).

In more detail, FMR-FE consists of two algorithms Gen and Rec. In turn, Rec makes calls to the helper algorithm Decode_t that decodes a random linear code with at most $t = O(\log(n))$ errors. Note that (n, m, q, χ) are chosen to ensure security from LWE and so that $m \geq 3n$.

The FMR-FE algorithm is as follows:

- $(p, r) \leftarrow \mathsf{Gen}(w)$:
  1. Sample $\mathbf{A} \in \mathbb{Z}_q^{m \times n}$ and $\boldsymbol{s} \in \mathbb{Z}_q^n$ uniformly.
  2. Let $p = (\mathbf{A}, \mathbf{A}\boldsymbol{s} + w)$.
  3. Let r be the first $n/2$ coordinates of the secret $\boldsymbol{s}$; that is, $r = \boldsymbol{s}_{1,\dots,n/2}$.
  4. Output (p, r).
- $r' \leftarrow \mathsf{Rec}(w', p)$:
  1. Parse p as $(\mathbf{A}, \boldsymbol{c})$; let $\boldsymbol{b} = \boldsymbol{c} - w'$.
  2. Compute $\boldsymbol{s}' = \mathsf{Decode}_t(\mathbf{A}, \boldsymbol{b})$.
  3. Output $r' = \boldsymbol{s}'_{1,\dots,n/2}$.
- $\boldsymbol{s}' \leftarrow \mathsf{Decode}_t(\mathbf{A}, \boldsymbol{b})$:
  1. Select $2n$ random rows without replacement $i_1, \dots, i_{2n} \leftarrow [1, m]$.
  2. Restrict $\mathbf{A}, \boldsymbol{b}$ to rows $i_1, \dots, i_{2n}$; denote these by $\mathbf{A}_{i_1,\dots,i_{2n}}, \boldsymbol{b}_{i_1,\dots,i_{2n}}$.
  3. Find n linearly independent rows of $\mathbf{A}_{i_1,\dots,i_{2n}}$. If no such rows exist, output $\perp$ and halt.
  4. Further restrict $\mathbf{A}_{i_1,\dots,i_{2n}}, \boldsymbol{b}_{i_1,\dots,i_{2n}}$ to these n rows; denote the result by $\mathbf{A}', \boldsymbol{b}'$.
  5. Compute $\boldsymbol{s}' = (\mathbf{A}')^{-1}\boldsymbol{b}'$.
  6. If $\boldsymbol{b} - \mathbf{A}\boldsymbol{s}'$ has more than t nonzero coordinates, restart at Step (1).
  7. Output $\boldsymbol{s}'$.

We remark that the correctness of FMR-FE depends primarily on the success of the Decode_t algorithm. [8] demonstrates that when w and w' differ by at most $t = O(\log(n))$ coordinates, the decoding succeeds. Intuitively, this is because $\boldsymbol{b}$ in Step (2) of Rec will have at most $O(\log(n))$ nonzero coordinates in its error vector, implying with good probability that Step (1) of Decode_t drops all the rows containing errors. When this occurs, the linear system becomes solvable with $(\mathbf{A}')^{-1}\boldsymbol{b}'$ being a solution to the linear system, and in particular, there is no difference in decoding output depending on whether w or w' was originally used. For the further details in the success probability and (expected) time complexity of Rec, see [8].

Standard FE security holds based on the LWE assumption, specifically by the hardness of LWE with uniform errors with appropriate parameters (e.g. a poly-size modulus) due to [7, Corollary 1] plus the hardcore bits lemma for LWE of Akavia et al. [1].

3.2 Vulnerabilities from Multiple Enrollments

Now we explore security issues that arise when the same input data (e.g., the same biometric) is enrolled (perhaps, with perturbations) at multiple servers. Intuitively, the running theme here will be that reusability can fail whenever a "second enrollment" (from running Gen twice) is as useful as a "second reading" (required to run Rec) for gaining access to the sensitive, extracted strings r.

Vulnerability when reusing the scheme "as-is." Consider a situation where the [8] scheme is used twice to enroll two (perhaps, noisy) versions of the same biometric w. That is, let $w_1 = w$ and $w_2 = w - \delta$, where perturbation δ has at most $t = O(\log(n))$ nonzero coordinates. Note that w_1 and w_2 are distance at most t from one another.

An adversary $\mathcal{A}$ who sees the public information obtains the public, helper strings $p_1 = (\mathbf{A}_1, \mathbf{A}_1 s_1 + w_1)$ and $p_2 = (\mathbf{A}_2, \mathbf{A}_2 s_2 + w_2)$. In order to violate security, $\mathcal{A}$ sets up the following system of linear equations:

$$p_1 - p_2 = \mathbf{A}_1 s_1 + w_1 - \mathbf{A}_2 s_2 - w_2$$

$$= (\mathbf{A}_1| - \mathbf{A}_2) \cdot \begin{bmatrix} s_1 \\ s_2 \end{bmatrix} + (w_1 - w_2)$$

$$= (\mathbf{A}_1| - \mathbf{A}_2) \cdot \begin{bmatrix} s_1 \\ s_2 \end{bmatrix} + \delta.$$

Next observe that $(\mathbf{A}_1| - \mathbf{A}_2) \cdot \begin{bmatrix} s_1 \\ s_2 \end{bmatrix}$ is a linear system with m rows and $2n$ columns and that δ has at most t nonzero coordinates. Let $N = 2n$. If $m \geq 3N$, then the verbatim decoding algorithm given by [8]—concretely, running $\mathsf{Decode}_t((\mathbf{A}_1| - \mathbf{A}_2), p_1 - p_2)$—recovers the secrets s_1, s_2 with good probability. Finally, the adversary $\mathcal{A}$ can recover the original biometric readings w_1 and w_2 by computing $w_1 = p_1 - (\mathbf{A}_1 s_1)$ and $w_2 = p_2 - (\mathbf{A}_2 s_2)$. This is a complete break.

The above attack on reusability of [8] succeeds as long as the algorithm $\mathsf{Decode}_t((\mathbf{A}_1| - \mathbf{A}_2), p_1 - p_2)$ is able to return a non-$\bot$ value. There are two technical issues to consider here: First, note that if $w_1 = w + \delta_1$ and $w_2 = w + \delta_2$, then w_1 and w_2 may be up to $2t$ distance from one another (rather than t). This is not a problem since $t = O(\log(n))$, so $2t = O(\log(n))$ as well. Second, note that we assumed $m \geq 3N = 6n$, whereas the [8] scheme only requires a setting of $m \geq 3n$. Is it possible that a setting of $3n \leq m \leq 6n$ avoids the attack? We argue informally that this is not the case. In particular, the above attacker $\mathcal{A}$ only needs to successfully guess *once* some set of $N = 2n$ rows that are linearly independent and are without errors. By a routine argument, over the independent and uniform choices of $\mathbf{A}_1$ and $\mathbf{A}_2$, such linearly-independent rows must exist with good probability. Since t is only $O(\log(n))$, the only difference will be the number of times the adversary must loop (and re-guess) until it succeeds, but this will still be poly-time.

3.3   A Partial Solution to Reusability (Weak Reusability)

In preparation for our upcoming reusable FE scheme in the next section, we introduce a slight change to the model, designed (to attempt) to thwart the previous attack on reusability. The resulting scheme will not yet be secure, but we then modify it in the sequel to achieve reusability.

Intuitively, the previous attack succeeded because the subtraction $p_1 - p_2$ of the public information formed a proper (noisy) linear system in dimension $2n$. In order to avoid such an attack, consider a case that all enrollment servers are given an access to a common global parameter. Specifically, assume that public matrix $\mathbf{A}$ is sampled uniformly at random from $\mathbb{Z}_q^{m \times n}$ once. Then, each invocation of Gen, Rec, and Decode$_t$ will use this common matrix $\mathbf{A}$.

Vulnerability when reusing the modified scheme (with public A). Consider a new situation where the [8] scheme is used twice to enroll two (perhaps, noisy) versions of the same biometric w – but where some uniform $\mathbf{A}$ is shared among all invocations of the fuzzy extractor algorithms.

An adversary $\mathcal{A}$ who sees the public information obtains the public matrix $\mathbf{A}$, plus the helper strings $p_1 = \mathbf{A}s_1 + w_1$ and $p_2 = \mathbf{A}s_2 + w_2$. Repeating the same format of attack as before, $\mathcal{A}$ sets up the following system of linear equations:

$$p_1 - p_2 = \mathbf{A}s_1 + w_1 - \mathbf{A}s_2 - w_2$$
$$= \mathbf{A}(s_1 - s_2) + w_1 - w_2$$
$$= \mathbf{A}(s_1 - s_2) + \delta.$$

Next observe that $\mathbf{A}(s_1 - s_2)$ is a linear system with m rows and n columns and that δ has at most t nonzero coordinates. Therefore, similar to before, $\mathcal{A}$ can compute Decode$_t(\mathbf{A}, p_1 - p_2)$, which outputs $s^* = s_1 - s_2$ with good probability.

On its own, s^* does not lead to an attack on security of the FE. Indeed, it is not hard to show that the above partial solution with public $\mathbf{A}$ is a weakly reusable FE as defined in Definition 2. For the completeness, we state the following theorem with its proof omitted.

Theorem 1. *Assume that a common parameter $\mathbf{A}$ is available such that $\mathbf{A}$ is sampled uniformly at random from $\mathbb{Z}_q^{m \times n}$. The FMR-FE is a computationally ϵ-weakly reusable as defined in Definition 2 where ϵ is a negligible in n.*

However, suppose that one of the two extracted strings, say r_1, is leaked to the adversary by some other means. Then, it becomes possible for $\mathcal{A}$ to additionally recover the second extracted string r_2 from s^*. To do so, $\mathcal{A}$ restricts s^* to its first $n/2$ coordinates to obtain $s^*_{1,\ldots,n/2}$. Recall that r_1 is set to be the restriction of s_1; that is, $r_1 = (s_1)_{1,\ldots,n/2}$. Further, observe that $s^*_{1,\ldots,n/2} = (s_1)_{1,\ldots,n/2} - (s_2)_{1,\ldots,n/2}$. Therefore, computing $r_1 - s^*_{1,\ldots,n/2}$ yields r_2, which is a break of reusability security for this scheme (conditioned on r_1 leaking to $\mathcal{A}$ directly, but not r_2). This attack indicates that the FMR-FE construction even with using the identical public random $\mathbf{A}$ is not able to satisfy the definition of reusable FE as in the Definition 3.

4 Upgrading Reusability

In this section, we show how to obtain a computational fully reusable FE when a common parameter is available in the random oracle model. As described in the previous section (Sect. 3.3), the main problem of partial solution is that the information leakage given by multiple public strings reveals linear relationships existing among the extracted random strings. This leads to a security vulnerability, preventing FMR-FE with common randomness from satisfying the full reusability with common randomness. A simple solution is to use a random oracle to break existing correlations (e.g., linear relation and any others) between extracted random strings, which we indeed show to be suffice to achieve the full reusability with common randomness.

4.1 Our Construction in the Random Oracle Model

Let $H : \mathbb{Z}_q^n \rightarrow \{0,1\}^\ell$ be a hash function modeled as a random oracle. We use notation G^H to denote algorithm G with oracle access to hash function H in the following. We present our random-oracle-based computational reusable FE $\Pi = (\mathsf{Gen}, \mathsf{Rec})$ as follows. Looking ahead, we assume that a public parameter is available in the form of a matrix $\mathbf{A}$ sampled uniformly at random from $\mathbb{Z}_q^{m \times n}$.

- $(p, r) \leftarrow \mathsf{Gen}^H(pp, w)$:
  1. Sample $s \in \mathbb{Z}_q^n$ uniformly.
  2. Parse pp as $\mathbf{A}$; let $p = \mathbf{A}s + w$.
  3. Let $r = H(s)$.
  4. Output (p, r).
- $r' \leftarrow \mathsf{Rec}^H(pp, w', p)$:
  1. Parse p as c; let $b = c - w'$.
  2. Parse pp as $\mathbf{A}$; compute $s' = \mathsf{Decode}_t(\mathbf{A}, b)$.
  3. Output $r' = H(s')$.

Remark on Decoding. Our new reusable FEs use [8]'s Decode_t verbatim and see Sect. 3.1 for the details.

The idea behind of upgrading the weak reusability to the full reusability is that we use the random oracle to extract the adaptive hardcore bits from the underlying secret. This essentially neutralizes the usefulness of any partial information revealed by multiple public helper strings.

Theorem 2. *Assume that a public common parameter $\mathbf{A}$ is available such that $\mathbf{A}$ is sampled uniformly at random from $\mathbb{Z}_q^{m \times n}$. The $\Pi = (\mathsf{Gen}^H, \mathsf{Rec}^H)$ is a computational reusable fuzzy extractor in the random-oracle model as defined in Definition 3.*

The approach of using the random-oracle to upgrade the weak reusability to the full reusability can be extended to a more generic case. Specifically, suppose that we have a weakly reusable fuzzy extractor $(\mathsf{Gen}, \mathsf{Rec})$ such that $(\mathsf{pub}, r) \leftarrow \mathsf{Gen}(w)$ and $r \leftarrow \mathsf{Rec}(\mathsf{pub}, w')$ satisfying Definition 2. Then, using random-oracle H, we construct a fully reusable extractor $(\mathsf{Gen}^*, \mathsf{Rec}^*)$ satisfying Definition 3 as follows: $(\mathsf{pub}^*, r^*) \leftarrow \mathsf{Gen}^*(w)$ such that $\mathsf{pub}^* = \mathsf{pub}$ and $r^* = H(r)$ where $(\mathsf{pub}, r) \leftarrow \mathsf{Gen}(w)$ and $r^* \leftarrow \mathsf{Rec}(\mathsf{pub}, w')$ such that $r^* = H(r')$ where $r' \leftarrow \mathsf{Rec}(\mathsf{pub}, w')$. In fact, the above approach is stronger as it transforms a fuzzy extractor satisfying the weak version of definition by Boyen [4, Definition 6] (see Sect. 2.2) where no extracted random string r is given to the adversary in its security game to a fully reusable extractor satisfying Definition 3. For completeness, we state our generic approach in the random oracle model below.

Theorem 3. *Suppose that $\Pi = (\mathsf{Gen}, \mathsf{Rec})$ is a weakly reusable fuzzy extractor satisfying the weaker version of definition by Boyen [4, Definition 6] such that no extracted random string r is given to an adversary in the security game. Then, there exists a fully reusable fuzzy extractor $\Pi^* = (\mathsf{Gen}^*, \mathsf{Rec}^*)$ satisfying Definition 3 in the random oracle model.*

The proof of Theorem 3 is a generalized version of the proof for Theorem 2 which is a special case. In this work, we rather omit the proof of Theorem 3 and provide the proof of Theorem 2 that can be easily extended to the general proof. We split the proof for Theorem 2 into two lemmas, Lemma 3 (correctness) and Lemma 4 (reusability).

Proof of Correctness. We observe that correctness of the above FE scheme follows from that of [8]:

Lemma 3. $\Pi = (\mathsf{Gen}^H, \mathsf{Rec}^H)$ *is correct.*

Proof. The only changes in terms of correctness for the above scheme, as compared to the [8] fuzzy extractor, is that we have a common parameter $\mathbf{A}$ sampled uniformly at random from $\mathbb{Z}_q^{m \times n}$ and we apply the random oracle H to s to obtain the extracted random string (instead of simply truncating s to its first $n/2$ coordinates). Therefore, our ROM scheme's correctness follows directly from [8].   □

4.2 Proof of Reusable Security in the ROM

Lemma 4. *Assume that $\mathsf{LWE}_{n,m,q,\mathcal{U}(-\beta,\beta)}$ for $\beta \in \mathsf{poly}(n)$ is hard and that H is a random oracle. Then, our fuzzy extractor scheme $\Pi = (\mathsf{Gen}^H, \mathsf{Rec}^H)$ is reusably-secure for uniform errors, as in Definition 3 in the ROM.*

Proof. To show reusable-security, we consider some PPT adversary $\mathcal{A}$ breaking the security of our fuzzy extractor scheme $(\mathsf{Gen}^H, \mathsf{Rec}^H)$ for the uniform error distribution $\mathcal{U}(-\beta, \beta)$ where $\beta \in \mathsf{poly}(n)$. Assuming that H is a random oracle, we will use $\mathcal{A}$ to construct a reduction $\mathcal{B}^{\mathcal{A}}$ that solves the search problem $\mathsf{LWE}_{n,m,q,\mathcal{U}(-\beta,\beta)}$.

The reduction $\mathcal{B}$ receives an $\mathsf{LWE}_{n,m,q,\mathcal{U}(-\beta,\beta)}$ search problem instance – either $\mathbf{A}, \boldsymbol{u}$ or $(\mathbf{A}, \boldsymbol{b} = \mathbf{A}\boldsymbol{s} + \boldsymbol{e})$ – where $\mathbf{A} \xleftarrow{\$} \mathbb{Z}_q^{m \times n}$, $\boldsymbol{u} \xleftarrow{\$} \mathbb{Z}_q^m$, $\boldsymbol{s} \xleftarrow{\$} \mathbb{Z}_q^n$, and $\boldsymbol{e} \xleftarrow{\$} \mathcal{U}(-\beta, \beta)$ for $\beta \in \mathsf{poly}(n)$.

The reduction $\mathcal{B}$ proceeds through the fuzzy extractor security experiment as follows. At the beginning, $\mathcal{B}$ chooses $\mathbf{A}$ from the LWE instance to be the public parameter pp. For the challenge (p^*, r^*), $\mathcal{B}$ chooses $p^* = \boldsymbol{b}$ and uniform r^*.

Next, $\mathcal{A}$ is invoked on $(\mathbf{A}, p^*)$ to (adaptively) produce a polynomially-long sequence of perturbations $\delta_1, ..., \delta_k$. After each time that $\mathcal{A}$ outputs a perturbation δ_i, $\mathcal{B}$ samples a *random vector-offset* $\Delta_i \in \mathbb{Z}_q^n$ and (using the additive key-homomorphism of LWE) sets $p_i = p^* + \mathbf{A}\Delta_i + \delta_i$; $\mathcal{B}$ chooses the corresponding extracted string r_i uniformly as well.

Observe that if the vector $\boldsymbol{b} = p^*$ is truly uniform then so are the p_i. On the other hand, if $\boldsymbol{b} = \mathbf{A}\boldsymbol{s} + \boldsymbol{e}$, then we have $p_i = \mathbf{A}\boldsymbol{s} + \boldsymbol{e} + \mathbf{A}\Delta_i = \mathbf{A}(\boldsymbol{s} + \Delta_i) + (\boldsymbol{e} + \delta_i)$.

Finally, in order for $\mathcal{A}$ to distinguish between this experiment at the original security experiment, $\mathcal{A}$ must query the random oracle H on either $\boldsymbol{s}$ or one of the $\boldsymbol{s} + \Delta_i$. Therefore, the reduction $\mathcal{B}$ can watch $\mathcal{A}$'s oracle queries, and attempt to use each to solve the LWE instance. If $\mathcal{A}$ distinguishes the experiments by such a query with probability ϵ, then the reduction $\mathcal{B}$ solves the challenge LWE instance with the same probability. This completes the proof. $\qquad\square$

5 A Reusable FE Without Random Oracles

In this section, we demonstrate a computational reusable FE based on the LWE assumption without relying on the random oracle under the assumption that a public parameter is available. More specifically, we demonstrate how to remove the random oracle H from the preceding scheme using a particular form of LWE-based symmetric encryption. Intuitively, by the results of [9] (cf. Lemma 2), symmetric-key LWE ciphertexts with appropriate parameterization may be viewed as *composable point functions*. This allows us to replace the previous scheme's use of H by a single lattice ciphertext.

We also note that it may be possible, in principle, to perform the upcoming proof strategy in a somewhat "black-box" manner using Lemma 2. However, in order to aid both the reader's understanding and future technical choices for implementations, we instead open up the proof structure as follows.

5.1 Construction of Our Reusable Fuzzy Extractor

Suppose that a public parameter pp is available such that $pp = \mathbf{A}$ is sampled uniformly at random from $\mathbb{Z}_q^{m \times n}$. We present our standard-model, reusable FE as follows:

- $(p, r) \leftarrow \mathsf{Gen}(pp, w)$:
  1. Sample $\boldsymbol{s} \in \mathbb{Z}_q^n$ uniformly.
  2. Parse pp as $\mathbf{A}$; let $\boldsymbol{c} = \mathbf{A}\boldsymbol{s} + w$.

3. Sample $r \in \{0,1\}^m, \mathbf{B} \in \mathbb{Z}_Q^{m \times n}$ uniformly, and sample $e \leftarrow \mathcal{D}_{\mathbb{Z},\alpha}^m$.
4. Let $\boldsymbol{h} = \mathbf{B}\boldsymbol{s} + \boldsymbol{e} + \frac{Q}{2}r$.
5. Let $p = (\boldsymbol{c}, \mathbf{B}, \boldsymbol{h})$.
6. Output (p, r).
- $r' \leftarrow \mathsf{Rec}(pp, w', p)$:
   1. Parse p as $(\boldsymbol{c}, \mathbf{B}, \boldsymbol{h})$; let $\boldsymbol{b} = \boldsymbol{c} - w'$.
   2. Parse pp as $\mathbf{A}$; compute $\boldsymbol{s}' = \mathsf{Decode}_t(\mathbf{A}, \boldsymbol{b})$.
   3. For each coordinate $i \in [m]$:
      (a) If the i-th coordinate of $\boldsymbol{h} - \mathbf{B}\boldsymbol{s} \in [\frac{3Q}{8}, \frac{5Q}{8}]$, then the i-th coordinate of r' is 1.
      (b) Else if the i-th coordinate of $\boldsymbol{h} - \mathbf{B}\boldsymbol{s}$ is less than $\frac{Q}{8}$ or greater than $\frac{7Q}{8}$, then the i-th coordinate of r' is 0.
      (c) Else, output $\perp$ and halt.
   4. Output the reconstructed string $r' \in \{0,1\}^m$.

Theorem 4. *Assume that a public common parameter $\mathbf{A}$ is available such that $\mathbf{A}$ is sampled uniformly at random from $\mathbb{Z}_q^{m \times n}$ where q is a super-polynomial in n. The $\Pi = (\mathsf{Gen}, \mathsf{Rec})$ is a computational reusable fuzzy extractor in the random-oracle model as defined in Definition 3.*

Similarly to the previous section, we prove the above theorem by proving two separate lemmas, Lemma 5 (correctness) and Lemma 6 (reusability).

Lemma 5. *$\Pi = (\mathsf{Gen}, \mathsf{Rec})$ is correct.*

Proof. The only change between this scheme and the prior scheme is that we have instantiated the oracle H using the LWE-based symmetric encryption scheme (respectively, multibit-output point obfuscator) of [9]. Therefore, correctness of $(\mathsf{Gen}, \mathsf{Rec})$ follows from that of $(\mathsf{Gen}^H, \mathsf{Rec}^H)$ and the correctness of [9]'s encryption scheme.   □

5.2   Proof of Security

Lemma 6. *Assume that $\mathsf{LWE}_{n,m,q,\mathcal{U}(-\beta,\beta)}$ and $\mathsf{LWE}_{\ell,m,Q,\mathcal{D}_{\mathbb{Z},\alpha}}$ for $n \geq 3m, \ell = \frac{k - \omega(\log(n))}{\log(q)}$, as well as appropriately parameterized $q, \beta \in \mathsf{poly}(n)$ and $Q, \alpha \in \mathsf{superpoly}(n)$, are hard. Then, our fuzzy extractor scheme $(\mathsf{Gen}, \mathsf{Rec})$ is reusably-secure, as in Definition 3.*

Proof. An adversary $\mathcal{A}$ that adaptively chooses perturbations $\{\delta_i\}_i$ should have a view of the form

$$\mathbf{A}, p^*, r^*, \{\delta_i, p_i, r_i\}_i$$

where each $p_i = (\boldsymbol{c}_i, \mathbf{B}_i, \boldsymbol{h}_i)$ (and $p^* = \boldsymbol{c}^*, \mathbf{B}^*, \boldsymbol{h}^*$) as in the original experiment.

We prove the lemma using a standard hybrid argument. Consider the following sequence of views:

View 1. The original security experiment, where in particular we have

$$c^* = \mathbf{A}s + w, \qquad c_i = \mathbf{A}s_i + w + \delta_i,$$

$$h^* = \mathbf{B}^*s + e + \frac{Q}{2}r^*, \qquad h_i = \mathbf{B}_is_i + e_i + \frac{Q}{2}r_i'.$$

View 2. Same as before, but we set $c_i = \mathbf{A}(s + \Delta_i) + w + \delta_i$, and also we set $h_i = \mathbf{B}_i(s + \Delta_i) + e_i + \frac{Q}{2}r_i'$.

View 3. Same as before, but we replace each $\mathbf{B}_i$ with $\mathbf{C}_i\mathbf{D}_i + \mathbf{E}_i$. That is, here we will have $h_i = (\mathbf{C}_i\mathbf{D}_i + \mathbf{E}_i)(s + \Delta_i) + e_i + \frac{Q}{2}r_i'$

View 4. Same as before, but we replace each $h_i = (\mathbf{C}_i\mathbf{D}_i)(s + \Delta_i) + \mathbf{E}_i(s + \Delta_i) + e_i + \frac{Q}{2}r_i'$ with instead $h_i = (\mathbf{C}_i\mathbf{D}_i)(s + \Delta_i) + e_i + \frac{Q}{2}r_i'$.

View 5. Same as before, but we (again) modify the h_i to $h_i = \mathbf{C}_iu_i + e_i + \frac{Q}{2}r_i'$, for i.i.d. $u_i \leftarrow \mathbb{Z}_q^n$.

View 6. Same as before, but – finally – we set the r_i' to be all-zeroes strings (rather than equal to the public r_i). In particular, the string h^* is now independent of the challenge extracted string r^*. (That is, the adversary has no advantage.)

We want to show that the distribution ensemble of View i is indistinguishable from the one of View $i + 1$ for all $i \in [5]$.

Claim. Views (1) and (2) are identically distributed.

Proof. This observation follows by the natural key-homomorphism of LWE samples. □

Claim. Views (2) and (3) are computationally indistinguishable under the hardness of decisional $\mathsf{LWE}_{n,m,q,\mathcal{U}(-\beta,\beta)}$.

Proof. This is a use of the LWE assumption "in reverse." That is, the random matrices $\mathbf{B}_i$ are replaced by pseudorandom matrices $\mathbf{C}_i\mathbf{D}_i + \mathbf{E}_i$ for uniform $\mathbf{C}_i \in \mathbb{Z}_Q^{m \times \ell}, \mathbf{D}_i \in \mathbb{Z}_Q^{\ell \times n}$ and m-by-n matrices $\mathbf{E}_i$ with entries from $\mathcal{U}(-\beta, \beta), \beta \in \mathsf{poly}(n)$. (We assume that the $\mathbf{C}_i, \mathbf{D}_i$ are known to the attacker.) It will be important in the sequel that the magnitude of the entries of $\mathbf{E}_i$ are "very short;" namely, that they are superpolynomially smaller than the magnitude of the e_i used to construct h_i. □

Claim. Views (3) and (4) are statistically indistinguishable.

Proof. This follows from the fact that the Discrete Gaussian samples e_i are superpolynomially larger than the norm of $\mathbf{E}_i(s + \Delta_i)$, since shifting a Gaussian by such a small value leads to a $\mathsf{negl}(n)$ statistical difference in the two distributions. □

Claim. Views (4) and (5) are computationally indistinguishable (assuming the randomized function $b^* = \mathbf{A}s + w \leftarrow f_{\mathbf{A}}(s)$ is 2^{-k}-hard to invert).

Proof. This uses a version of the Goldreich-Levin theorem for $\mathbb{Z}_Q$. Here, we are using a computationally leakage-resilient form of the leftover hash lemma to "re-randomize" the key s used to construct the vectors h^* and h_i to some independent and uniform u. In fact, this claim follows from Lemma 2 in the presence of the LWE instance $b^* = As + w$, assuming that it's 2^{-k} hard to invert. To be more concrete, since matrix multiplication (or for that matter, any universal hash function) is a strong randomness extractor, we have that

$$(\mathbf{D}_i, \mathbf{D}_i s, \mathbf{A}s + w) \stackrel{\text{stat}}{\approx} (\mathbf{D}_i, u_i, \mathbf{A}s + w),$$

which gives the lemma. □

Claim. Views (5) and (6) are computationally indistinguishable under the hardness of decisional $\mathsf{LWE}_{\ell,m,Q,\mathcal{D}_{\mathbb{Z},\alpha}}$.

Proof. Since the key vector $u_i = u + \Delta_i$ is now independent of the c^* component's secret vector s, we may apply the $\mathsf{LWE}_{\ell,m,Q,\mathcal{D}_{\mathbb{Z},\alpha}}$ assumption to switch h^* and all of the h_i to uniform, then back to encodings in the same form of $r_i' = \mathbf{0}$. □

Combining the above claims of indistinguishability between the views, we complete the proof of the Lemma 6. □

6   Practical Comparison of Reusable Fuzzy Extractors

In this section, we briefly overview the previous reusable FE of [5] and compare its concrete efficiency to our proposed FEs.

Reusable fuzzy extractor based on digital lockers. Canetti et al. [5] proposed a construction of reusable FE by relying on a strong cryptographic primitive, called digital locker. A digital locker is a symmetric-key cryptographic primitive of pair of algorithms denoted by (lock, unlock) such that it locks a secret message s under a secret password key k. That is, $\mathsf{lock}(k, s) = c$ and $\mathsf{unlock}(c, k) = s$. Its security guarantees that guessing the hidden message is as hard as guessing the secret password.

In practice, the digital locker can be constructed by using a very strong hash function H. Lynn et al. [12] showed a simple construction of a digital locker by using the (programmable) random oracle, which is briefly outlined as follows. Let r be a secret random string to hide under a key k for appropriate lengths respectively. Then, the digital locker for security parameter κ can be constructed as $\mathsf{lock}(r, k) = \mathsf{H}(s\|k) \oplus (m\|0^\kappa)$ where s is a nonce and "$\|$" denotes concatenation.

To see the actual size of reusable FE in practice, consider the following specific case: let $n = 100$ denote the length of biometric template in the binary representation, let $t = 10$ be the maximum number of errors in a biometric reading, and let $\delta = 1/1000$ (e.g., the authentication fails on a correct fingerprint with at most 0.1%). be our allowable error probability for the correct system. Then, the reusable FE generates a public string that consists of l digital lockers where $l = 100^{\frac{10 \cdot 90}{100 \cdot \ln 100}} \log(2 \cdot 1000) \approx 100^2 \cdot 11 \approx 110000$. Assuming that H is instantiated with SHA-256 hash function, the size of public string is 3.5 MByte.

The time complexity of reproduction algorithm is involved with the computation of $l/2$ hash values on average. Therefore, the decoding takes approximately 2 min on average assuming that the computation of SHA-256 takes 1 ms.

Concrete efficiency of our reusable fuzzy extractor. In order to compare the above against the cost of our FE, we need to estimate concrete sizes of LWE-based ciphertexts. For a fair comparison against the ROM-based FE of [5], we evaluate our ROM-based construction, which is effectively bounded by the cost of the FMR-FE. (Certainly, our standard-model variant will be somewhat more expensive.)

Our concrete calculations are based on the work of Albrecht, Player, and Scott [3]. Note that for Regev-style LWE ciphertexts, as in our ROM-based FE, we set lattice dimension n according to the desired level of real-world security, then choose the number of LWE samples m and the modulus q accordingly. Namely, we will have $m = 3n$ and $q = n^2$.

In order to achieve approximately 128 bits of security, the lattice dimension should be at least 256 [3]. The public parameter pp is an $n \times m$ matrix of $\mathbb{Z}_q$ integers. This has size $n \cdot m \cdot \log_2(q) = 256 \cdot (3 \cdot 256) \cdot 16 = 3,145,728$ *bits*, which is about 0.39 MBytes. The public strings per-enrollment, i.e. the p, are then 256 times smaller, or about 12 KBytes each.

Finally, we mention that replacing "plain" LWE with either Ring-LWE or Module-LWE may lead to further efficiency improvements, and leave experimenting with those alternative choices as an interesting problem for future work.

References

1. Akavia, A., Goldwasser, S., Vaikuntanathan, V.: Simultaneous hardcore bits and cryptography against memory attacks. In: Reingold, O. (ed.) TCC 2009. LNCS, vol. 5444, pp. 474–495. Springer, Heidelberg (2009). doi:10.1007/978-3-642-00457-5_28
2. Alamélou, Q., Berthier, P.-E., Cauchie, S., Fuller, B., Gaborit, P.: Reusable fuzzy extractors for the set difference metric and adaptive fuzzy extractors (2016). http://eprint.iacr.org/2016/1100
3. Albrecht, M.R., Player, R., Scott, S.: On the concrete hardness of learning with errors. J. Math. Cryptol. **9**(3), 169–203 (2015)
4. Boyen, X.: Reusable cryptographic fuzzy extractors. In: 11th ACM Conference on Computer and Communications Security, pp. 82–91. ACM Press (2004)
5. Canetti, R., Fuller, B., Paneth, O., Reyzin, L., Smith, A.: Reusable fuzzy extractors for low-entropy distributions. In: Fischlin, M., Coron, J.-S. (eds.) EUROCRYPT 2016. LNCS, vol. 9665, pp. 117–146. Springer, Heidelberg (2016). doi:10.1007/978-3-662-49890-3_5
6. Dodis, Y., Reyzin, L., Smith, A.: Fuzzy extractors: how to generate strong keys from biometrics and other noisy data. In: Cachin, C., Camenisch, J.L. (eds.) EUROCRYPT 2004. LNCS, vol. 3027, pp. 523–540. Springer, Heidelberg (2004). doi:10.1007/978-3-540-24676-3_31
7. Döttling, N., Müller-Quade, J.: Lossy codes and a new variant of the learning-with-errors problem. In: Johansson, T., Nguyen, P.Q. (eds.) EUROCRYPT 2013. LNCS, vol. 7881, pp. 18–34. Springer, Heidelberg (2013). doi:10.1007/978-3-642-38348-9_2

8. Fuller, B., Meng, X., Reyzin, L.: Computational fuzzy extractors. In: Sako, K., Sarkar, P. (eds.) ASIACRYPT 2013. LNCS, vol. 8269, pp. 174–193. Springer, Heidelberg (2013). doi:10.1007/978-3-642-42033-7_10

9. Goldwasser, S., Kalai, Y.T., Peikert, C., Vaikuntanathan, V.: Robustness of the learning with errors assumption. In: 1st Innovations in Computer Science, ICS 2010, pp. 230–240. Tsinghua University Press (2010)

10. Huth, C., Becker, D., Guajardo, J., Duplys, P., Güneysu, T.: Securing systems with scarce entropy: LWE-based lossless computational fuzzy extractor for the IoT (2016). http://eprint.iacr.org/2016/982

11. Kamp, J., Zuckerman, D.: Deterministic extractors for bit-fixing sources and exposure-resilient cryptography. In: 44th Annual Symposium on Foundations of Computer Science (FOCS), pp. 92–101. IEEE, October 2003

12. Lynn, B., Prabhakaran, M., Sahai, A.: Positive results and techniques for obfuscation. In: Cachin, C., Camenisch, J.L. (eds.) EUROCRYPT 2004. LNCS, vol. 3027, pp. 20–39. Springer, Heidelberg (2004). doi:10.1007/978-3-540-24676-3_2

13. Regev, O.: On lattices, learning with errors, random linear codes, and cryptography. In: Gabow, H.N., Fagin, R. (eds.) 37th Annual ACM Symposium on Theory of Computing (STOC), pp. 84–93. ACM Press, May 2005

GENFACE: Improving Cyber Security Using Realistic Synthetic Face Generation

Margarita Osadchy[1]($\boxtimes$), Yan Wang[2], Orr Dunkelman[1], Stuart Gibson[2],
Julio Hernandez-Castro[3], and Christopher Solomon[2]

[1] Computer Science Department, University of Haifa, Haifa, Israel
{rita,orrd}@cs.haifa.ac.il
[2] School of Physical Sciences, University of Kent, Canterbury, UK
{s.j.gibson,c.j.solomon}@kent.ac.uk
[3] School of Computing, University of Kent, Canterbury, UK
J.C.Hernandez-Castro@kent.ac.uk

Abstract. Recent advances in face recognition technology render face-based authentication very attractive due to the high accuracy and ease of use. However, the increased use of biometrics (such as faces) triggered a lot of research on the protection biometric data in the fields of computer security and cryptography.

Unfortunately, most of the face-based systems, and most notably the privacy-preserving mechanisms, are evaluated on small data sets or assume ideal distributions of the faces (that could differ significantly from the real data). At the same time, acquiring large biometric data sets for evaluation purposes is time consuming, expensive, and complicated due to legal/ethical considerations related to the privacy of the test subjects. In this work, we present GENFACE, the first publicly available system for generating synthetic facial images. GENFACE can generate sets of large number of facial images, solving the aforementioned problem. Such sets can be used for testing and evaluating face-based authentication systems. Such test sets can also be used in balancing the ROC curves of such systems with the error correction codes used in authentication systems employing *secure sketch* or *fuzzy extractors*. Another application is the use of these test sets in the evaluation of privacy-preserving biometric protocols such as GSHADE, which can now enjoy a large number of synthetic examples which follow a real-life distribution of biometrics. As a case study, we show how to use GENFACE in evaluating SecureFace, a face-based authentication system that offers end-to-end authentication and privacy.

Keywords: Synthetic face generation · GENFACE · SecureFace · Biometrics · Face-based authentication · Face verification

1 Introduction

Biometric systems have become prevalent in many computer security applications, most notably authentication systems which use different types of biometrics, such as fingerprints, iris codes, and facial images.

© Springer International Publishing AG 2017
S. Dolev and S. Lodha (Eds.): CSCML 2017, LNCS 10332, pp. 19–33, 2017.
DOI: 10.1007/978-3-319-60080-2_2

A central problem in developing reliable and secure biometric systems is the fuzziness of biometric samples. This fuzziness is caused by the sampling process and by the natural changes in appearance. For example, two images of the same face are never identical and could change significantly due to illumination, pose, facial expression, etc. To reduce these undesirable variations in practical applications, images are usually converted to lower-dimensional representations by a feature extraction process. Unfortunately, no representation preserves the original variation in identity while cancelling the variation due to other factors. As a result, there is always a trade-off between the robustness of the representation (consistency of recognition) and its discriminating power (differentiating between different individuals).

Devising robust features requires a significant number of training samples—the most successful systems use several million images (see [19] for a more detailed discussion). Most of these samples are used for training, while evaluation (testing) of the system is typically done on a relatively small dataset. Since such systems may need to support large sets of many users (such as border control systems or national biometric databases including the entire population[1]), a small-scale evaluation is insufficient when testing the system's consistency in large-scale applications.

The solution seems to be the testing of the system on a large dataset. However, acquiring large data sets is difficult due to several reasons: The process of collecting the biometric samples is very time consuming (as it requires many participants, possibly at different times) and probably expensive (e.g., due to paying the participants or the need to annotate the samples). Moreover, as biometric data is extremely private and sensitive (e.g., biometric data cannot be easily changed), collecting (and storing) a large set of samples may face legal and ethical constraints and regulations.

A possible mitigation to these problems is working with synthetic biometrics—"biometrics" synthetically generated from some underlaying model. Given a reasonable model, which approximates the real-life biometric trait, offering efficient sampling procedures, with sufficiently many "possible samples", designers of biometric systems can query the model, rather than collecting real data. This saves both the time and the effort needed for constructing the real dataset. It has the additional advantage of freeing the designer from legal and ethical constraints. Moreover, if the model is indeed "rich" enough, one can obtain an enormous amount of synthetic biometric samples for testing and evaluate large-scale systems.

In the case of fingerprints, the SFinGe system [4] offers the ability to produce synthetic fingerprints. SFinGe can generate a database of such synthetic fingerprints, which can be used for the training and evaluation of fingerprints' biometric systems. Fingerprint recognition algorithms have been shown to perform equally well on the outputs of SFinGe and on real fingerprint databases.

In the case of faces, some preliminary results were discussed in [24]. The aim of [24] was to transform one sensitive data set (which contains real users) into a

---

[1] Such as Aadhaar, the Indian biometric database of the full Indian population.

fixed, secondary data set of synthetic faces of the same size. The transformation was based on locating close faces and "merging" them into new synthetic faces. This approach has strong limitations, both in the case of security (as one should perform a very thorough security analysis) as well as usability (as there is only a single, fixed size, possible data set). Finally, [24] does not offer a general purpose face generation system and it is not publicly available.

In this paper, we introduce GENFACE, system that generates synthetic faces and is publicly available. GENFACE allows the generation of many synthetic facial images: the user picks how many "identities" are needed, as well as how many images per "identity" to generate. Then, using the active appearance model [7], the system generates the required amount of synthetic samples. Another important feature of our system is the ability to control the "natural" fuzziness in the generated data.[2]

Synthetic faces created by GENFACE can be used in evaluating and testing face-based biometric systems. For example, using GENFACE it is possible to efficiently evaluate and determine the threshold that face recognition systems should set in distinguishing between same identity and different ones. By varying this threshold, one can obtain the ROC curve of the system's performance, which is of great importance in studying the security of an authentication system.

Similarly, a large volume of work in privacy-preserving techniques for biometrics (such as fuzzy commitment [14] and fuzzy extractors [9]) treat the fuzziness using error correction codes. Such solutions and most notably, systems using such solutions (e.g., [5,6,10,26]) and their efficiency rely on the parameter choices. Our system offers face generation with a controlled level of fuzziness, which allows for a more systematic evaluation of these protection mechanisms and systems.

We note that GENFACE can also be used in other contexts within computer security. Protocols such as GSHADE [3], for privacy-preserving biometric identification, should be evaluated with data which is similar to real biometrics. By using the output of GENFACE, such protocols could easily obtain a large amount of synthetic samples 'at the press of a button', which would allow for more realistic and accurate simulation of real data, without the need for collecting and annotating the dataset.

Finally, GENFACE can be used in evaluating large scale face-based biometric systems. For example, consider representations (also called templates) which are too short to represent a large set of users without collisions, but are sufficiently long to represent a small set of users without such collisions. It is of course better to identify such problems *before* the collection of millions of biometric samples, which can be easily done using synthetic biometrics.

This paper is organized as follows: Sect. 2 introduces GENFACE and how it produces synthetic faces. In Sect. 3 we show that a real system, *SecureFace* [10], reaches similar accuracy results using synthetic faces and using real faces and its evaluation on a large-scale data set, generated by GENFACE. Section 4 concludes our contribution and discusses future research directions.

---

[2] Changes in viewing conditions require the use of a 3D model and will be considered in future work.

2   GenFace System for Synthetic Faces

The main function of the GenFace System is generating a large number of face images of the same or different identities. It generates a (user) specified number of random, different, facial identities (face images) which we refer to as *seed points*. A set of faces, comprising subtle variations of a seed face can then be generated. The difference (distance to the seed point) in appearance between the images comprising the set can be controlled by the user. Below a certain distance threshold the differences will be sufficiently small such that faces in the set, which we refer to later as *offspring* faces, have the same identity as the seed.

In this section, we first describe the generative model that we use for sampling synthetic faces. We then introduce our specific methods for sampling seed faces and for sampling offspring faces. We then present the GenFace user interface and some guidelines for using our system.

2.1   Model for Representing Facial Appearance

Different models have been proposed for generating and representing faces including, active appearance models [7], 3D deformable models [2], and convolutional neural networks [18,30]. Here we use an appearance model (AM) [7] due to its capacity for generating photo-realistic faces (e.g. [12,23]). The representation of faces within an AM is consistent with human visual perception and hence also compatible with the notion of face-space [28].[3] In particular, the perceptual similarity of faces is correlated with distance in the AM space [17].

AMs describe the variation contained within the training set of faces, used for its construction. Given that this set spans all variations associated with identity changes, the AM provides a good approximation to any desired face. The distribution of AM coefficients (that encode facial identity) of faces belonging to the same ethnicity are well approximated by an independent, multivariate, Gaussian probability density function [20,27,29].

We follow the procedure for AM construction, described in [12]. The training set of facial images, taken under the same viewing conditions, is annotated using a point model that delineates the face shape and the internal facial features. In this process, 22 landmarks are manually placed on each facial image. Based on these points, 190 points of the complete model are determined (see [12] for details). For each face, landmark coordinates are concatenated to form a shape vector, x. The data is then centered by subtracting the mean face shape, $\bar{x}$, from each observation. The shape principle components P_s are derived from the set of mean subtracted observations (arranged as columns) using principal components analysis (PCA). The synthesis of a face shape, denoted by $\hat{x}$, from the *shape model* is achieved as follows,

$$\hat{x} = P_s b_s + \bar{x}, \tag{1}$$

---

[3] Hereafter we use the term face-space to mean the space spanned by a set of principal components, derived from a set of training face images.

where b_s is a vector in which the first m elements are normally distributed parameters that determine the linear combination of shape principal components and the remaining elements are equal to zero. We refer to b_s as the *shape coefficients*.

Before deriving the texture component of the AM, training images must be put into correspondence using non-rigid shape alignment procedure. Each shape normalized and centered RGB image of a training face is then rearranged as a vector g. Such vectors for all training faces form a matrix which is used to compute the texture principle components, P_g, by applying PCA. A face texture, denoted by $\hat{g}$, is reconstructed from the *texture model* as follows,

$$\hat{g} = P_g b_g + \bar{g}, \tag{2}$$

where b_g are the *texture coefficients* which are also normally distributed and $\bar{g}$ is the mean texture.

The final model is obtained by a PCA on the concatenated shape and texture parameter vectors. Let Q denote the principal components of the concatenated space. The AM coefficients, c, are obtained from the corresponding shape, x, and texture, g, as follows,

$$c = Q^T \begin{bmatrix} rb_s \\ b_g \end{bmatrix} = Q^T \begin{bmatrix} wP_s^T(x - \bar{x}) \\ P_g^T(g - \bar{g}) \end{bmatrix} \tag{3}$$

where w is a scalar that determines the weight of shape relative to texture.

Generating new instances of facial appearance from the model requires a sampling method for AM coefficients, c, which is described in detail in Sect. 2.2. The shape and texture coefficients are then obtained from the AM coefficients as follows: $b_s = Q_s c$ and $b_g = Q_g c$, where $[Q_s^T Q_g^T]^T$ is the AM basis. The texture and shape of the face are obtained via Eq. (1) and (2) respectively. Finally, the texture $\hat{g}$ is warped onto the shape $\hat{x}$, resulting in a face image.

The identity change in the face-space is a slowly changing function. Thus, there is no clear border between different identities. However, given the success of numerous face recognition methods, which rely on the Euclidean distance in the face-space to make recognition decisions (e.g., [1,11,25]), we can conclude that Euclidean distance in the face-space is correlated with the dissimilarity of the corresponding facial images. Another conclusion that can be drawn from the success of the face-space recognition methods, is that face-space is isotropic.[4] Based on these two observations, we suggest simulating the variation in appearance of the same person by sampling points in a close proximity to each other. Specifically, we define an identity as a random seed in the face-space and generate different images of this identity by sampling points in the face-space at a distance s from the seed point. Here s is a parameter of the system which is evaluated in our experiments (in Sect. 3).

2.2 Sampling AM Coefficients for Seed Faces

Since AM coefficients follow a multivariate Gaussian distribution, most of the faces in AM representation are concentrated near the hyper-ellipsoid with radii

---

[4] Our experiments, reported in Sect. 3.2 verify these assumptions.

approximately equal to $\sqrt{d}$ standard deviation units, where d is the dimension of the face-space. This follows from the fact that in high dimensional space, the distance between the samples of a multivariate Gaussian distribution to its mean follows the chi-distribution with an average of about $\sqrt{d}$ and variance that saturates at 0.7 (for $d > 10$). Since the standard deviation of the corresponding chi-distribution is narrow, sampling on the surface of the hyper-ellipsoid retains most of the probabilistic volume (i.e., most of the "plausible" faces in this face-space[5]). Also, it prevents the caricature effect associated with the displacement of a face in the direction directly away from the origin [13,16].

Let $N(0, \sigma)$ denote the distribution of AM coefficients where $\sigma = [\sigma_1, \sigma_2, \cdots, \sigma_d]$ are the standard deviations of the face-space. Let V denote the hyper-ellipsoid in R^d with radii $\sqrt{d}\sigma_i$ and ∂V denote its surface. To ensure the diversity of synthesized faces with respect to identity change, we should sample seed faces on ∂V uniformly at random. To this end, we implemented an algorithm from [21] that offers a method for uniform sampling from a manifold enclosing a convex body (summarized in Algorithm 1). Given a confidence level $1 - \epsilon$, the algorithm first draws a point x from the volume V uniformly at random, and then simulates a local diffusion process by drawing a random point p from a small spherical Gaussian centered at x. If p is located outside V, the intersection between ∂V and a line segment linking p with x will be a valid sample, the distribution of which has a variation distance $O(\epsilon)$ from the uniform distribution on ∂V. It is essential to draw uniform samples efficiently from within V, thus we have also implemented an efficient sampler for this purpose, as proposed in [8].

2.3 Sampling AM Coefficients for Offspring Faces

Once a seed face has been uniformly sampled from ∂V, a set of offspring faces can be sampled from a local patch of ∂V around the seed face. Given a sampling distance s, the system can randomly pick up an offspring face located on ∂V which is at a distance s away from the seed face. If s is sufficiently small, all offspring faces will correspond to the same identity as the seed face.

Before describing the algorithm for sampling offspring face coefficients, we introduce the following notations: Let $P : \{p|p \neq \mathbf{0}\} \to \{\hat{p}|p \in \partial V\}$ be a mapping function that projects any point in the face-space, except the origin, onto ∂V by rescaling: $\hat{p} = \frac{k}{\|M^{-1}p\|}p$, where M is a $d \times d$ diagonal matrix of standard deviations σ_i $(i = 1, .., d)$ and $k \approx \sqrt{d}$.

To provide diversity among offspring faces, we sample them along random directions. Given a seed face x and a distance s, the algorithm (summarized in Algorithm 2) repeatedly samples a face $\hat{x}$ on ∂V at random until its distance to x exceeds s. The vector $\hat{x} - x$ defines a sampling direction for an offspring face. The algorithm proceeds by finding a point in this direction such that its

---

[5] It is unclear what should be the training size of a face-space that models all possible faces. However, we note that a face which is not "plausible" in some face-space, i.e., is very far from the surface of the face-space is likely to not "work" properly in a system which relies on the face-space.

Algorithm 1. Uniform sampling on the surface of a hyper-ellipsoid

---

1: **function** UNIFORMSAMPLE$(V, \partial V, d, S, \epsilon)$ ▷ V a d-dimensional
    hyper-ellipsoid, ∂V: surface of V, S: a uniform sampler on V, ϵ: variation distance
    from the uniform distribution.
2:     Draw N points uniformly from V using S,
3:     Estimate κ – the smallest eigenvalue of the Inertia matrix $E[(\boldsymbol{x} - \bar{\boldsymbol{x}})(\boldsymbol{x} - \bar{\boldsymbol{x}})^T]$
    ▷ $N = O(dlog^2(d)log\frac{1}{\epsilon})$
4:     $\sqrt{t} \leftarrow \frac{\epsilon\sqrt{\kappa}}{32d}$
5:     $p \leftarrow \varnothing$
6:     **while** $p == \varnothing$ **do**
7:         $x \in_R S$
8:         $y \in_R Gaussian(x, 2tI_d)$ ▷ y follows a normal distribution.
9:         **if** $y \notin V$ **then**
10:            $p \leftarrow \overrightarrow{\boldsymbol{xy}} \bigcup \partial V$
11:         **end if**
12:     **end while**
13:     **return** p
14: **end function**

---

projection into ∂V is at a distance s away from $\boldsymbol{x}$. We sample a large number of points in ∂V off-line to speed up the search for a sampling direction.

2.4   User Interface of GenFace System

The user interface of GenFace is shown in Fig. 1. The user can set the sampling distance by entering a number into the DISTANCE text box in the PARAME-TERS panel. She has options for either sampling offspring faces within, or on the perimeter of a region, centered at the seed face. By pressing the GENERATE button in the SEED panel, a seed face and its eight offspring faces are generated and displayed in the OFFSPRING panel with the seed face positioned in the center. The distance from each offspring face to the seed will be displayed on the top of each offspring face.[6] The user can select any face and save the face image and coefficients by pressing the SAVE AS button. Alternatively, the user can generate a large number of faces using a batch mode: The user may input the number of seed faces and the number of offspring in the BATCH panel. By pressing GENERATE button in the panel, all faces will be generated and saved automatically into a set of sequentially named folders with each folder containing a seed and its offsprings. A progress bar will be shown until all faces have been generated. If the generation procedure is interrupted, the user can resume it by specifying the starting number of the seed face and pressing the GENERATE button. The user is allowed to load saved data into the system and the face image will be displayed in the center of the OFFSPRING panel.

Version 1.0.0 of GenFace is implemented in Matlab and compiled for Windows. Generated images are saved as JPEG files and the corresponding meta

---

[6] This feature is more relevant to the "sample within" option, as the distance from each offspring image to the seed could be different.

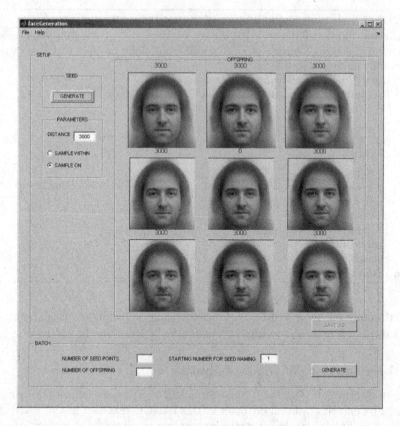

Fig. 1. The user interface of the GENFACE system

Algorithm 2. Sampling an offspring face

1: **function** SAMPLEOFFSPRING$(V, \partial V, d, \boldsymbol{x}, s, \epsilon, k, M)$ $\triangleright$ V: a d-dimensional hyper-ellipsoidal body, ∂V: the surface of V, $\boldsymbol{x}$: a seed face, s: a sampling distance, ϵ: relative error tolerance of sampling distance, k: the normalized radius of V, M: a diagonal matrix defined as $diag(\sigma_1, \sigma_2, \cdots, \sigma_d)$
2:     **repeat**
3:        Sample a random face $\hat{\boldsymbol{x}}$ uniformly on ∂V
4:     **until** $\|\hat{\boldsymbol{x}} - \boldsymbol{x}\| \geq s$
5:     $t \leftarrow s, \hat{t} \leftarrow 0$
6:     **repeat**
7:        $\boldsymbol{p} \leftarrow \boldsymbol{x} + t\boldsymbol{v}$
8:        $\hat{\boldsymbol{p}} = \frac{k}{\|M^{-1}\boldsymbol{p}\|}\boldsymbol{p}$ $\triangleright$ Project $\boldsymbol{p}$ onto ∂V
9:        $\hat{t} \leftarrow \|\hat{\boldsymbol{p}} - \boldsymbol{x}\|$
10:    $t \leftarrow \frac{s}{t}t$
11:    **until** $|\frac{\hat{t}}{s} - 1| \leq \epsilon$
12:    **return** $\hat{\boldsymbol{p}}$
13: **end function**

---

data is saved in a text file with the same name. The installation package of the GENFACE is publicly available at https://crypto.cs.haifa.ac.il/GenFace/.[7]

3   Testing the GENFACE System with SecureFace

In this section we show an example of using synthetic facial images generated by GENFACE for evaluating the recently introduced SecureFace system [10] for key derivation from facial images. We start with a brief description of the SecureFace system and then turn to a comparison of its results on real and synthetic data sets of the same size. We also show that using GENFACE we can generate data with different levels of fuzziness (simulating the inherent fuzziness of biometric samples) and that this fuzziness directly affects the success rate of the SecureFace system. A similar procedure can be used to choose the parameters for face generation to fit the expected level of fuzziness in the real data. Finally, we test the scalability of the SecureFace system using a much larger set of synthetic faces with the fuzziness parameters that approximate the variation in the real data used in our experiments.

3.1   The SecureFace System

The SecureFace system [10] derives high-entropy cryptographic keys from frontal facial images while fully protecting the privacy of the biometric data, including the training phase. SecureFace (as most of the biometric cryptosystems) offers a two-stage template generation process. Before this process, an input face is aligned to the canonical pose by applying an affine transformation on physical landmarks found by a landmark detector (e.g., [15]).

The first stage of SecureFace converts input images into real-valued representations that suppress the fuzziness in images of the same person due to viewing conditions (such as pose, illumination, camera noise, small deformations etc.). A very large volume of work exists on this topic in the computer vision community. The SecureFace system uses a combination of standard local features that do not require user-dependent training, specifically, *Local Binary Patterns* (LBPs), *Histogram of Oriented Gradients* (HoG), and *Scale Invariant Feature Transform* (SIFT). The extracted features are reduced in dimensionality (by PCA), then concatenated and whitened to remove correlations.

The second phase of the processing transforms real-valued representations (obtained by the first step) to binary strings with the following properties: consistency/discriminability, efficiency (high entropy), and zero privacy loss. Let $x \in R^D$ be a data point (real-valued feature vector) and w_k be a projection vector. The transformation to a binary string is done by computing $1/2(\mathbf{sgn}(w_k^T x) + 1)$. Vectors w_k form the embedding matrix $W = [w_1, \ldots, w_K]$. W is obtained by a learning process with the goal of mapping the templates

---

[7]   GENFACE does not require full Matlab, but the installation package will install the "MATLAB Component Runtime".

of the same person close to each other in the Hamming space and templates of different people far from each other. The resulting binary strings have almost full entropy, specifically, each bit has $\approx 50\%$ chance of being one or zero, and different bits are independent of each other. The embedding, W, is learned on a different (public) set of people, thus it does not reveal any information regarding system's users. Finally, W is generated only once and can be used with any data set of faces without any re-training.

In the acquisition stage a generated template, $t = 1/2(\mathbf{sgn}(W^T x)+1)$, is used as the input to the fuzzy commitment scheme [14] which chooses a codeword C and computes $s = C + t$. The helper data s is released to the user. For a re-sampling, a new image of a user is converted to a binary string using the two-stage template generation process (described above) resulting in a template $t' = 1/2(\mathbf{sgn}(W^T x) + 1)$. The user supplies s and along with t' the system computes $C' = EC(s+t')$, where $EC()$ is the error-correction function. If x and x' are close (in terms of Hamming distance), then $C' = C$.

3.2 Experiments

Our experiments include two parts. In the first part, we compare the performance of the SecureFace system on real and synthetic sets of equal sizes, varying the level of fuzziness in the synthetics data. This experiment allows us to choose (if possible) the level of fuzziness that fits the real data. In the second part we test the scalability of the SecureFace system on larger sets of synthetic data using the selected parameters. Acquiring real facial data of that size (5,000–25,000 subjects) is problematic as discussed earlier.

Matching Size Experiment: We used a subset of the in-house dataset of real people, which contains 508 subjects, with 2.36 images per subject on average. All images were collected in the same room while the participants were sitting at the same distance from the camera. The lighting conditions were kept the same during collection, and the subjects were told to keep a neutral expression. The top row of Table 1 illustrates two subjects with 3 images per each from the dataset. Even though, the variation in viewing conditions in this set were kept to a minimum, some variation in lighting (due to the different height of people), pose, and facial expressions is still present. We choose this particular set of real faces in our experiments, as it was used as a test-bed for the SecureFace system and because it focuses on "natural" variation in identity (as opposed to mix of all variations) which the proposed GENFACE can offer.

For the synthetic datasets, we generated 5 sets using different parameterization of GENFACE. Each set contained 500 seeds with 5 offsprings per seed, which can be viewed as 500 subjects with 5 images per subject. Rows 2–5 of Table 1 demonstrate examples of generated images for different distance parameters (each row corresponds to a different distance parameter and shows two subjects with 3 images per each).

We ran the acquisition stage of the SecureFace system for all 508 subjects. Then we simulated 1,200 genuine attempts and 1,665 imposter attempts (all

Table 1. Example of images from the real set (first row) and from the synthetic set with different levels of fuzziness. Each row shows two different subjects with 3 images per subject.

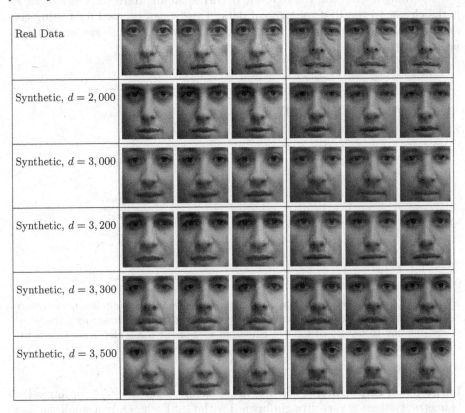

queries in the genuine attempts were with 5 reference subjects chosen uniformly at random from 508). The corresponding ROC curve is depicted in Fig. 2. We note that in the context of SecureFace, a False Positive happens when the face of an imposter "succeeds" to unlock the key of a real user.

Our hypothesis is that the distance parameter controls the fuzziness of the generated data. To test this hypothesis, we tested GenFace on each set, using all generated subjects along with 2,500 genuine and 2,500 imposter attempts (again, all queries in the genuine attempts were with 5 reference subjects chosen uniformly at random from 500). The ROC curves of these experiments are shown in Fig. 2, showing that data sets with higher distances between the offsprings to the corresponding seed result in worse ROC curves. This can happen either due to the increased fuzziness in the images of the same seed (subject) or due to decreased distances between different seeds. We compared the distributions of distances among different subjects between all tested sets and found them well aligned with each other. Thus, we can conclude that the fuzziness of the same person indeed grows with the distance parameter.

According to the plot, the ROC curve corresponding to the distance of 3,000 is the closest to the ROC of the real data. Thus it can be used for testing the system in the large-scale experiment. The third row of Fig. 2 shows examples of facial images for this distance.

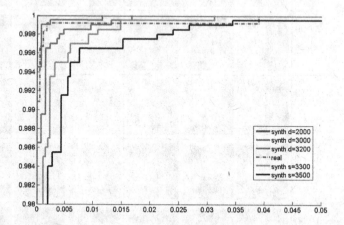

Fig. 2. ROC curves for the "Matching Size" experiment showing the performance SecureFace on real data and on three generated sets with different distance parameters (corresponding to levels of fuzziness among images of the same identity). This figure is best viewed in color. (Color figure online)

Large-Scale Experiment: Testing the scalability of SecureFace (or any other biometric system) is very straightforward with GENFACE. Only a small number of parameters need to be set: the distance to the value that approximates the variation in the real data, the number of seeds (subjects), and the number of offspring (images per subject). After that, the required data can be generated in a relatively short time. For example, generating a seed face on a 32-bit Windows XP with an Intel Core i7, 2.67 Hz, 3.24 GB RAM using a code written and compiled in Matlab R2007b (7.5.0.342) takes 0.9 s on average, and generating an offspring for a given seed takes 0.636 s on average. The process can be further optimized and adjusted to run in parallel.

The results of testing the SecureFace using generated sets of different sizes (from 500 to 25,000 subjects) with the distance parameter of 3,000 are shown in Fig. 3. The ROC curves show very plausible behavior that we believe closely approximates that of real data with a similar level of fuzziness.

4   Conclusions

We proposed a system for generating plausible images of faces belonging to the same and different identities. We showed that our system could easily generate a

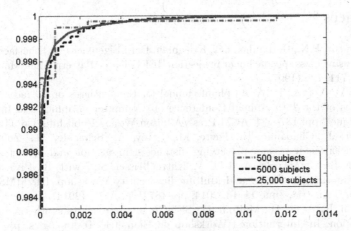

Fig. 3. ROC curves for the "Large-Scale" experiment, showing the performance of SecureFace on synthetic data of different sizes (from 500 to 25,000 subjects) and with a distance of 3,000 (between offsprings and the corresponding seed). The ROC curve corresponding to 500 subjects (d = 3,000) is copied from Fig. 2 for the reference. This figure is best viewed in color. (Color figure online)

very large number of facial images while controlling their fuzziness. We suggested that our system can be used for systematic evaluation of biometric systems, instead of real faces. We showed the merit of this approach using the recently introduced SecureFace.

Future research directions include: Evaluating the merits of using GENFACE's images for training purposes. Evaluating facial landmark detectors by comparing their results with GENFACE's output (that can contain landmarks). This aspect can be used for improving future landmark detectors, which have various applications throughout computer vision beyond biometrics. In the current work, we addressed only the "natural variation" in facial appearance. Future work will include integrating variations resulting from a change in viewing conditions. This will require rendering facial appearances, using a face 3D model (e.g., such as in [22]). Additionally, studying the difference in the behavior of other biometric systems (e.g., Amazon cognitive systems) when using synthetic faces rather than real ones is left to future works.

Finally, we plan to extend the platform support, e.g., introduce Linux support. We also plan on offering a parallel process for the sampling of faces and offsprings.

Acknowledgements. This research was supported by UK Engineering and Physical Sciences Research Council project EP/M013375/1 and by the Israeli Ministry of Science and Technology project 3-11858. We thank Mahmood Sharif for his support in experiments using SecureFace. We thank the anonymous reviewers of this paper for their ideas and suggestions.

References

1. Belhumeur, P.N., Hespanha, J.P., Kriegman, D.J.: Eigenfaces vs. fisherfaces: recognition using class specific linear projection. IEEE Trans. Pattern Anal. Mach. Intell. **19**(7), 711–720 (1997)
2. Blanz, V., Vetter, T.: A morphable model for the synthesis of 3D faces. In: Proceedings of the 26th Annual Conference on Computer Graphics and Interactive Techniques, pp. 187–194. ACM Press/Addison-Wesley Publishing Co. (1999)
3. Bringer, J., Chabanne, H., Favre, M., Patey, A., Schneider, T., Zohner, M.: GSHADE: faster privacy-preserving distance computation and biometric identification. In: Unterweger, A., Uhl, A., Katzenbeisser, S., Kwitt, R., Piva, A. (eds.) ACM Information Hiding and Multimedia Security Workshop, IH&MMSec 2014, Salzburg, Austria, June 11–13, 2014, pp. 187–198. ACM (2014)
4. Cappelli, R., Maio, D., Maltoni, D.: SFinGe: an approach to synthetic fingerprint generation. In: International Workshop on Biometric Technologies, pp. 147–154 (2004)
5. Chang, Y., Zhang, W., Chen, T.: Biometrics-based cryptographic key generation. In: IEEE International Conference on Multimedia and Expo (ICME), pp. 2203–2206 (2004)
6. Chen, C., Veldhuis, R., Kevenaar, T., Akkermans, A.: Biometric binary string generation with detection rate optimized bit allocation. In: CVPR Workshop on Biometrics, pp. 1–7 (2008)
7. Cootes, T.F., Edwards, G.J., Taylor, C.J.: Active appearance models. IEEE Trans. Pattern Anal. Mach. Intell. **23**(6), 681–685 (2001)
8. Dezert, J., Musso, C.: An efficient method for generating points uniformly distributed in hyperellipsoids. In: The Workshop on Estimation, Tracking and Fusion: A Tribute to Yaakov Bar-Shalom (2001)
9. Dodis, Y., Ostrovsky, R., Reyzin, L., Smith, A.D.: Fuzzy extractors: how to generate strong keys from biometrics and other noisy data. SIAM J. Comput. **38**(1), 97–139 (2008)
10. Dunkelman, O., Osadchy, M., Sharif, M.: Secure authentication from facial attributes with no privacy loss. In: Sadeghi, A., Gligor, V.D., Yung, M. (eds.) 2013 ACM SIGSAC Conference on Computer and Communications Security, CCS 2013, Berlin, Germany, November 4–8, 2013, pp. 1403–1406. ACM (2013)
11. Edwards, G.J., Cootes, T.F., Taylor, C.J.: Face recognition using active appearance models. In: Burkhardt, H., Neumann, B. (eds.) ECCV 1998. LNCS, vol. 1407, pp. 581–595. Springer, Heidelberg (1998). doi:10.1007/BFb0054766
12. Gibson, S.J., Solomon, C.J., Bejarano, A.P.: Synthesis of photographic quality facial composites using evolutionary algorithms. In: Proceedings on British Machine Vision Conference, BMVC 2003, Norwich, UK, pp. 1–10, September 2003 (2003)
13. Gibson, S.J., Solomon, C.J., Pallares-Bejarano, A.: Nonlinear, near photo-realisticcaricatures using a parametric facial appearance model. Behav. Res. Methods **37**(1), 170–181 (2005). http://dx.doi.org/10.3758/BF03206412
14. Juels, A., Wattenberg, M.: A fuzzy commitment scheme. In: Motiwalla, J., Tsudik, G. (eds.) CCS 1999, Proceedings of the 6th ACM Conference on Computer and Communications Security, Singapore, November 1–4, 1999, pp. 28–36. ACM (1999)
15. Kazemi, V., Sullivan, J.: One millisecond face alignment with an ensemble of regression trees. In: 2014 IEEE Conference on Computer Vision and Pattern Recognition, CVPR 2014, Columbus, OH, USA, June 23–28, 2014, pp. 1867–1874 (2014)

16. Lee, K., Byatt, G., Rhodes, G.: Caricature effects, distinctiveness, and identification: testing the face-space framework. Psychol. Sci. **11**(5), 379–385 (2000)
17. Lewis, M.: Face-space-R: towards a unified account of face recognition. Vis. Cogn. **11**(1), 29–69 (2004)
18. Li, M., Zuo, W., Zhang, D.: Convolutional network for attribute-driven and identity-preserving human face generation. arXiv preprint arXiv:1608.06434 (2016)
19. Masi, I., Tran, A.T., Hassner, T., Leksut, J.T., Medioni, G.: Do we really need to collect millions of faces for effective face recognition? In: Leibe, B., Matas, J., Sebe, N., Welling, M. (eds.) ECCV 2016. LNCS, vol. 9909, pp. 579–596. Springer, Cham (2016). doi:10.1007/978-3-319-46454-1_35
20. Matthews, I., Baker, S.: Active appearance models revisited. Int. J. Comput. Vis. **60**(2), 135–164 (2004)
21. Narayanan, H., Niyogi, P.: Sampling hypersurfaces through diffusion. In: Goel, A., Jansen, K., Rolim, J.D.P., Rubinfeld, R. (eds.) APPROX/RANDOM - 2008. LNCS, vol. 5171, pp. 535–548. Springer, Heidelberg (2008). doi:10.1007/978-3-540-85363-3_42
22. Paysan, P., Knothe, R., Amberg, B., Romdhani, S., Vetter, T.: A 3D face model for pose and illumination invariant face recognition. In: Tubaro, S., Dugelay, J. (eds.) Sixth IEEE International Conference on Advanced Video and Signal Based Surveillance, AVSS 2009, 2–4 September 2009, Genova, Italy, pp. 296–301. IEEE Computer Society (2009).,
23. Solomon, C.J., Gibson, S.J., Mist, J.J.: Interactive evolutionary generation of facial composites for locating suspects in criminal investigations. Appl. Soft Comput. **13**(7), 3298–3306 (2013)
24. Sumi, K., Liu, C., Matsuyama, T.: Study on synthetic face database for performance evaluation. In: Zhang, D., Jain, A.K. (eds.) ICB 2006. LNCS, vol. 3832, pp. 598–604. Springer, Heidelberg (2005). doi:10.1007/11608288_79
25. Turk, M., Pentland, A.: Eigenfaces for recognition. J. Cogn. Neurosci. **3**(1), 71–86 (1991)
26. Tuyls, P., Akkermans, A.H.M., Kevenaar, T.A.M., Schrijen, G.-J., Bazen, A.M., Veldhuis, R.N.J.: Practical biometric authentication with template protection. In: Kanade, T., Jain, A., Ratha, N.K. (eds.) AVBPA 2005. LNCS, vol. 3546, pp. 436–446. Springer, Heidelberg (2005). doi:10.1007/11527923_45
27. Tzimiropoulos, G., Pantic, M.: Optimization problems for fast AAM fitting in-the-wild. In: IEEE International Conference on Computer Vision, ICCV, pp. 593–600 (2013)
28. Valentine, T.: A unified account of the effects of distinctiveness, inversion, and race in face recognition. Q. J. Exp. Psychol. **43**(2), 161–204 (1991)
29. Wu, H., Liu, X., Doretto, G.: Face alignment via boosted ranking model. In: 2008 IEEE Computer Society Conference on Computer Vision and Pattern Recognition (CVPR 2008), 24–26 June 2008, Anchorage, Alaska, USA (2008)
30. Zhang, L., Lin, L., Wu, X., Ding, S., Zhang, L.: End-to-end photo-sketch generation via fully convolutional representation learning. In: Proceedings of the 5th ACM on International Conference on Multimedia Retrieval, pp. 627–634. ACM (2015)

Supervised Detection of Infected Machines Using Anti-virus Induced Labels

(Extended Abstract)

Tomer Cohen[1], Danny Hendler[1(✉)], and Dennis Potashnik[2]

[1] Department of Computer Science, Ben-Gurion University of the Negev,
Beer Sheva, Israel
hendlerd@cs.bgu.ac.il
[2] IBM Cyber Center of Excellence, Beer Sheva, Israel

Abstract. Traditional antivirus software relies on signatures to uniquely identify malicious files. Malware writers, on the other hand, have responded by developing obfuscation techniques with the goal of evading content-based detection. A consequence of this arms race is that numerous new malware instances are generated every day, thus limiting the effectiveness of static detection approaches. For effective and timely malware detection, signature-based mechanisms must be augmented with detection approaches that are harder to evade.

We introduce a novel detector that uses the information gathered by IBM's QRadar SIEM (Security Information and Event Management) system and leverages anti-virus reports for automatically generating a labelled training set for identifying malware. Using this training set, our detector is able to automatically detect complex and dynamic patterns of suspicious machine behavior and issue high-quality security alerts. We believe that our approach can be used for providing a detection scheme that complements signature-based detection and is harder to circumvent.

1 Introduction

With tutorials for writing sophisticated malicious code, as well as malicious source code and tools for malware generation freely available on the Internet in general and the dark web in particular [10,18], developing new malware is becoming easier and sharing malicious code and exploits between different cyber-criminal projects becomes common.

Moreover, polymorphic and metamorphic malware that utilize dead-code insertion, subroutine reordering, encryption, and additional obfuscation techniques, are able to automatically alter a file's content while retaining its functionality [15,23]. As a consequence, the number of new malicious files is growing quickly. Indeed, the number of new malware released to the wild on January 2017 alone is estimated as approx. 13 million and the number of *known* malware is estimated as approx. 600 million [21]!

© Springer International Publishing AG 2017
S. Dolev and S. Lodha (Eds.): CSCML 2017, LNCS 10332, pp. 34–49, 2017.
DOI: 10.1007/978-3-319-60080-2_3

A direct implication of this high rate of new malware is that anti-virus products, which rely heavily on signatures based on file-contents for identifying malware, must be complemented by detection approaches that are harder to evade. One such approach is to attempt to identify the behavioral patterns exhibited by compromised machines instead of relying only on the signatures of the files they download. This approach is more robust to most malware obfuscation techniques, since new malware variants typically have new signatures but exhibit the same behavioral patterns.

Many organizations employ SIEM (Security Information and Event Management) systems, which are software products and services that consolidate information streams generated by multiple hardware and software sources within the organization. SIEM systems facilitate the real-time analysis of gathered information in terms of its security implications. They are able to enforce enterprise security policies and to generate events and alarms when (customizable) static rules are triggered. The rich data available to contemporary SIEM systems holds the potential of allowing to distinguish between the behavior of benign and compromised machines.

This is the approach we take in this work. Our study is based on data collected by IBM® Security QRadar® system [9]. QRadar collects, normalizes and stores the data, received from various networks and security devices, and also enriches it with additional analytics, such as new events generated by its Custom Rule Engine (CRE). We have developed a detector for compromised machines that automatically mines these log files in a big-data environment, in order to extract and compute numerous features for identifying malicious machine behavior. A key challenge is that of generating a *labelled* set of training examples, each representing the activity of a specific machine during a specific time interval, on which supervised machine learning algorithms can be invoked. Analyzing numerous activity logs manually in order to determine whether or not they are benign is expensive and time-consuming. Moreover, since the behavior of infected machines may vary over time, the process of manual labeling should be repeated sufficiently often, which is unrealistic.

Instead, our detector leverages the alerts issued by the anti-virus running on the client machine in order to automatically identify periods of time in which the machine is likely infected and label them as "black" instances. "White" training instances are generated based on the activity of machines for which no AV alerts were issued for an extended period of time. This allows our detector to automatically and periodically generate labelled training sets. Our work thus makes the following contributions:

- Our experiments prove that malicious behavior of infected machines can be accurately detected based on their externally observed behavior.
- We devise a novel detector for infected machines in big data SIEM environments, that uses anti-virus induced labels for supervised classification.
- We present the results of extensive evaluation of our detector, conducted based on more than 6 terabytes of QRadar logs collected in a real production environment of a large enterprise, over the period of 3.5 months between

1.12.15–16.3.2016. Our evaluation shows that our detector identifies security incidents that trigger AV alerts with high accuracy and indicates that it is also able to alert on suspicious behavior that is unobserved by the AV.

2 Related Work

Some previous work used anti-virus (AV) labels for categorizing malicious executables into malware categories such as bots, Trojan horses, viruses, worms, etc. Perdisci et al. [20] proposed an unsupervised system to classify malware based on its network activities. Nari and Ghorbani [17] proposed a scheme of building network activity graphs. Graph nodes represent communication flows generated by the malware and are marked by the type of the flow (DNS, HTTP, SSL, etc.) and edges represent the dependencies between the flows. Bekerman et al. propose a machine-learning based system for detecting malicious executables. The data set consists of malware network traces, that were tagged based on the detections of an IDS system, and traces of benign flows. Whereas the goal of the aforementioned studies is to detect malicious executables or to categorize them to malware families, the goal of our system is to detect infected machines based on their network behavior.

We now describe several previous works that proposed systems for identifying infected machines. Narang et al. [16] propose a method for detecting a P2P botnet in its dormant phase (in standby for receiving a command from the $C\&C$) by collecting network traces from benign P2P application and P2P botnets and extracting features from the traces. Several studies [3,11] attempt to detect the life-cycle phase when a newly infected bot searches for its $C\&C$, by leveraging the fact that this search often generates a large number of DNS query failures. Another approach [5,7] is to leverage the fact that different bots send similar messages to the $C\&C$. Unlike these works, our system does not look for a specific phase in the malware's life-cycle, nor does it limit itself to the detection of specific malware types.

Some previous work examines the behavior of hosts in a time window, extracting features based on the events associated with the host in the time windows and training a machine learning classifier based on these features. Bocchi et al. [4] use tagged data set based on a single day of network traces, captured by a commercial ISP. They use time windows of 30 min when a host is inspected and an instance is generated based on all the events that occurred during the time window. Unlike their detector which operates on offline data, our system is integrated within the QRadar SIEM system. It can therefore use a multitude of *log sources* (described in Sect. 4), reporting about different aspects of the computer's network's behavior. Moreover, out study was conducted based on data collected during a significantly longer period of 3.5 months.

Gu et al. [6] present a system called Botminer, that uses clustering algorithms for detecting infected hosts. A key difference between our system and theirs is that, whereas our detector is designed to detect any type of malware, Botminer is limited to bot detection. Yen et al. [22] propose an unsupervised machine-learning based system to detect anomalous behavior of machines in an enterprise environment. The data is collected using a SIEM system used by EMC.

The authors define a time window of a single day to inspect a host and create an instance based on all the events relating to this host that occurred during this time window.

3   The QRadar Environment

Our detector is based on data collected by an IBM® Security QRadar® SIEM (Security Information and Event Management) system [9], which we shortly describe in this section. SIEM systems are organization-level software products and services that consolidate information streams generated by multiple hardware and software sources and facilitate the real-time analysis of this data in terms of its security implications. The devices that send their reports to the instance of QRadar which we used in this study are the following: Symantec endpoint protection solution, network firewalls, personal firewall, IDS, routers, DNS servers, DHCP servers and authentication servers.

Devices that send streams of information to QRadar are called *log sources*. Some of these devices (e.g. firewalls and AVs) send streams of events. Network devices (e.g. routers and switches), on the other hand, generate streams of *flows* and are therefore called *flow sources*. Events include security-related data, such as firewall denials, ssh unexpected messages, teardown UDP connection, teardown TCP connection, etc. QRadar also enriches event streams by adding new events, generated by its CRE, such as port scanning, excessive firewall denials across multiple internal hosts from a remote host, local windows scanner detected, etc. Flows, on the other hand, report on network traffic. Table 1 lists key flow fields.

Table 1. Flow fields

Field name	Description
Source IP	The IP address of the machine that initiated the flow
Destination IP	The IP address of the destination machine
Source port	The port used by the machine that initiated the flow
Destination port	The port used by the destination machine
First packet time	The time of the first packet
Incoming packets	The number of packets sent by the source
Outgoing packets	The number of packets sent by the destination
Incoming bytes	The total number of bytes sent by the source
Outgoing bytes	The total number of bytes sent by the destination
Direction	Indicates who initiated the flow, a machine that belongs to the enterprise or a machine outside the enterprise
Source IP location	The geographical location of the source IP
Destination IP location	The geographical location of the destination IP
TCP flags	The TCP flags used in the flow session

QRadar's event collector collects the logs from the devices, parses, and normalizes them to unified QRadar events. The normalized stream of data is then passed to an event processor, which enriches the data and generates additional events, according to the custom rules defined in the system. Some events may designate a security threat and trigger an alert, called an *offense*, that is sent to a security analyst. For example, an AV report can trigger such an offense. Another example is a collection of firewall denials generated by a single remote host that may trigger an "excessive firewall denials across multiple hosts from a remote host" offense.

The QRadar instance we worked with can locally save up to one month of events and flows. In order to aggregate more data, normalized data is forwarded to a remote HDFS [1] cluster. The data gathered in the HDFS cluster sums up to more than 2 TB of data each month, from which approximately 1.6 TB are events and the rest are flows. On an average day, more than 2000 unique enterprise user IPs are seen and more than 32 M events and 6.5 M flows relating to these users are collected.

Gathered data is partitioned to hourly files: each hour is represented by two files – one aggregating all flows reported during that hour, and the other aggregating all events reported during that hour. Each event is comprised of fields such as event name, low and high level categories, source and destination IPs, event time, and event payload. The event name designates the event's type. Each event type (identified by the event's event name field) belongs to a single low-level category, which serves to group together several related event types. Similarly, low-level categories are grouped together to more generic high-level categories. Table 2 lists a few events, specifying for each its low- and high-level categories.

We implemented our detector on Spark [2]. Spark is an open source cluster computing framework that is able to interface with Hadoop Distributed File System (HDFS) [1]. Since our feature extraction is done using Spark, our detector is scalable and can cope with big data.

Table 2. QRadar sample events

Event name	Low-level category	High-level category
Virus Detected, Actual action: Left alone	Virus Detected	Malware
Virus Detected, Actual action: Detail pending	Virus Detected	Malware
Teardown UDP connection	Firewall session closed	Access
Firewall allow	Firewall permit	Access
Built TCP connection	Firewall session opened	Access

4 The Detector

Our detector leverages the alerts issued by the anti-virus running on client machines in order to automatically identify time windows in which the machine is likely infected and label them as "black" instances. "White" training instances are generated based on the activity of machines for which no AV alerts were issued for an extended period of time. Each such time window includes events and flows pertaining to the machine under consideration. Before providing a detailed description of our detector and the features it uses, we define more precisely what we mean by the terms flows, black instances, and white instances.

A *flow* is an aggregation of packets that share the same source address, source port, destination address, and destination port. The stopping condition for aggregation differs between udp and tcp. Aggregation of udp packets stops when no new packets arrive for a predefined time. Aggregation of tcp packets stops either when no new packets arrive for a predefined time or when a packet with a set FIN flag has been sent or received.

A *black instance* is a positive example that is provided to the ML algorithm used by our detector. It describes a period of time in which a host is assumed to be infected. The generation of a black instance is triggered upon the occurrence of certain types of anti-virus (AV) events on a host. The black instance encapsulates all the events/flows from/to the host before and after the AV event occurred on it. We have set the length of the time window to 7 h, 6 h before the AV event and 1 h after it.

Events of the following two types trigger the generation of a black instance: "Virus Detected, Actual action: Left alone" and "Virus Detected, Actual action: Details pending". The reason for basing the generation of black instances on these AV events is that their semantics guarantees that a malware is indeed active on the host when the event is generated for it. In both these event types, a malware was detected on the machine but was not deleted by the AV, hence it is likely that the malware was active on the host some time before the AV event and possibly remained active for at least some period of time after the AV event.

The AV's configuration in our system is such that it attempts to delete all detected malware. Nevertheless, in some circumstances, the malware cannot be deleted, either because it is the image of a running process or because the AV lacks permissions for doing so.[1]

A *white instance* is a negative example that is provided to our detector. It describes a period of time in which it is assumed that no malware operates within the host. It encapsulates the events and flows pertaining to a clean host within a time window of 7 h. By clean, we mean that no significant AV event for that machine was reported for 7 days before and after the white instance.

---

[1] Newer AV versions have the capability of stopping the process and then deleting the file.

Features. As mentioned previously, our data consists of network flows (provided in netflow form) and security-related events. We have defined more that 7000 features to be computed based on a time window of data. The features can be divided to the following groups.

1. Event name #: number of events with the same specific name (that is, events of the same specific type) in the time window.
2. Event name distribution: number of events with the same name divided by the number of all events in the time window.
2. Low level category #: number of events with the same low level category in the time window.
4. Low level category distribution: number of events with the same low level category divided by the number of all events in the time window.
5. High level category #: number of events with the same high level category in the time window.
6. High level category distribution: number of events with the same high level category divided by the number of all events in the time window.
7. Flows on port: the number of flows with the specific port number (split into outgoing ports and incoming ports).
8. Flow port distribution: the number of flows with the specific port number divided by the number of all flows in the time window.
9. Flow statistics: average, max, min, variance and deviation of the aggregation fields of the flows, for example: number of packets in/out, number of bytes in/out, etc.

In order to lower dimensionality, we used well-known feature selection algorithm. We applied 6 algorithms and selected the top 40 features output by each, thus reducing the number of features to 135. The algorithms we applied are: Information Gain [12], Information Gain Ratio [8], Chi Squared, Relief [13], Filter [24] and SVM-based feature selection. Each algorithm contributed approx. 20 unique features and the rest of the features were shared between one or more algorithms.

Constructing the Machine Learning Model. The training set is labelled based on AV reports. To construct black instances, we find all significant AV reports (see the definition of a black instance earlier in this section) available from the data collected by QRadar. Then, *raw black instances* are built around each such event (6 h before and 1 h after the event). Raw instances are the aggregation of all events and flows pertaining to the IP of the machine on which the AV event occurred and fall within the time window. Then, we compute the features, normalize the instance (we provide more details on normalization later) and label the instance as black.

Raw white instances are created based on sampled clean periods of machines. Such instances are normalized similarly and are labeled as white. Based on the set of these training instances, a machine learning model is built. The data set we worked with exhibits imbalance: there were far more white instances than black instances. In order to alleviate this problem, we employed under-sampling. After experimenting with a few under-sampling ratios, we have set the ratio between the black and the white instances in our training set to 1:10 (i.e., there are 10 times more white instances than black instances). This is consistent with the findings of Moskovitch et al. [14], who investigated how the ratio between black and white instances in the training set affects model accuracy in the malware detection domain.

The required number of white instances is obtained by random sampling of a large pool of white instances. This pool is created by generating, per every machine IP appearing in the data, 8 instances per day (centered around hours 24am, 3am, 6am, 9am, 12pm, 15pm, 18pm and 21pm) for each day in the training period.

We invoked the detector on a test set of the instances. The construction of the training set and the test set[2] is done on the HDFS cluster using map-reduce code. The computation of feature selection, model construction and classification are currently done on a stand-alone machine, since these ML algorithms were not available on our HDFS cluster.

The input to our detector are QRadar's events and flows. These are accumulated and aggregated within predetermined time windows, per every machine under consideration. Then, features are extracted and their values are normalized, resulting in instances to be classified. Next, an ML algorithm is invoked on these instances, using the most recent model available, resulting in classification of the instances, based on which detection warnings are issued when necessary.

Data Filtering. The QRadar data we received consists of events and flows for all the machines in the enterprise network. These include servers (e.g. DNS servers and mail servers), endpoint computers, and smartphones. The AV reports which we use for labelling, however, are only generated on endpoint computers. Moreover, the network behavior of servers and smartphones is substantially different from that of endpoint computers in terms of flow and event type distributions. In our evaluation we therefore focused on modelling and detecting infected endpoint computers (both desktops and laptops) and filtered out data from servers and smartphones.

Some of the training instances created by the algorithm had little or no activity, which is not helpful for differentiating between malicious and benign activity. Therefore, when constructing the training set, we filtered out all instances whose level of activity (in terms of numbers of flows and events) was below a certain threshold.

Some events always accompany significant AV event of the types "Virus Detected, Actual action: Left alone" or "Virus Detected, Actual action: Details pending". We had to eliminate features based on these events in order to avoid over-fitting that would detect AV events rather than suspicious network behavior.

---

[2] More details on the test set are provided in Sect. 5.

Normalization of Data Values. If the resolution of feature values is too fine, then over-fitting may result due to the fact that the model may classify two instances differently based on very small differences in field values. To mitigate this potential problem, we normalized feature values in the following manner.

- Features that count the number of events/flows from any specific type were normalized by applying the log function and then rounding the result to the closest integer value.
- Distribution-features (whose values represent fractions) were rounded up to the nearest second decimal place.

Setting Time Window Length. A key issue affecting detection accuracy is that of setting the length of the time windows that define training and detection instances. As described in Sect. 4, black instances are built around significant AV events indicating that a malware is running on the machine for which the event is issued. In general, however, we do not know for how long the malware is active before and after the AV event and the extent to which its activity is reflected in the network behavior of the infected host. The following questions have to be addressed in this regard.

1. How long should the time window extend before the AV event?
2. How long should the time window extend after the AV event?

Setting the length of the time window before and after the AV event involves striking a good balance in the following inherent tradeoff: an overly small window may imply losing significant information characterizing malicious activity, whereas an overly large window may imply incorporating data that is irrelevant. Let us call the period within the time window that precedes the AV event the *earlier period* and the period following the AV event the *later period*. We conducted several experiments in order to optimize these periods. Based on the results of these experiments, we chose to define time windows of 7 h–6 h before the AV event and 1 h after it.

5 Evaluation

In this section we report on the results of our evaluation. We created a training set based on QRadar data from 1.12.2015–29.2.2016. We use undersampling with a ratio of 1:10 between black and white instances to overcome the imbalance exhibited by our dataset. We then created a classification model based on this training set and evaluated its accuracy on the rest of the data, spanning the period 1.3.2016–16.3.2016 (henceforth referred to as the *evaluation period*). We conducted a few experiments, described in the following.

Instance-Based Experiment. We created a test set that consists of all the time windows built around significant AV events that were reported on the first half of March 2016 (there were 59 such instances) and all *white instances* during this period. White instances are created in a manner similar to that described in Sect. 4. More precisely, 8 instances are constructed per IP per day starting 24am, every 3 h. Those instances whose time interval does not contain any significant AV event are designated as white. The total number of machines that appear in the test set is 4987 and the total number of instances in the test set is 285,494.

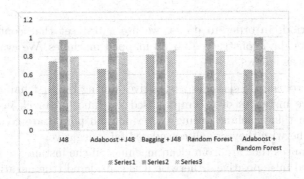

Fig. 1. Detection results for instance-based experiment.

We evaluated the following machine learning algorithm implemented in WEKA [19]: Random Forest, Decision tree (J48), Adaboost+J48, Bagging+J48, Random Forest, and Adaboost+Random Forest. The results are presented in Fig. 1. Best results were obtained by the Bagging+J48 algorithm, which yields a TPR of 0.81 on black instances and a TPR of 0.996 on white instances, with ROC area 0.86.

Compromised Machine Detection Experiment. As mentioned in Sect. 4, events and flows are bound to an endpoint IP. In the enterprise network, however, the IP of a machine changes on a regular basis. Consequently, we cannot track the activity of a machine over an extended period of time solely based on the IPs associated with events/flows.

In order to address this problem, our code constructs a mapping between IP/time pairs and machine MAC addresses. This mapping is constructed using DHCP events and personal firewall events reported to Q-Radar. Unfortunately, DHCP and firewall events were not available for all machines in the course of the first half of March, but we were able to construct the mapping for 10 machines on which a total of 32 black instances occurred (which we henceforth refer to as *compromised machines*) and for additional 2110 machines on which no such events occurred (which we henceforth refer to as *clean machines*).

In the compromised machine detection experiment, we evaluate the extent to which the classifier alerts on compromised machines during the

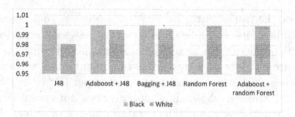

Fig. 2. Detection results for the compromised machine detection experiment.

evaluation period. In order to do so, we use a test set that contains all the instances (8 per day) constructed for all mapped machines. We evaluate classification results as follows.

- If the detector alerts (classifies as positive) during the first half of March on one or more instances of a compromised machine, all black instances that occurred on that machine during the evaluation period are considered true positive, otherwise they are all considered false negative.
- For all instances that are from clean machines, if the instance is alerted on it is considered false positive, otherwise it is considered true negative.

The results of this experiment are shown by Fig. 2. In terms of true positive rate, 3 algorithms (J48, Bagging+J48, Adaboost+J48) alert on all compromised machines (hence all 32 black instances are true positive), whereas the other two algorithms (random forest and Adaboost+random forest) alert on 9 out 10 compromised machines and only miss a single machine, on which a single black instance occurred during the evaluation period (hence 31 black instances are true positives).

All algorithms exhibit excellent detection rate on white instances but the random forest algorithms (with and without Adaboost) are the clear winners, both exceeding true positive rate of 0.9997 (!). In absolute numbers, Adaboost+random forest has only 29 false positives and random forest only has 31 (out of a total of approx. 135,000 white instances), all of which occurred on a set of 14 clean machines.

5.1 Early Detection Experiment

In this section, we evaluate the extent to which our detector is able to alert on suspicious activity before the AV does. In order to be able to track machine activity across day boundaries, we conduct also this evaluation for the set of machines for which a mapping exists throughout the evaluation period. In this experiment, we consider a black instance as a true positive, if our detector alerted on the machine on which it occurred within ±3 days of the event.

Similarly to the compromised machine detection experiment, we use a test set that contains all the instances (8 per day) constructed for all mapped machines. First, we compute true positive and false positive rates as follows.

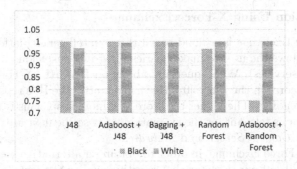

Fig. 3. True positive/negative results for a ±3 days alert period.

- For every black instance, if the detector alerts (classifies an instance as positive) on that machine within ±3 days of the black instance, it is considered true positive, otherwise it is considered false negative.
- For all instances that are at least 3 days afar from all black instances (if any) on their machine, if the instance is alerted on it is considered false positive, otherwise it is considered true negative.

The results are shown in Fig. 3. Comparing with Fig. 2, true positive results are identical for all algorithms except for Adaboost+random forest, for which the number of true positives is now down to 24 (as compared with 31 in the compromised machine detection experiment). True negative results are also similar to those presented in Fig. 2. The random forest algorithms are the winners and both exceed true negative rate of 0.9997. In absolute numbers, Adaboost+random forest has 31 false positives and random forest has 38 (out of a total of approx. 135,000 white instances), all of which occurred on a set of 14 clean machines.

As we described above, a black instance is considered a true positive if the detector alerts on that machine within ±3 days. We say that the detector provides *early detection* for a black instance, if it alerts on the same machine on some instance that precedes the black instance by at most 3 days. If the algorithm provides early detection for an algorithm, then the *early detection period* is the length of time between the black instance and the earliest alert on that machine that precedes it by at most 3 days.

Clearly, an instance for which the detector provides early detection is a true positive. Based on the results presented by Figs. 2 and 3, we conclude that the Random Forest algorithm strikes the best balance between true positive and true negative rates. In the early detection experiment, we evaluated its capability of providing early detection.

Overall, out of the 31 true positives, the Random Forest algorithm provides early detection for 25. The average early detection period is approx. 22 h. Figure 4 presents a histogram of the early detection period achieved by the algorithm.

5.2   Evaluation Using X-Force Exchange

IBM's X-Force Exchange is a cloud-based threat intelligence platform. It offers an API that allows the user to make queries regarding suspicious IPs and URLs (among other services). When querying about an IP/URL, X-Force Exchange returns a few scores in the range $[0 - 10]$ that quantify the IP's reputation w.r.t. several risk categories. The higher the score, the more suspicious is the IP/URL. The risk categories relevant for the evaluation we describe next are: *Bots*, *Botnet Command and Control Server*, and *Malware*.

We used X-Force Exchange in order to obtain an alternative method of evaluating the quality of our detector that is orthogonal to the features it uses. We did this by querying X-Force Exchange on IPs and URLs that are communicated with during instances in our test set. Our hypothesis was that black instances, as well as instances alerted on by our detector, would possess statistically-significant X-Force Exchange lower reputation levels (and therefore higher risk scores).

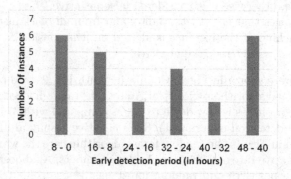

Fig. 4. Early detection histogram for the Random Forest algorithm.

Table 3 presents the average scores obtained for suspected (alerted), black, and white instances in the 3 relevant categories. As expected, the average score of black instances is significantly higher than that of white instances for all categories and the gap is especially large for the Botnet C&C category. Suspected instances on which our detector alerts obtain scores that are also significantly higher than those of white instances and, for the Bots and Malware categories, even higher than those of black instances.

Table 3. X-Force average scores

Category	Suspected	Black	White
Bots	4.211538	3.246154	1.969231
Malware	5.812	5.7	4.24
Botnet command and control	3.015385	4.6	1.323077

Table 4. X-Force p-values

Category	Black vs. Suspectet	Black vs. White	Suspectet vs. White
Bots	0.223193	0.046219	0.014266
Malware	0.161492	8.28E−08	1.42E−08
Botnet command and control	0.031646	1.04E−06	0.024139

We have also checked the statistical significance of the differences in score by computing their p-values and show the results in Table 4. As anticipated, the differences between the scores of black and white instances, as well as those of suspected and white instances, are statistically significant. On the other hand, the differences between black and suspected instances in the Bots and malware categories are insignificant, but they are significant for the botnet C&C category, where the grade of black instances is higher. Collectively, these results provide a clear indication that at least part of the presumed false positives of our detector do indicate suspicious host behavior that is not alerted by the AV.

6   Discussion

In this work, we presented a novel detector for infected machines in big data SIEM environments, that uses anti-virus induced labels for supervised classification. Our detection uses features that were selected out of more that 7000 features, computed based on time windows of QRadar data, containing reports on machines' flows and events.

We also present the results of extensive evaluation of our detector, conducted based on more than 6 terabytes of QRadar logs collected in a real production environment of a large enterprise, over the period of 3.5 months. Our evaluation shows that our detector identifies security incidents that trigger AV alerts with high accuracy. Moreover, it is able to provide early detection for a majority of these events. Our evaluation also indicates that our detector is able to alert on suspicious behavior that is unobserved by the AV.

One direction for future work is to find additional features that can further improve the accuracy of our detection approach. One possible way of doing this is to use the reputation of IPs/URLs that are communicated with during a time-window. These were used by us for evaluation, but not for constructing a detection model.

Another interesting direction is to conduct an experiment for observing the rate of concept drift exhibited by the models learnt using our approach and leverage its results for optimizing the duration and frequency in which models are learnt. Finally, it would also be interesting to check the extent by which models learnt on one QRadar system are applicable to other QRadar systems.

Acknowledgments. This research was supported by IBM's Cyber Center of Excellence in Beer Sheva and by the Cyber Security Research Center and the Lynne and William Frankel Center for Computing Science at Ben-Gurion University. We thank Yaron Wolfshtal from IBM for allowing Tomer to use IBM's facilities, for providing us the data on which this research is based, and for many helpful discussions.

References

1. Hadoop distributed file system. http://hadoop.apache.org/
2. Spark cluster computing. http://spark.apache.org/
3. Antonakakis, M., Perdisci, R., Nadji, Y., Vasiloglou II, N., Abu-Nimeh, S., Lee, W., Dagon, D.: From throw-away traffic to bots: Detecting the rise of DGA-based malware. In: USENIX Security Symposium, vol.12 (2012)
4. Bocchi, E., Grimaudo, L., Mellia, M., Baralis, E., Saha, S., Miskovic, S., Modelo-Howard, G., Lee, S.-J.: Magma network behavior classifier for malware traffic. Comput. Netw. **109**, 142–156 (2016)
5. Dietrich, C.J., Rossow, C., Pohlmann, N.: CoCoSpot: clustering and recognizing botnet command and control channels using traffic analysis. Comput. Netw. **57**(2), 475–486 (2013)
6. Gu, G., Perdisci, R., Zhang, J., Lee, W., et al.: BotMiner: clustering analysis of network traffic for protocol-and structure-independent botnet detection. In: USENIX Security Symposium, vol. 5, pp. 139–154 (2008)
7. Gu, G., Zhang, J., Lee, W.: BotSniffer: detecting botnet command and control channels in network traffic (2008)
8. Hall, M.A., Smith, L.A.: Practical feature subset selection for machine learning (1998)
9. IBM: IBM Security QRadar SIEM. http://www-03.ibm.com/software/products/en/qradar-siem/
10. iicybersecurity: International institute of cyber security. https://iicybersecurity.wordpress.com
11. Jiang, N., Cao, J., Jin, Y., Li, L.E., Zhang, Z.-L.: Identifying suspicious activities through DNS failure graph analysis. In: 2010 18th IEEE International Conference on Network Protocols (ICNP), pp. 144–153. IEEE (2010)
12. Kent, J.T.: Information gain and a general measure of correlation. Biometrika **70**(1), 163–173 (1983)
13. Kira, K., Rendell, L.A.: The feature selection problem: traditional methods and a new algorithm. In: AAAI, vol. 2, pp. 129–134 (1992)
14. Moskovitch, R., Stopel, D., Feher, C., Nissim, N., Elovici, Y.: Unknown malcode detection via text categorization and the imbalance problem. In: IEEE International Conference on Intelligence and Security Informatics, ISI 2008, pp. 156–161. IEEE (2008)
15. Musale, M., Austin, T.H., Stamp, M.: Hunting for metamorphic JavaScript malware. J. Comput. Virol. Hacking Tech. **11**(2), 89–102 (2015)
16. Narang, P., Ray, S., Hota, C., Venkatakrishnan, V.: PeerShark: detecting peer-to-peer botnets by tracking conversations. In: 2014 IEEE Security and Privacy Workshops (SPW), pp. 108–115. IEEE (2014)
17. Nari, S., Ghorbani, A.A.: Automated malware classification based on network behavior. In: 2013 International Conference on Computing, Networking and Communications (ICNC), pp. 642–647. IEEE (2013)

18. Deep Web News. https://darkwebnews.com
19. Weka 3: Data mining software in Java. University of Waikato. http://www.cs.
    waikato.ac.nz/ml/weka/
20. Perdisci, R., Lee, W., Feamster, N.: Behavioral clustering of http-based malware
    and signature generation using malicious network traces. In: NSDI, vol. 10, p. 14
    (2010)
21. AV TEST: The independent it-security institute. https://www.av-test.org/en/
    statistics/malware/
22. Yen, T.-F., Oprea, A., Onarlioglu, K., Leetham, T., Robertson, W., Juels, A.,
    Kirda, E.: Beehive: Large-scale log analysis for detecting suspicious activity in
    enterprise networks. In: Proceedings of the 29th Annual Computer Security Appli-
    cations Conference, pp. 199–208. ACM (2013)
23. You, I., Yim, K.: Malware obfuscation techniques: a brief survey. In: 2010 Interna-
    tional Conference on Broadband, Wireless Computing, Communication and Appli-
    cations (BWCCA), pp. 297–300. IEEE (2010)
24. Yu, L., Liu, H.: Feature selection for high-dimensional data: a fast correlation-based
    filter solution. In: ICML, vol. 3, pp. 856–863 (2003)

Building Regular Registers with Rational Malicious Servers and Anonymous Clients

Antonella Del Pozzo[1]([✉]), Silvia Bonomi[1], Riccardo Lazzeretti[1],
and Roberto Baldoni[1,2]

[1] Department of Computer and System Sciences "Antonio Ruberti",
Research Center of Cyber Intelligence and Information Security (CIS),
Sapienza Università di Roma, Rome, Italy
{delpozzo,bonomi,lazzeretti,baldoni}@dis.uniroma1.it
[2] CINI Cybersecurity National Laboratory, Rome, Italy

Abstract. The paper addresses the problem of emulating a regular register in a synchronous distributed system where clients invoking read() and write() operations are anonymous while server processes maintaining the state of the register may be compromised by rational adversaries (i.e., a server might behave as *rational malicious Byzantine* process). We first model our problem as a Bayesian game between a client and a rational malicious server where the equilibrium depends on the decisions of the malicious server (behave correctly and not be detected by clients vs returning a wrong register value to clients with the risk of being detected and then excluded by the computation). We prove such equilibrium exists and finally we design a protocol implementing the regular register that forces the rational malicious server to behave correctly.

Keywords: Regular register · Rational malicious processes · Anonymity · Bayesian game

1 Introduction

To ensure high service availability, storage services are usually realized by replicating data at multiple locations and maintaining such data consistent. Thus, replicated servers represent today an attractive target for attackers that may try to compromise replicas correctness for different purposes, such as gaining access to protected data, interfering with the service provisioning (e.g. by delaying operations or by compromising the integrity of the service), reducing service availability with the final aim to damage the service provider (reducing its reputation or letting it pay for the violation of service level agreements), etc. A compromised replica is usually modeled trough an arbitrary failure (i.e. a Byzantine failure) that is made transparent to clients by employing Byzantine Fault Tolerance (BFT) techniques. Common approaches to BFT are based on the deployment of a sufficiently large number of replicas to tolerate an estimated number f of compromised servers (i.e. BFT replication). However, this approach has a strong

© Springer International Publishing AG 2017
S. Dolev and S. Lodha (Eds.): CSCML 2017, LNCS 10332, pp. 50–67, 2017.
DOI: 10.1007/978-3-319-60080-2_4

limitation: a smart adversary may be able to compromise more than f replicas in long executions and may get access to the entire system when the attack is sufficiently long. To overcome this issue, Sousa et al. designed the *proactive-reactive recovery* mechanism [22]. The basic idea is to periodically reconfigure the set of replicas to rejuvenate servers that may be under attack (proactive mode) and/or when a failure is detected (reactive mode).

This approach is effective in long executions but requires a fine tuning of the replication parameters (upper bound f on the number of possible compromised replicas in a given period, rejuvenation window, time required by the state transfer, etc.) and the presence of secure components in the system. In addition, it is extremely costly during good periods (i.e. periods of normal execution) as a high number of replicas must be deployed independently from their real need. In other words, the system pays the cost of an attack even if the attack never takes place.

In this paper, *we want to investigate the possibility to implement a distributed shared variable (i.e. a register) without making any assumption on the knowledge of the number of possible compromised replicas*, i.e. without relating the total number of replicas n to the number of possible compromised ones f. To overcome the impossibility result of [5,19], we assume that (i) clients preserve their privacy and do not disclose their identifiers while interacting with server replicas (i.e. anonymous clients) and (ii) at least one server is always alive and never compromised by the attacker. We first model our protocol as a game between two parties, a client and a rational malicious server (i.e. a server controlled by rational adversaries) where each rational malicious server gets benefit by two conflicting goals: (i) it wants to have continuous access to the current value of the register and, (ii) it wants to compromise the validity of the register returning a fake value to a client. However, if the rational malicious server tries to accomplish goal (ii) it could be detected by a client and it could be excluded from the computation, precluding it to achieve its first goal. We prove that, under some constraints, an equilibrium exists for such game. In addition, we design some distributed protocols implementing the register and reaching such equilibrium when rational malicious servers privilege goal (i) with respect to goal (ii). As a consequence, rational malicious servers return correct values to clients to avoid to be detected by clients and excluded by the computation and the register implementation is proved to be correct.

The rest of the paper is organized as follows: Sect. 2 discusses related works, Sects. 3 and 4 introduce respectively the system model and the problem statement. In Sect. 5 we model the problem as a Bayesian game and in Sect. 6 we provide a protocol matching the Bayesian Nash Equilibrium that works under some limited constraints, while in Sect. 7 we presents two variants of the protocol that relax the constraints, at the expense of some additional communications between the clients or protocol complexity increase. Finally, Sect. 8 presents a discussion and future work.

2   Related Work

Building a distributed storage able to resist arbitrary failures (i.e. Byzantine) is a widely investigated research topic. The Byzantine failure model captures the most general type of failure as no assumption is made on the behavior of faulty processes. Traditional solutions to build a Byzantine tolerant storage service can be divided into two categories: *replicated state machines* [20] and *Byzantine quorum systems* [5,17–19]. Both the approaches are based on the idea that the state of the storage is replicated among processes and the main difference is in the number of replicas involved simultaneously in the state maintenance protocol. Replicated state machines approach requires that every non-faulty replica receives every request and processes requests in the same order before returning to the client [20] (i.e. it assumes that processes are able to totally order requests and execute them according to such order). Given the upper bound on the number of failures f, the replicated state machine approach requires only $2f + 1$ replicas in order to provide a correct register implementation. Otherwise, Byzantine quorum systems need just a sub-set of the replicas (i.e. *quorum*) to be involved simultaneously. The basic idea is that each operation is executed by a quorum and any two quorums must intersect (i.e. members of the quorum intersection act as witnesses for the correct execution of both the operations). Given the number of failures f, the quorum-based approach requires at least $3f + 1$ replicas in order to provide a correct register implementation in a fully asynchronous system [19]. Let us note that, in both the approaches, the knowledge of the upper bound on faulty servers f is required to provide deterministic correctness guarantees. In this paper, we follow an orthogonal approach. We are going to consider a particular case of byzantine failures and we study the cost, in terms of number of honest servers, of building a distributed storage (i.e. a register) when clients are anonymous and have no information about the number of faulty servers (i.e. they do not know the bound f). In particular, the byzantine processes here considered deviate from the protocol by following a strategy that brings them to optimize their own benefits (i.e., they are *rational*) and such strategy has the final aim to compromise the correctness of the storage (i.e., they are *malicious*). In [16], the authors presented Depot, a cloud storage system able to tolerate any number of Byzantine clients or servers, at the cost of a weak consistency semantics called *Fork-Join-Causal consistency* (i.e., a weak form of causal consistency).

Another different solution can rely on Proactive Secret Sharing [26]. Secret Sharing [27] guarantees that a secret shared by a client among n parties (servers) cannot be obtained by an adversary corrupting no more than f servers. Moreover, if no more than f servers are Byzantines, the client can correctly recover the secret from the shares provided by any $f + 1$ servers. Recent Proactive Secret Sharing protocols, e.g. [28], show that Secret Sharing can be applied also to synchronous networks. Even if Proactive Secret Sharing can guarantee the privacy of the data (this is out of the scope of the paper) against up to $f = n - 2$ passive adversaries, the solution has some limitations. First of all, clients are not able to verify whether the number of Byzantines exceeds f and hence understand if the message obtained is correct. Secondly, Secret Sharing protocols operating in a synchronous distributed system with Byzantines (active adversaries) correctly work with a small number of Byzantines and have high complexity ($f < n/2 - 1$ and $\mathcal{O}(n^4)$ in [28]).

In [3], the authors introduced the *BAR (Byzantine, Altruistic, Rational) model* to represent distributed systems with heterogeneous entities like peer-to-peer networks. This model allows to distinguish between Byzantine processes (arbitrarily deviating from the protocol, without any known strategy), altruistic processes (honestly following the protocol) and rational processes (may decide to follow or not the protocol, according to their individual utility). Under the BAR model, several problems have been investigated (e.g. reliable broadcast [7], data stream gossip [14], backup service through state machine replication [3]). Let us note that in the BAR model the utility of a process is measured through the cost sustained to run the protocol. In particular, each step of the algorithm (especially sending messages) has a cost and the objective of any rational process is to minimize its global cost. As a consequence, rational *selfish* processes deviate from the protocol just by skipping to send messages, if not properly encouraged by some reward. In contrast with the BAR model, in this paper we consider malicious rational servers that can deviate from the protocol with different objectives, benefiting from preventing the correct protocol execution rather than from saving messages.

More recently, classical one-shot problems as leader election [1,2], renaming and consensus [2] have been studied under the assumption of rational agents (or rational processes). The authors provide algorithms implementing such basic building blocks, both for synchronous and asynchronous networks, under the so called *solution preference* assumption i.e., agents gain if the algorithm succeeds in its execution while they have zero profit if the algorithm fails. As a consequence, processes will not deviate from the algorithm if such deviation interferes with its correctness. Conversely, the model of rational malicious processes considered in this paper removes implicitly this assumption as they are governed by adversaries that get benefit when the algorithm fails while in [1,2] rational processes get benefit from the correct termination of the protocol (i.e. they are selfish according with the BAR model).

Finally, the model considered here can be seen as a particular case of BAR where rational servers take malicious actions, with the application similar to the one considered in [3]. However, in contrast to [3], we do not assume any trusted third party to identify users, we assume that clients are anonymous (e.g., they are connected through the Tor anonymous network [23]), and we investigate the impact of this assumption together with the rational model. To the best of our knowledge, this is the first paper that analyzes how the anonymity can help in managing rational malicious behaviors.

3 System Model

The distributed system is composed by a set of n servers implementing a distributed shared memory abstraction and by an arbitrary large but finite set of clients $\mathcal{C}$. Servers are fully identified (i.e. they have associated a unique identifier $s_1, s_2 \ldots s_n$) while clients are anonymous, i.e. they share the same identifier.

Communication model and timing assumptions. Processes can communicate only by exchanging messages through *reliable* communication primitives, i.e. messages are not created, duplicated or dropped. The system is synchronous in the following sense: all the communication primitives used to exchange messages guarantee a timely delivery property. In particular, we assume that clients communicate with servers trough a *timely* reliable broadcast primitive (i.e., there exists an integer δ, known by clients, such that if a client broadcasts a message m at time t and a server s_i delivers m, then all the servers s_j deliver m by time $t + \delta$). Servers-client and client-client communications are done through "point-to-point" *anonymous timely* channels (a particular case of the communication model presented in [10] for the most general case of homonyms). Considering that clients are identified by the same identifier ℓ, when a process sends a point-to-point message m to an identifier ℓ, all the clients will deliver m. More formally, there exists an integer $\delta' \leq \delta$, known by processes, such that if s_i sends a message m to a client identified by an identifier ℓ at time t, then all the clients identified by ℓ receive m by time $t + \delta'$ (for simplicity in the paper we assume $\delta = \delta'$).

We assume that channels are *authenticated* ("oral" model), i.e. when a process identified by j receives a message m from a process identified by i, then p_j knows that m has been generated by a process having identifier i.

Failure model. Servers are partitioned into two disjoint sub-sets: *honest* servers and *malicious* servers (*attackers*). Honest servers behave according to the protocol executed in the distributed system (discussed in Sect. 6) while malicious servers represent entities compromised by an adversary that may deviate from the protocol by dropping messages (omission failures), changing the content of a message, creating spurious messages, exchanging information outside the protocol, etc. Malicious servers are *rational*, i.e. they deviate from the protocol by following a strategy that aims at increasing their own benefit (usually performing actions that may prevent the correct execution of the protocol). We assume that rational malicious servers act independently, i.e. they do not form a coalition and each of them acts for its individual gain.

Servers may also fail by crashing and we identify as *alive* the set of non crashed servers[1]. However, we assume that at least one honest alive server always exists in the distributed system.

4   Regular Registers

A register is a shared variable accessed by a set of processes, i.e. clients, through two operations, namely read() and write(). Informally, the write() operation updates the value stored in the shared variable while the read() obtains the value contained in the variable (i.e. the last written value). Every operation issued on

---

[1] Alive servers may be both honest or malicious.

a register is, generally, not instantaneous and it can be characterized by two events occurring at its boundary: an *invocation* event and a *reply* event. These events occur at two time instants (invocation time and reply time) according to the fictional global time.

An operation *op* is *complete* if both the invocation event and the reply event occur (i.e. the process executing the operation does not crash between the invocation and the reply). Contrary, an operation *op* is said to be *failed* if it is invoked by a process that crashes before the reply event occurs. According to these time instants, it is possible to state when two operations are concurrent with respect to the real time execution. For ease of presentation we assume the existence of a fictional global clock and the invocation time and response time of operations are defined with respect to this fictional clock.

Given two operations *op* and *op'*, and their invocation event and reply event times ($t_B(op)$ and $t_B(op')$) and return times ($t_E(op)$ and $t_E(op')$), we say that *op precedes op'* ($op \prec op'$) iff $t_E(op) < t_B(op')$. If *op* does not precede *op'* and *op'* does not precede *op*, then *op* and *op'* are *concurrent* ($op||op'$). Given a write(v) operation, the value v is said to be written when the operation is complete.

In case of concurrency while accessing the shared variable, the meaning of *last written value* becomes ambiguous. Depending on the semantics of the operations, three types of register have been defined by Lamport [15]: *safe*, *regular* and *atomic*. In this paper, we consider a regular register which is specified as follows:

- **Termination**: If an alive client invokes an operation, it eventually returns from that operation.
- **Validity**: A read operation returns the last value written before its invocation, or a value written by a write operation concurrent with it.

Interestingly, safe, regular and atomic registers have the same computational power. This means that it is possible to implement a multi-writer/multi-reader atomic register from single-writer/single-reader safe registers. There are several papers in the literature discussing such transformations (e.g., [6,12,21,24,25] to cite a few). In this paper, we assume that the register is single writer in the sense that no two write() operations may be executed concurrently. However, any client in the system may issue a write() operation. This is not a limiting assumption as clients may use an access token to serialize their writes[2]. We will discuss in Sect. 8 how this assumption can be relaxed.

5 Modeling the Register Protocol as a Game

In a distributed system where clients are completely disjoint from servers, it is possible to abstract any register protocol as a sequence of requests made by clients (e.g. a request to get the value or a request to update the value) and responses (or replies) provided by servers, plus some local computation.

---

[2] Let us recall that we are in a synchronous system and the mutual exclusion problem can be easily solved also in presence of failures.

If all servers are honest, clients will always receive the expected replies and all replies will always provide the right information needed by the client to correctly terminate the protocol. Otherwise, a compromised server can, according to its strategy, omit to send a reply or can provide bad information to prevent the client from terminating correctly. In this case, in order to guarantee a correct execution, the client tries to detect such misbehavior, react and punish the server. Thus, a distributed protocol implementing a register in presence of rational malicious servers can be modeled as a two-party game between a client and each of the servers maintaining a copy of the register: the client wants to correctly access the register while the server wants to prevent the correct execution of a read() without being punished.

Players. The two players are respectively the client and the server. Each player can play with a different role: servers can be divided into *honest* servers and *malicious* servers while clients can be divided in those asking a *risky request* (i.e., clients able to detect misbehaviors and punish server[3]) and those asking for a *risk-less request* (i.e., clients unable to punish servers).

Strategies. Players' strategies are represented by all the possible actions that a process may take. Clients have just one strategy, identified by $\mathcal{R}$, that is *request information to servers*. Contrarily, servers have different strategies depending on their failure state:

- malicious servers have three possible strategies: (i) $\mathcal{A}$, i.e. *attack the client* by sending back wrong information (it can reply with a wrong value, with a wrong timestamp or both), (ii) $\mathcal{NA}$, i.e. *not attack the client* behaving according to the protocol and (iii) $\mathcal{S}$, i.e. *be silent* omitting the answer to client's requests;
- honest servers have just the $\mathcal{NA}$ strategy.

Let us note that the game between a honest client and a honest server is trivial as they have just one strategy that is to follow the protocol. Thus, in the following we are going to skip this case and we will consider only the game between a client and a rational malicious server.

Utility functions and extensive form of the game. Clients and servers have opposite utility functions. In particular:

- every client increases its utility when it is able to read a correct value from the register and it wants to maximize the number of successful read() operations;
- every server increases its utility when it succeeds to prevent the client from reading a correct value, while it loses when it is detected by the client and it is punished.

In the following, we will denote as G_c the gain obtained by the client when it succeeds in reading, G_s the gain obtained by the server when it succeeds in

---

[3] Notice that the client ability to detect a server misbehaviors depends on the specific protocol.

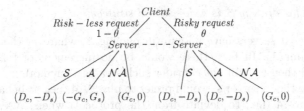

Fig. 1. Extensive form of the game. Dashed line represents the unknown nature of requests from the risk point of view. Outcome pairs refer to client and server gains respectively.

preventing the client from reading and as D_c the gain of the client when detecting the server and as D_s the loss of the server when it is detected. Such parameters are characteristic of every server and describe its behavior in terms of subjective gains/losses they are able to tolerate. Without loss of generality, we assume that G_c, G_s, D_c and D_s are all greater than 0, that all the servers have the same G_s and D_s[4] and that all the clients have the same G_c and D_c. Figure 1 shows the extensive form of the game.

The game we are considering is a Bayesian game [11] as servers do not have knowledge about the client role but they can estimate the probability of receiving a risky request or a risk-less request i.e., they have a *belief* about the client role.

We denote as θ (with $\theta \in [0, 1]$) the server belief of receiving a risky request (i.e. the client may detect that the server is misbehaving) and with $1 - \theta$ the server belief of receiving a risk-less request (i.e. the client is not be able to detect that the server is misbehaving).

Analysis of the Bayesian Game. In the following, we are going to analyze the existence (if any) of a *Bayesian Nash Equilibrium* i.e., a Nash Equilibrium[5] computed by considering the players' belief.

Let us note that in our game, clients have just one strategy. Thus, the existence of the equilibrium depends only on the decisions taken by servers according to their utility parameters G_s, D_s and their belief about the nature of a request (i.e., its evaluation of θ).

Let us now compute the expected gain $E()$ of a server s_i while selecting strategies $\mathcal{S}$, $\mathcal{NA}$ and $\mathcal{A}$:

$$E(\mathcal{S}) = (-D_s \times (1 - \theta)) + (-D_s \times \theta) = -D_s \tag{1}$$
$$E(\mathcal{NA}) = ((1 - \theta) \times 0) + (\theta \times 0) = 0 \tag{2}$$
$$E(\mathcal{A}) = ((1 - \theta) \times G_s) - (\theta \times D_s) \tag{3}$$

---

[4] Let us note that if two servers have different values for G_s and D_s, the analysis shown in the following is simply repeated for each server.

[5] Let us recall that a Nash Equilibrium exists when each player selects a strategy and none of the players increases its utility by changing strategy.

Lemma 1. *The strategy $\mathcal{S}$ is a dominated strategy.*

It follows that servers have no gain in playing $\mathcal{S}$, whatever the other player does (cf. Lemma 1). In fact, there would be no increment of their utility by playing $\mathcal{S}$ and then we will not consider such strategy anymore.

Let us note that a server s_i would prefer to play $\mathcal{NA}$ (i.e., to behave honestly) with respect to $\mathcal{A}$ (i.e., to deviate from the protocol) when $E(\mathcal{NA}) > E(\mathcal{A})$. Combining Eqs. (2) and (3) we have that a s_i would prefer to play $\mathcal{NA}$ when

$$\frac{G_s}{(G_s + D_s)} > \theta. \tag{4}$$

The parameters G_s and D_s are strictly dependent on the attackers profile (i.e., an attacker for which is more important to stay in the system rather than subvert it or vice versa), thus we can not directly work on them. In the remaining part of the work we propose protocols to tune the θ parameter in such a way that the inequality (4) holds. To this purpose, we derive the following Lemmas:

Lemma 2. *Let s_i be a rational malicious server. If $D_s < G_s$ and $\theta < \frac{1}{2}$ then the best response of s_i is to play strategy $\mathcal{A}$ (i.e. $\mathcal{NA}$ is a dominated strategy).*

Lemma 3. *Let s_i be a rational malicious server. If $D_s > G_s$ and $\theta \geq \frac{1}{2}$ then the best response of s_i is to never play strategy $\mathcal{A}$ (i.e. $\mathcal{NA}$ is a dominant strategy).*

Due to the lack of space, proofs of the previous Lemmas can be found in [9].

6 A Protocol $\mathcal{P}$ for a Regular Register when $D_s \gg G_s$

In this section, we propose a protocol $\mathcal{P}$ implementing a regular register in a synchronous distributed system with anonymous clients and up to $n-1$ malicious rational servers. The protocol works under the assumption that the server loss D_s in case of detection is much higher than its gain G_s obtained when the client fails during a read (i.e. $D_s \gg G_s$[6]). This assumption models a situation where the attacker is much more interested in having access to data stored in the register and occasionally interfere with the server rather than causing a reduction of the availability (e.g., no termination or validity violation). We will relax this assumption to the simple case $D_s > G_s$ in the next section extending $\mathcal{P}$ in two different ways.

Our protocol $\mathcal{P}$ follows the classical quorum-based approach. When a client wants to write, it sends the new value together with its timestamp to servers and waits for acknowledgments. Similarly, when it wants to read, it asks for values and corresponding timestamps and then it tries to select a value among the received ones. Let us note that, due to the absence of knowledge on the upper bound of malicious processes, it could be impossible for a reader to select

---

[6] More precisely, $\mathcal{P}$ works when $D_s > cG_s$ where c is the estimated number of clients in the system.

a value among those reported by servers and, in addition, the reader may be unable to distinguish well behaving servers from malicious ones. To overcome this issue we leverage on the following observation: the last client c_w writing a value v is able to recognize such value while reading after its write (as long as no other updates have been performed). This makes the writer c_w the only one able to understand which server s_i is reporting a wrong value $v_i \neq v$, detect it as malicious and punish it by excluding s_i from the computation. Thus, the basic idea behind the protocol is to exploit the synchrony of the system and the anonymity of clients to makes the writer indistinguishable from readers and "force" malicious servers to behave correctly.

Let us note that anonymity itself is not enough to make the writer indistinguishable from other clients. In fact, if we consider a naive solution where we add anonymity to a register implementation (e.g., to the one given by Attiya, Bar-Noy and Dolev [4]), we have that servers may exploit the synchrony of the channels to estimate when the end of the write operation occurs and to infer whether a read request may arrive from the writer or from a different client (e.g., when it is received too close to a write request and before the expected end of the write). To this aim, we added in the write() operation implementation some *dummy* read requests. These messages are actually needed to generate message patterns that make impossible to servers to distinguish messages coming from the writer from messages arriving from a different client. As a consequence, received a read request, a server s_i is not able to distinguish if such request is risky (i.e. it comes from the writer) or is risk-less (i.e. it comes from a generic client).

In addition, we added a detection procedure that is executed both during read() and write() operations by any client. In particular, such procedure checks that every server answered to a request and that the reported information are "coherent" with its knowledge (e.g., timestamps are not too old or too new). The detection is done first locally, by exploiting the information that clients collect during the protocol execution, and then, when a client detects a server s_j, it disseminates its detection so that the malicious server is permanently removed from the computation (collaborative detection).

Finally, the timestamp used to label a new written value is updated by leveraging acknowledgments sent by servers at the end of the preceding write() operation. In particular, during each write() operation, servers must acknowledge the write of the value by sending back the corresponding timestamp. This is done on the anonymous channels that deliver such message to all the clients that will update their local timestamp accordingly. As a consequence, any rational server is inhibited from deviating from the protocol, unless it accepts the high risk to be detected as faulty and removed from the system.

In the following, we provide a detailed description of the protocol $\mathcal{P}$ shown in Figs. 2, 3 and 4.

The read() operation (Fig. 2). When a client wants to read, it first checks if the *last_ts* variable is still equal to 0. If so, then there is no write() operation terminated before the invocation of the read() and the client returns the

```
Init:
(01)  replies ← ∅; my_last_val ← ⊥; my_last_ts ← 0; last_ts ← 0;
(02)  ack ← ∅; honest ← {s₁, s₂ ... sₙ}; writing ← false;

operation read():
(03)  if (last_ts = 0)
(04)    then return ⊥;
(05)    else replies ← ∅;
(06)      broadcast READ();
(07)      wait (2δ);
(08)      if (∀ sᵢ ∈ honest, ∃ < −, ts, val > ∈ replies)
(09)        then broadcast READACK();
(10)          return val;
(11)        else wait (δ);
(12)          if (∀ sᵢ ∈ honest, ∃ < −, ts, val > ∈ replies)
(13)            then broadcast READACK();
(14)              return val;
(15)            else execute detection(repliesᵢ, R)
(16)              broadcast READACK();
(17)              if (∀ sᵢ ∈ honest, ∃ < −, ts, val > ∈ replies)
(18)                then return val;
(19)                else abort ;
(20)              endif
(21)          endif
(22)      endif
(23)  endif

when REPLY(< j, ts, v, ots, ov >) is delivered:
(24)  replies ← replies ∪ {< j, ts, v >};
(25)  replies ← replies ∪ {< j, ots, ov >};

when DETECTED(sⱼ) is delivered:
(26) honest ← honest \ {sⱼ};
```

(a) Client Protocol

```
Init:
(01)  valᵢ ← ∅; tsᵢ ← 0;
(02)  old_valᵢ ← ⊥; old_tsᵢ ← 0; readingᵢ ← 0;

when READ() is delivered:
(03)  readingᵢ ← readingᵢ + 1;
(04)  send REPLY (< i, tsᵢ, valᵢ, old_tsᵢ, old_valᵢ >);

when READACK() is delivered:
(05)  readingᵢ ← readingᵢ − 1;
```

(b) Server Protocol

Fig. 2. The read() protocol for a synchronous system.

default value $\perp$ (line 04, Fig. 2(a)). Otherwise, c_i queries the servers to get the last value of the register by sending a READ() message (line 06, Fig. 2(a)) and remains waiting for 2δ times, i.e. the maximum round trip message delay (line 07, Fig. 2(a)).

When a server s_i delivers a READ() message, the $reading_i$ counter is increased by one and then s_i sends a REPLY($<i, ts_i, val_i, old\_ts_i, old\_val_i>$) message containing the current and old values and timestamp stored locally (lines 03–04, Fig. 2(b)).

When the reading client delivers a REPLY($<j, ts, val, ots, ov>$) message, it stores locally the reply in two tuples containing respectively the current and the old triples with server id, timestamp and corresponding value (lines 24–25, Fig. 2(a)). When the reader client is unblocked from the wait statement, it checks if there exists a pair $<ts, val>$ in the $replies$ set that has been reported by all servers it believes honest (line 08, Fig. 2(a)) and, in this case, it sends a READ_ACK() message (line 09, Fig. 2(a)) and it returns the corresponding value (line 10, Fig. 2(a)). Received the READ_ACK() message, a server s_i just decreases by one its $reading_i$ counter (line 05, Fig. 2(b)). Otherwise, a write() operation may be in progress. To check if it is the case, the client keeps waiting for other δ time units and then checks again if a good value exists (lines 11–12, Fig. 2(a)). If, after this period, the value is not yet found, it means that some of the servers behaved maliciously. Therefore, the client executes the detection() procedure to understand who is misbehaving (cfr. Fig. 4). Let us note that such procedure cleans up the set of honest servers when they are detected to be malicious. Therefore, after the execution of the procedure, the reader checks for the last time if a good value exists in its $replies$ set and, if so, it returns such value (line 18, Fig. 2(a)); otherwise the special value abort is returned (line 19, Fig. 2(a)). In any case, a READ_ACK() is sent to block the forwarding of new values at the server side (line 16, Fig. 2(a)).

The write() operation (Fig. 3). When a client wants to write, it first sets its *writing* flag to true, stores locally the value and the corresponding timestamp, obtained incrementing by one the current timestamp stored in $last\_ts$ variable (lines 0102, Fig. 3(a)), sends a WRITE() message to servers, containing the value to be written and the corresponding timestamp (line 03, Fig. 3(a)), and remains waiting for δ time units.

When a server s_i delivers a WRITE(v, ts) message, it checks if the received timestamp is greater than the one stored in the ts_i variable. If so, s_i updates its local variables keeping the current value and timestamp as old and storing the received ones as current (lines 02–05, Fig. 3(b)). Contrarily, s_i checks if the timestamp is the same stored locally in ts_i. If this happens, it just adds the new value to the set val_i (line 06, Fig. 3(b)). In any case, s_i sends back an ACK() message with the received timestamp (lines 08, Fig. 3(b)) and forwards the new value if some read() operation is in progress (lines 09, Fig. 3(b)). Delivering an ACK() message, the writer client checks if the timestamp is greater equal than its my_last_ts and, if so, it adds a tuple $<j, ts, ->$ to its ack set (line 16, Fig. 3(a)).

When the writer is unblocked from the wait statement, it sends a READ() message, waits for δ time units and sends another READ() message (lines 06–08, Fig. 3(a)). This message has two main objectives: (i) create a message pattern that makes impossible to malicious servers to distinguish a real reader from the writer and (ii) collect values to detect misbehaving servers. In this way, a rational malicious server, that aims at remaining in the system, is inhibited from misbehaving as it could be detected from the writer and removed from the computation. The writer, in fact, executes the detection() procedure both on the ack set and on the $replies$ set collected during the write() (lines 09–11, Fig. 3(a)). Finally, the writer sends two READ_ACK() messages to block the forwarding of replies, resets its *writing* flag to false and returns from the operation (lines 12–15, Fig. 3(a)).

```
operation write(v):
(01) writing ← true; ack ← ∅;
(02) my_last_ts ← last_ts + 1; my_last_val ← v;
(03) broadcast WRITE(< my_last_val, my_last_ts >);
(04) wait(δ);
(05) replies ← ∅;
(06) broadcast READ();
(07) wait(δ);
(08) broadcast READ();
(09) execute detection(ack, A);
(10) wait(δ);
(11) execute detection(replies_i, R);
(12) broadcast READACK();
(13) broadcast READACK();
(14) writing ← false;
(15) return(ok).
─────────────────────────────────────────────
when WRITE_ACK(ts, s_j) is delivered:
(16)  if (ts ≥ my_last_ts) then ack ← ack ∪ {< j, ts, − >} endif
─────────────────────────────────────────────
when ∃ ts such that S = {j | ∃ < j, ts', − >∈ ack} ∧ S ⊇ honest:
(17)  if (ts ≥ last_ts) then last_ts ← ts endif
(18)  for each < j, ts', − > ∈ ack such that ts' = ts do ack ← ack\ < j, ts', − > endFor.
```

(a) Client Protocol

```
when WRITE(< val, ts >) is delivered:
(01) if (ts > ts_i)
(02)    then old_ts_i ← ts_i;
(03)         old_val_i ← val_i;
(04)         ts_i ← ts;
(05)         val_i ← {val};
(06)    else if (ts_i = ts) then val_i ← val_i ∪ {val}; endif
(07) endIf
(08) send WRITE_ACK(ts, i);
(09) if (reading_i > 0) then send REPLY (< i, ts_i, val_i, old_ts_i, old_val_i >) endif.
```

(b) Server Protocol

Fig. 3. write() protocol for a synchronous system.

Let us note that, the execution of a write() operation triggers the update of the $last\_ts$ variable at any client. This happens when in the ack set there exists a timestamp reported by any honest server (lines 17–18, Fig. 3(a)).

The detection() procedure (Fig 4). This procedure is used by clients to detect servers misbehaviors during the execution of read() and write() operations. It takes as parameter a set (that can be the $replies$ set or the ack set) and a flag that identifies the type of the set (i.e. A for ack, R for replies). In both cases, the client checks if it has received at least one message from any server it saw honest and detects as faulty all the servers omitting a message (lines 01–08).

If the set to be checked is a set of ACK() messages, the client (writer) just checks if some server s_j acknowledged a timestamp that is different from the one it is using in the current write() and, if so, s_j is detected as malicious (lines 38–42). Otherwise, if the set is the $replies$ set (flagged as R), the client checks if it is running the procedure while it is writing or reading (line 10). If the client is writing, it just updated the state of the register. Thus, the writer checks that all servers sent back the pair $<v, ts>$ corresponding to the one stored locally in

```
procedure detection(replies_set, set_type):
(01)  S = {j|∃ < j, −, − >∈ replies_set};
(02)  if (honest ⊄ S)
(03)    then for each s_j ∈ (honest_i \ S) do
(04)              trigger detect(s_j);
(05)              honest_i ← honest_i \ {s_j};
(06)              broadcast DETECTED(s_j);
(07)         endFor
(08)  endif
(09)  if (set_type = R)
(10)    then if (writing)
(11)       then R = {j|∃ < j, my_last_val, my_last_ts >∈ replies_set};
(12)          if (honest ⊄ R)
(13)            then for each s_j ∈ (honest_i \ R) do
(14)                     trigger detect(s_j);
(15)                     honest_i ← honest_i \ {s_j};
(16)                     broadcast DETECTED(s_j);
(17)                 endFor
(18)          endIf
(19)       else for each < j, ts, − >∈ replies_set such that ts < last_ts − 1 do
(20)                 trigger detect(s_j);
(21)                 honest ← honest \ {s_j};
(22)                 broadcast DETECTED(s_j);
(23)          endFor
(24)          for each < j, ts, val >∈ replies_set such that ts = my_last_ts do
(25)                 D_i = {v | (∃ < j, ts, val >∈ replies_set) ∧ (ts = my_last_ts)};
(26)                 if ((my_last_val ≠ ⊥) ∧ (my_last_ts = last_ts) ∧ (last_val ∉ D_i))
(27)                     then trigger detect(s_j);
(28)                          honest ← honest \ {s_j};
(29)                          broadcast DETECTED(s_j);
(30)                 endif
(31)          endFor
(32)          for each < j, ts, val >∈ replies_set such that ts > last_ts + 1 do
(33)                 trigger detect(s_j);
(34)                 honest_i ← honest_i \ {s_j};
(35)                 broadcast DETECTED(s_j);
(36)          endFor
(37)       endif
(38)    else for each < j, ts, − >∈ replies_set such that ts ≠ my_last_ts do
(39)              trigger detect(s_j);
(40)              honest ← honest \ {s_j};
(41)              broadcast DETECTED(s_j);
(42)         endFor
(43)  endif.
```

Fig. 4. detection() function invoked by an anonymous client for a synchronous system.

the variables my_last_val and my_last_ts. If someone reported a bad value or timestamp, it is detected as misbehaving (lines 11–18). If the client is reading, it is able to detect servers sending back timestamps that are too old (lines 19–23) or too new to be correct (lines 32–36) or servers sending back the right timestamp but with a wrong value (lines 24–31).

Due to the lack of space, the correctness proofs of $\mathcal{P}$ are reported in [9].

7 $\mathcal{P}_{cv}$ and $\mathcal{P}_{hash}$ Protocols for a Regular Register when $D_s \geq G_s$

In the following, we show how to modify the protocol to get $\theta \geq \frac{1}{2}$, when $D_s \geq G_s$. In particular, we propose two possible extensions: the first using a probabilistic collaborative detection at the client side (introducing a cost in terms of number of messages needed to run the detection) and the second using a kind of fingerprint to prevent servers misbehavior (introducing a computational cost).

A collaborative detection protocol $\mathcal{P}_{cv}$. The collaborative detection involves all the clients in the detection process and exploits the fact that the last writer remains in the system and it is always able to identify a faulty server. The basic idea is to look for a write witness (i.e., the writer) each time that a reader is not able to decide about the correctness of a value. This solution allows to identify malicious server and to decide and return always a correct value. However, considering that (i) we want to decouple as much as possible servers and client, (ii) this collaborative approach has a cost in terms of messages and (iii) to force rational servers to behave correctly it is sufficient to get $\theta \geq \frac{1}{2}$ (according to Lemma 3), then we use this collaborative approach only with a given probability.

More in details, in $\mathcal{P}_{cv}$ protocol, when a reader does not collect the same value from all servers it flips a coin to decide if running the collaborative detection or not. If the outcome is 1, then it broadcasts to all the other clients the timestamps collected during the read operation and waits that some writer acknowledge them. When a client receives a check timestamp request, it checks if it corresponds to its last written value and if so, it replies with such a value so that the reader can double-check information provided by servers. If there is no match between values and timestamps, then clients are able to detect a faulty server and exclude it from the computation.

The introduction of this probabilistic step in the protocol increases the value of θ to $\frac{1}{2}$. As a consequence, following Lemma 3, any rational server will decide to behave correctly to avoid to be detected.

A fingerprint-based detection protocol $\mathcal{P}_{hash}$. Let us recall that the basic idea behind the detection process is to include inside reply messages (i.e., write acknowledgements or read replies) "enough" information to verify the correctness of the provided information. In particular, in protocol $\mathcal{P}$, servers are required to acknowledge write operations by sending back the corresponding timestamp so that each client is always aware about it and the writer is able to verify that no bad timestamps are sent to clients.

In protocol $\mathcal{P}_{hash}$, the basic idea is to extend $\mathcal{P}$ by including another information i.e., a fingerprint of the value and its timestamp (e.g., its hash), in the write message and in its acknowledgement so that it is always possible for a client to check that servers are replying correctly. More in details, when a client writes, it computes the hash of the value and its corresponding timestamp and attaches such fingerprint to the message. In such way (as for $\mathcal{P}$) when servers acknowledge a write, they send back the correct fingerprint to all clients. Having such information, all clients are potentially able to detect locally if values collected during a read operation are never written values (this can be simply done by computing the hash of the message and compare it with the one received during the last write). However, as in the case of $\mathcal{P}_{cv}$, this detection has a cost and, to get $\theta \geq \frac{1}{2}$ it is sufficient that this is done with a certain probability. Thus, when

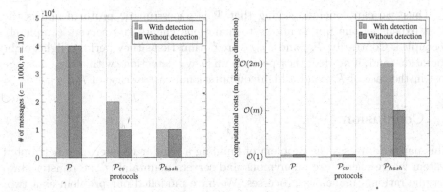

Fig. 5. Qualitative analysis of protocols with respect to their message complexity (left figure) and computational complexity (right figure). For the message complexity we consider a system where the number of servers is $n = 10$ and the number of clients is $c = 1000$. For the computational complexity we consider the cost with respect to the message size m.

a reader does not collect the same value from all servers, it flips a coin and if the outcome is 1 then it computes the hash of the messages it delivered and compares them with the hashes it knows to be associated to a specific timestamp. The introduction of this step is enough to get $\theta = \frac{1}{2}$ and to prevent rational servers deviating from the protocol. Notice that, as for $\mathcal{P}_{cv}$, the employment of the random coin has a twofold purpose: (i) to provide a solution for $D_s \geq G_s$, for which it is enough to have $\theta \geq \frac{1}{2}$ and (ii) to avoid to always perform the costly detection operation.

Due to the lack of space, proofs for the correctness of $\mathcal{P}_{cv}$ and $\mathcal{P}_{hash}$ protocols are sketched in the [9].

Trade offs. Figure 5 shows a qualitative comparison of the three proposed protocols in terms of message complexity and computational cost. In particular, we compare the cost of the protocols both in presence and absence of a server attack (i.e., when the detection is necessary or not). As we can see, $\mathcal{P}$ requires the highest number of messages and such number does not depend on the real need of doing detection but it is rather required to mask the type of operation that a client is doing and to make indistinguishable real read messages from dummy ones. Concerning its computational cost, it is constant since it does not depend on the message size.

In $\mathcal{P}_{cv}$ it is possible to save the dummy read messages as we do not need anymore to mask the message pattern but we need to pay the cost of the collaborative detection, if it is needed. In fact, if a reader is not able to decide a value, it needs to send messages to contact all the other clients (higher message complexity in case of server misbehaviour). Concerning the computational cost, it is not affected by the detection. Conversely, $\mathcal{P}_{hash}$ exhibits the dual behaviour: message complexity is not affected by server misbehaviour but the computational cost is impacted by the need of detection.

Thus, we can conclude saying that $\mathcal{P}$ is a pessimistic protocol and is the most expensive one but it allows to maintains clients and servers completely decoupled. Contrarily, $\mathcal{P}_{cv}$ and $\mathcal{P}_{hash}$ are optimistic as they perform lightweight operations and, if needed, they perform an heavy detection (with a high message cost in the case of $\mathcal{P}_{cv}$ and a high computational cost in case of $\mathcal{P}_{hash}$).

8   Conclusion

This paper addresses the problem of building a regular register in a distributed system where clients are anonymous and servers maintaining the register state may be rational malicious processes. We have modelled our problem as a two-parties Bayesian game and we designed distributed protocols able to reach the Bayesian Nash Equilibrium and to emulate a regular register when the loss in case of detection is greater than the gain obtained from the deviation (i.e. $D_s > G_s$). To the best of our knowledge, our protocols are the first register protocols working in the absence of knowledge on the number of compromised replicas.

The protocols rely on the following assumptions: (i) rational malicious servers act independently and do not form a coalition, (ii) the system is synchronous, (iii) clients are anonymous and (iv) write operations are serialised.

As future works, we are investigating how to solve the same problem under weaker synchrony assumption or in the case an attacker controls a coalition of processes. Addressing these points is actually far from be trivial. Considering a fully asynchronous system, in fact, makes impossible to use our punishment mechanism as clients are not able to distinguish alive but silent servers from those crashed. Additionally, when the attacker is able to compromise and control a coalition of processes, the model provided in this paper is no more adequate and we are studying if and how it is possible to define a *Bayesian Coalitional Game* [13] for our problem and if an equilibrium can be reached in this case.

Acknowledgments. This present work has been partially supported by the EURA-SIA project, and CINI Cybersecurity National Laboratory within the project Filiera-Sicura: Securing the Supply Chain of Domestic Critical Infrastructures from Cyber Attacks (www.filierasicura.it) funded by CISCO Systems Inc. and Leonardo SpA.

References

1. Abraham, I., Dolev, D., Halpern, J. Y. Distributed protocols for leader election: a game-theoretic perspective. In: DISC 2013, pp. 61–75 (2003)
2. Afek, Y., Ginzberg, Y., Feibish, S.L., Sulamy, M.: Distributed computing building blocks for rational agents. In: PODC 2014, pp. 406–415 (2014)
3. Aiyer, A.S., Alvisi, L., Clement, A., Dahlin, M., Martin, J.P., Porth, C.: BAR fault tolerance for cooperative services. In: ACM SIGOPS Operating Systems Review. ACM, pp. 45–58 (2005)
4. Attiya, H., Bar-Noy, A., Dolev, D.: Sharing memory robustly in message-passing systems. J. ACM **42**(1), 124–142 (1995)

5. Bazzi, R.A.: Synchronous Byzantine Quorum systems. Distrib. Comput. **13**(1), 45–52 (2000)
6. Chaudhuri, S., Kosa, M.J., Welch, J.: One-write algorithms for multivalued regular and atomic registers. Acta Informatica **37**, 161–192 (2000)
7. Clement, A., Li, H.C., Napper, J., Martin, J., Alvisi, L., Dahlin, M.: BAR primer. In: DSN, M. (ed.), pp. 287–296 (2008)
8. Clement, A., Napper, J., Li, H., Martin, J.P., Alvisi, L., Dahlin, M.: Theory of BAR games. In: PODC 2007, pp. 358–359 (2007)
9. Del Pozzo, A., Bonomi, S., Lazzeretti, R., Baldoni, R.: Building regular registers with rational malicious servers and anonymous clients – extended version (2017). Available online on arXiv
10. Delporte-Gallet, C., Fauconnier, H., Tran-The, H.: Uniform consensus with homonyms and omission failures. In: ICDCN 2013, pp. 61–175 (2013)
11. Fudenberg, D., Tirole, J.: Game Theory. Massachusetts, Cambridge (1991)
12. Haldar, S., Vidyasankar, K.: Constructing 1-writer multireader multivalued atomic variables from regular variables. JACM **42**(1), 186–203 (1995)
13. Ieong, S., Shoham, Y.: Bayesian coalitional games. In: AAAI 2008, pp. 95–100 (2008)
14. Li, H.C., Clement, A., Wong, E.L., Napper, J., Roy, I., Alvisi, L., Dahlin, M.: BAR gossip. In: OSDI 2006, pp. 191–204 (2006)
15. Lamport, L.: On interprocess communication, part 1: models, part 2: algorirhms. Distrib. Comput. **1**(2), 77–101 (1986)
16. Mahajan, P., Setty, S., Lee, S., Clement, A., Alvisi, L., Dahlin, M., Walfish, M.: Depot : cloud storage with minimal trust. ACM TOCS **29**(4), 12 (2011)
17. Malkhi, D., Reiter, M.K.: Byzantine Quorum systems. Distrib. Comput. **11**(4), 203–213 (1998)
18. Martin J., Alvisi L., Dahlin M.: Small Byzantine Quorum systems. In: DSN 2002, pp. 374–388 (2002)
19. Martin J., Alvisi L., Dahlin M.: Minimal Byzantine storage. In: DISC (2002)
20. Schneider, F.B.: Implementing fault-tolerant services using the state machine approach: a tutorial. ACM Comput. Surv. **22**(4), 299–319 (1990)
21. Singh, A.K., Anderson, J.H., Gouda, M.: The Elusive atomic register. JACM **41**(2), 331–334 (1994)
22. Sousa, P., Bessani, A.N., Correia, M., Neves, N.F., Verissimo, P.: Highly available intrusion-tolerant services with proactive-reactive recovery. IEEE TPDS **21**(4), 452–465 (2010)
23. The Tor project. https://www.torproject.org
24. Vidyasankar, K.: Converting Lamport's regular register to atomic register. IPL **28**(6), 287–290 (1988)
25. Vityani, P., Awerbuch, B.: Atomic shared register access by asynchronous hardware. In: FOCS 1987, pp. 223–243 (1987)
26. Ostrovsky, R., Yung, M.: How to withstand mobile virus attacks. In: PODC 1991, pp. 51–59 (1991)
27. Shamir, A.: How to share a secret. Comm. ACM **22**(11), 612–613 (1979)
28. Dolev, S., ElDefrawy, K., Lampkins, J., Ostrovsky, R., Yung, M.: Proactive secret sharing with a dishonest majority. In: SCN 2016, pp. 529–548 (2016)
29. Cramer, R., Damgard, I.B.: Secure Multiparty Computation. Cambridge University Press, New York (2015)

On the Optimality of the Exponential Mechanism

Francesco Aldà[(⊠)] and Hans Ulrich Simon

Faculty of Mathematics, Horst Görtz Institute for IT Security,
Ruhr-Universität Bochum, Universitätsstraße 150, 44801 Bochum, Germany
{francesco.alda,hans.simon}@rub.de

Abstract. In this work, we investigate one of the most renowned tools used in differential privacy, namely the exponential mechanism. We first study the optimality of the error introduced by the exponential mechanism in the average-case scenario, when the input/output universe of the mechanism can be modeled as a graph where each node is associated with a database. By leveraging linear programming theory, we provide some regularity conditions on the graph structure under which the exponential mechanism minimizes the average error. Moreover, we give a toy example in which the optimality is preserved (up to a constant factor) even if these regularity conditions hold only to a certain extent. Finally, we prove the worst-case optimality of the exponential mechanism when it is used to release the output of a sorting function.

1 Introduction

Differential privacy [8] is a highly popular paradigm for privacy-preserving statistical analysis. It ensures privacy by limiting the influence of an individual input datum on the released information. In addition to the rigorous privacy guarantees provided, the recognition of this framework can be traced back to some crucial factors: the composition property which permits to combine differentially private mechanisms while controlling privacy degradation, and the existence of very simple tools which easily endorse its adoption. The Laplace [8] and the exponential [19] mechanism represent the perfect example. While the Laplace mechanism provides differential privacy to vector-valued functions, the exponential mechanism is intentionally designed for applications where the response set can be arbitrary and possibly non-numeric [19]. If we ignore the efficiency issues that this algorithm inherently has, it has proved extremely successful in a number of applications, from privately generating synthetic databases that can accurately answer a large class of queries [4], to private PAC-learning [16]. Moreover, it has been shown to outperform the accuracy guarantees provided by the Laplace mechanism in some numeric settings [3].

In this paper, we first investigate under which conditions the exponential mechanism is optimal in terms of the average-case error. We consider the setting where the input and output universe of a privacy mechanism coincide and can be modeled as a graph, where each node is associated with a database,

S. Dolev and S. Lodha (Eds.): CSCML 2017, LNCS 10332, pp. 68–85, 2017.
DOI: 10.1007/978-3-319-60080-2_5

and adjacent nodes correspond to neighboring databases. The optimal privacy mechanism can then be expressed as the solution of a linear program, where we seek to minimize the average error introduced by the mechanism subject to the constraints induced by differential privacy. We show that, if the induced graph has a transitive automorphism group and a so-called regular layer sequence, then the exponential mechanism is actually optimal, i.e., its solution coincides with that of the optimal mechanism. We then provide a toy example in which this result holds (up to a constant factor) even if the aforementioned conditions are met only to a large extent. Finally, we introduce the sorting function and show that the error introduced by the exponential mechanism is actually optimal in the worst-case. We underline that this last result carries over and extends the analysis discussed in a work currently under review [1].

Related Work. A general upper bound on the error introduced by the exponential mechanism is given by McSherry and Talwar [19]. Lower bounds in differential privacy have been extensively studied and a range of techniques for proving lower bounds have been introduced [1,6,7,12,13,18]. The optimality of differentially private mechanisms has been the subject of recent studies. Kairouz et al. [15] introduce a family of mechanisms which contains a utility-maximizer under the local model of privacy. Koufogiannis et al. [17] investigate the optimality of the Laplace mechanism under the Lipschitz privacy framework. In particular, they show that the Laplace mechanism is optimal for identity queries in terms of the mean-squared error, when privacy is guaranteed with respect to the L_1-norm. Geng et al. [10] show that the mean-squared error introduced by the staircase mechanism is optimal for low-dimensional queries. Linear programming theory can be leveraged to show lower bounds on the error needed for achieving any meaningful privacy guarantee [6,9]. Hsu et al. [14] investigate how to solve a linear program under differential privacy. Hardt and Talwar [13] exploit linear programming theory to show tight upper and lower bounds on the amount of noise needed to provide differential privacy for r linear queries on databases in $\mathbb{R}^N$. Our contribution is mostly related to the work of Ghosh et al. [11] and Brenner and Nissim [5]. In their paper, Ghosh et al. [11] consider Bayesian *information consumers* that wish to compute the number of entries in a database satisfying a given predicate. An information consumer is characterized by a prior belief and a loss-function, which quantify the consumer's side knowledge and the quality of the answer. Introducing a linear program modeling the privacy constraints, they show that a discrete variant of the Laplace mechanism enables optimality (after a deterministic post-processing of the output) for *all* Bayesian information consumers. Such a mechanism is usually referred to as universally optimal. In a follow up work, Brenner and Nissim [5] show that universally optimal mechanisms for Bayesian consumers are extremely rare, proving that they essentially exist only for a single count query. Their proof makes use of a so-called *privacy constraint graph*, where the vertices correspond to the values of the output space, and the edges correspond to pairs of values resulting by applying the query function to neighboring databases. In contrast to [11] and [5], we restrict our attention to a single information consumer who has a uniform prior over the input/output space and measures the loss in terms of the record-exchange metric. We then study under which conditions on the structure of the privacy constraint graph the solution of the optimal differentially private mechanism (modeled by a linear program similar to the one introduced by Ghosh et al. [11]) coincides with the solution that the exponential mechanism delivers.

2   Preliminaries

Let $\mathcal{X}$ be a domain. A database $\mathcal{D}$ is an N-dimensional vector over $\mathcal{X}$, i.e. $\mathcal{D} \in \mathcal{X}^N$. N is referred to as the size of the database $\mathcal{D}$. Two databases $\mathcal{D}, \mathcal{D}'$ are said to be *neighboring*, denoted $\mathcal{D} \sim \mathcal{D}'$, if they can be obtained from each other by a single record exchange.

Definition 1 ([8]). *Let $\mathcal{X}$ be a domain and $\mathcal{R}$ be a (possibly infinite) set of responses. A random mechanism $\mathcal{M} : \mathcal{X}^N \to \mathcal{R}$ is said to provide ε-differential privacy for $\varepsilon > 0$ if, for every pair $(\mathcal{D}, \mathcal{D}')$ of neighboring databases and for every measurable $S \subseteq \mathcal{R}$, we have*

$$\Pr[\mathcal{M}(\mathcal{D}) \in S] \le e^\varepsilon \cdot \Pr[\mathcal{M}(\mathcal{D}') \in S].$$

The exponential mechanism [19] is a well-known tool for achieving differential privacy. Let $u \colon \mathcal{X}^N \times \mathcal{R} \to \mathbb{R}$ be a *utility* function, mapping a database/output pair to a score. Given a database $\mathcal{D} \in \mathcal{X}^N$, the exponential mechanism defines a probability distribution over $\mathcal{R}$ weighted according to the utility function $u(\mathcal{D}, \cdot)$.

Definition 2 ([19]). *Let $u \colon \mathcal{X}^N \times \mathcal{R} \to \mathbb{R}$ and $\varepsilon > 0$. The exponential mechanism $\mathcal{M}_{exp} \colon \mathcal{X}^N \to \mathcal{R}$ assigns to $s \in \mathcal{R}$ a probability density proportional to*

$$\exp\left(\frac{\varepsilon \cdot u(\mathcal{D}, s)}{S(u)}\right), \tag{1}$$

where $S(u) = \sup_{s \in \mathcal{R}} \sup_{\mathcal{D} \sim \mathcal{D}'} |u(\mathcal{D}, s) - u(\mathcal{D}', s)|$ is the sensitivity of u. It then returns a value sampled from such distribution.

We briefly note that the definition in [19] is slightly more general and has an additional factor $\mu(s)$ in (1), which represents a prior distribution on $\mathcal{R}$. In this paper, we deal with a uniform prior and have therefore omitted $\mu(s)$ from Definition 2.

Lemma 1 ([19]). *The exponential mechanism provides 2ε-differential privacy.*

In several cases (see for example the unit demand auction setting in [19]) the factor 2 in the statement of Lemma 1 can be removed, strengthening the privacy guarantees to ε-differential privacy.

3   Optimal Mechanisms and Linear Programming

Let $G = (\mathcal{K}, E)$ denote a graph with $K = |\mathcal{K}|$ nodes and diameter D. Intuitively, we should think of each node $x \in \mathcal{K}$ as a piece of information associated with a database $\mathcal{D}$. Moreover, we should think of adjacent nodes in G as nodes whose underlying databases are neighbored in the sense that they can be obtained from each other by a single record exchange. Hence a node y has distance d from another node x iff d is the smallest number of record exchanges which transforms the database underlying y into the database underlying x. We consider the following special situation:

- The (randomized) mechanisms $\mathcal{M}$ under investigation should provide ε-differential privacy and, given a node $x \in \mathcal{K}$, they should return another node in $\mathcal{K}$ (the choice of which depends on $\mathcal{M}$'s internal randomization).
- The cost (= negated utility) of an output y, given input x, is defined as the distance between x and y in G, which is denoted as $d(x, y)$. We will refer to this distance measure as the *record-exchange metric*. Note that $|d(x, y) - d(x', y)| \leq 1$ holds for all $x, x', y \in \mathcal{K}$ such that x and x' (resp. their underlying databases) are neighbored. Thus $-d$ (viewed as a utility function) has sensitivity 1.

Note that the record-exchange metric coincides with what is called "geodesic distance" w.r.t the graph G in some papers.

We consider two examples where, in both cases, the record-exchange metric coincides with $1/2$ times the L_1-metric.

Example 1. Suppose that the nodes in G represent histograms with N users and T types of records (briefly called (N, T)-histograms hereafter), i.e., we may identify a node $x \in \mathcal{K}$ with a vector $(v_1, \ldots, v_T)$ such that $N = \sum_{t=1}^{T} v_t$ and v_t is the number of users whose record is of type t. Note that the record-exchange metric satisfies $d(x, y) = \frac{1}{2}\|y - x\|_1$ because each record-exchange can decrease the L_1-distance between two histograms by an amount of 2 (but not more).

Example 2. Suppose that the nodes in G represent sorted (N, T)-histograms, i.e., we may identify a node $x \in \mathcal{K}$ with a sorted sequence $v_1 \geq \ldots \geq v_T$ such that $\sum_{t=1}^{T} v_t = N$. Here v_1 (resp. v_2 and so on) denotes the number of users whose record occurs most often (resp. 2nd most often and so on) in the database. Alternatively, we may be interested in the $r \leq T$ largest values of $v_1 \geq \ldots \geq v_T$ only, i.e., we identify a node $x \in \mathcal{K}$ with the initial segment $v_1 \geq \ldots, \geq v_r$ of the full sequence $v_1 \geq \ldots \geq v_T$.

In this section, we investigate under which conditions the exponential mechanism is optimal in the sense of incurring the smallest possible expected error (measured in terms of the record-exchange metric) where expectation is taken over the (uniformly distributed) inputs $x \in_R \mathcal{K}$ and over the internal randomization of the mechanism. We start by introducing several linear programs. The optimal solution of the first linear program we consider, denoted LP[1] below, corresponds to the solution of the optimal ε-differentially private mechanism.

Another linear program, denoted LP[3] below, has an optimal solution which coincides with the one given by the exponential mechanism. We then provide some regularity conditions on the graph G under which an optimal solution of LP[3] also optimizes LP[1] (so that the exponential mechanism is optimal whenever the regularity conditions are valid).

We can now continue with our general discussion. Note that a (randomized) mechanism $\mathcal{M}$ with inputs and outputs taken from $\mathcal{K}$ is formally given by probability parameters $p(y|x)$ denoting the probability of returning $y \in \mathcal{K}$ when given $x \in \mathcal{K}$ as input. Since, for each x, $p(y|x)$ is a distribution on $\mathcal{K}$, we have

$$(\forall x, y \in \mathcal{K} : \ p(y|x) \geq 0) \wedge \left(\forall x \in \mathcal{K} : \ \sum_{y \in \mathcal{K}} p(y|x) = 1 \right). \tag{2}$$

Moreover, if $\mathcal{M}$ provides ε-differential privacy, we have

$$\forall y \in \mathcal{K}, \forall \{x, x'\} \in E : \ p(y|x') \geq e^{-\varepsilon} \cdot p(y|x). \tag{3}$$

Conversely, every choice of these probability parameters that satisfies (2) and (3) represents a mechanism that provides ε-differential privacy.

Suppose that $\mathcal{M}$ is given by its probability parameters $p = (p(y|x))$ as described above. The average distance between $x \in \mathcal{K}$ and the output $y \in \mathcal{K}$, returned by $\mathcal{M}$ when given x as input, is then given as follows:

$$f^G(p) = \frac{1}{K} \cdot \sum_{x \in \mathcal{K}} \sum_{y \in \mathcal{K}} p(y|x) d(x, y). \tag{4}$$

Let $S_d = S_d(y)$ denote the set of all nodes in $\mathcal{K}$ with distance d to y (the d-th layer of G w.r.t. start node y). Then

$$f^G(p) = \frac{1}{K} \cdot \sum_{y \in \mathcal{K}} f_y^G(p) \ \text{ for } \ f_y^G(p) = \sum_{d=0}^{D} \sum_{x \in S_d(y)} p(y|x) d. \tag{5}$$

We pursue the goal to find an ε-differentially private mechanism $\mathcal{M}$ that minimizes $d(x, y)$ on the average. For this reason, we say that a mechanism $\mathcal{M}_*$ with probability parameters p_* is *optimal* w.r.t. G if p_* is a minimizer of $f^G(p)$ among all p that satisfy (2) and (3). It is obvious from our discussion that the probability parameters $p_*(y|x)$ representing an optimal mechanism w.r.t. G are obtained by solving the following linear program:

$$\text{LP}[1] : \min_{p = (p(y|x))_{x,y \in \mathcal{K}}} \ f^G(p) \ \text{ s.t. } (2) \text{ and } (3).$$

We will refer to this linear program as $\text{LP}^G[1]$ whenever we want to stress the dependence on the underlying graph G. We now bring into play the following modifications of the condition (2):

$$(\forall x, y \in \mathcal{K} : p(y|x) \geq 0) \, \wedge \, \left(\sum_{x \in \mathcal{K}} \sum_{y \in \mathcal{K}} p(y|x) = K \right). \qquad (6)$$

$$(\forall x, y \in \mathcal{K} : p(y|x) \geq 0) \, \wedge \, \left(\forall y \in \mathcal{K} : \sum_{x \in \mathcal{K}} p(y|x) = 1 \right). \qquad (7)$$

Note that (7) implies (6). Consider the following relatives of $\mathrm{LP}^G[1]$:

$$\mathrm{LP}[2] : \min_{p=(p(y|x))_{x,y \in \mathcal{K}}} f^G(p) \text{ s.t. (6) and (3)};$$

$$\mathrm{LP}[3] : \min_{p=(p(y|x))_{x,y \in \mathcal{K}}} f^G(p) \text{ s.t. (7) and (3)}.$$

As for $\mathrm{LP}^G[1]$, we will use the notations $\mathrm{LP}^G[2]$ and $\mathrm{LP}^G[3]$ to stress the dependence on the underlying graph G. Given a graph $G = (\mathcal{K}, E)$, a permutation σ of $\mathcal{K}$ is called *automorphism* if, for all $x, y \in \mathcal{K}$, $\{x, y\} \in E \Leftrightarrow \{\sigma(x), \sigma(y)\} \in E$. The set of all automorphisms of $\mathcal{K}$, under the operation of composition of functions, forms a group called the automorphism group of G. Such a group is called *transitive* if, for every $x, y \in \mathcal{K}$, there exists an automorphism σ of $\mathcal{K}$ such that $\sigma(x) = y$.

Lemma 2. *Suppose that the graph G has a transitive automorphism group. Then every feasible solution p for $\mathrm{LP}^G[2]$ can be transformed into another feasible solution p' such that $f^G(p') \leq f^G(p)$ and p' satisfies (7).*

Proof. Let p be any feasible solution for $\mathrm{LP}^G[2]$. For every $y \in \mathcal{K}$, let $K_y(p) = \sum_{x \in \mathcal{K}} p(y|x)$. According to (6), we have $\sum_{y \in \mathcal{K}} K_y(p) = K$. Define

$$\bar{p}(y|x) = \frac{1}{K_y(p)} p(y|x) \quad \text{and} \quad \bar{f}_y(p) = \sum_{d=0}^{D} \sum_{x \in S_d(y)} \bar{p}(y|x) d$$

and note that $\bar{p}$ satisfies (3) and (7). We may now write $f^G(p)$ as follows:

$$f^G(p) = \sum_{y \in \mathcal{K}} \frac{K_y(p)}{K} \cdot \bar{f}_y(p).$$

Thus $f^G(p)$ can be interpreted as the average of the cost terms $f_y(p)$ where the term $f_y(p)$ is chosen with probability $K_y(p)/K$. According to the pigeonhole principle, there exists $y^* \in \mathcal{K}$ such that $\bar{f}_{y^*}(p) \leq f^G(p)$. Our strategy is to use the automorphism of G for building a new (and superior) feasible solution p' whose components contain K duplicates of the parameter collection $(\bar{p}(y^*|x)_{x \in \mathcal{K}})$. To this end, let σ_y be the automorphism which maps y to y^* and define

$$p'(y|x) = \bar{p}(y^*|\sigma_y(x)).$$

Note that $x \in S_d(y)$ if and only if $\sigma_y(x) \in S_d(y^*)$. Obviously, $p' \geq 0$ and, for every $y \in \mathcal{K}$, we have

$$K_y(p') = \sum_{x \in \mathcal{K}} p'(y|x) = \sum_{x \in \mathcal{K}} \bar{p}(y^*|\sigma_y(x)) = \sum_{x \in \mathcal{K}} \bar{p}(y^*|x) = 1.$$

This shows that p' satisfies (7). Moreover, p' satisfies (3) since, for every $y \in \mathcal{K}$ and every $\{x, x'\} \in E$, we have

$$e^{-\varepsilon} \cdot p'(y|x) = e^{-\varepsilon} \cdot \bar{p}(y^*|\sigma_y(x)) \le \bar{p}(y^*|\sigma_y(x')) = p'(y|x'),$$

where the inequality follows from the fact that $\bar{p}$ satisfies (3) and σ_y is an automorphism. The following calculation shows that $f_y(p') = \bar{f}_{y^*}(p)$ holds for every $y \in \mathcal{K}$:

$$f_y(p') = \sum_{d=0}^{D} \sum_{x \in S_d(y)} p'(y|x)d = \sum_{d=0}^{D} \sum_{x \in S_d(y)} \bar{p}(y^*|\sigma_y(x))d$$

$$= \sum_{d=0}^{D} \sum_{x \in S_d(y^*)} \bar{p}(y^*|x)d = \bar{f}_{y^*}(p).$$

We now obtain

$$f^G(p') = \frac{1}{K} \cdot \sum_{y \in \mathcal{K}} f_y(p') = \bar{f}_{y^*}(p) \le f^G(p),$$

which concludes the proof.   □

The following result is an immediate consequence of Lemma 2.

Corollary 1. *The optimal values of the problems* LP[2] *and* LP[3] *coincide. Moreover, every optimal solution for* LP[3] *is an optimal solution for* LP[2].

We say that the graph G has a *regular layer sequence w.r.t.* $y \in \mathcal{K}$ if, for all d and for all $x, x' \in S_d(y)$, the nodes x and x' have the same number of neighbors in $S_{d-1}(y)$ and the same number of neighbors in $S_{d+1}(y)$. Let $E[y] = E \cap (S_d(y) \times S_{d+1}(y))$, i.e., $E[y]$ contains the edges in E which connect two nodes in subsequent layers (but excludes the edges which connect two nodes in the same layer).

Lemma 3. *Suppose that the graph* $G = (\mathcal{K}, E)$ *has a transitive automorphism group and a regular layer sequence w.r.t. any* $y \in \mathcal{K}$. *Then the problems* $\mathrm{LP}^G[2]$ *and* $\mathrm{LP}^G[3]$ *have an optimal solution that satisfies*

$$\forall y \in \mathcal{K}, \forall (x, x') \in E[y] : p(y|x') \ge e^{-\varepsilon} \cdot p(y|x) \tag{8}$$

with equality.

Proof. The problem LP[3] decomposes into $K = |\mathcal{K}|$ independent subproblems, one subproblem LP(y) for each fixed choice of $y \in \mathcal{K}$:

$$\mathrm{LP}(y) : \min_{p=(p(y|x))_{x \in \mathcal{K}}} f_y^G(p) = \sum_{d=0}^{D} \left(\sum_{x \in S_d(y)} p(y|x) \right) d$$

$$\text{s.t. } (p \ge 0) \wedge \left(\sum_{x \in \mathcal{K}} p(y|x) = 1 \right) \wedge \left(\forall \{x, x'\} \in E : p(y|x') \ge e^{-\varepsilon} \cdot p(y|x) \right).$$

Let LP[5] (the numbering will become clear in Sect. 4) be the linear program that is obtained from LP(y) by substituting the weaker constraint

$$\forall(x, x') \in E[y] : p(y|x') \geq e^{-\varepsilon} \cdot p(y|x)$$

for

$$\forall\{x, x'\} \in E : p(y|x') \geq e^{-\varepsilon} \cdot p(y|x).$$

In Sect. 4 we will prove the following result:

Claim 1. *If $G(y)$ has a regular layer sequence, then LP[5] has an optimal solution with the following properties:*

1. *The parameter vector $(p(y|x))_{x \in \mathcal{K}}$ (with a fixed choice of y) assigns the same probability mass to all nodes x taken from the same layer.*
2. *For every $(x, x') \in E[y]$, it satisfies the constraint $p(y|x') \geq e^{-\varepsilon} \cdot p(y|x)$ with equality.*

It immediately follows that this optimal solution is also an optimal solution for LP(y), which completes the proof. □

The proof of Claim 1 is lengthy and will therefore be given later. See Lemma 5 in Sect. 4. Recall that $d(x, y)$ denotes the distance between x and y w.r.t. the record-exchange metric. Here comes the main result of this section which essentially states that the exponential mechanism is optimal under the assumptions made in Lemma 3.

Theorem 1. *Under the same assumptions as in Lemma 3, the following holds. An optimal mechanism for $\mathrm{LP}^G[1]$ (and even for $\mathrm{LP}^G[2]$ and for $\mathrm{LP}^G[3]$) is obtained by setting*

$$\forall x, y \in \mathcal{K} : \quad p(y|x) \propto \exp(-\varepsilon \cdot d(x, y)).$$

Proof. Let p be the optimal solution for $\mathrm{LP}^G[2]$ and $\mathrm{LP}^G[3]$ that satisfies (8) with equality so that

$$\forall y \in \mathcal{K}, \forall(x, x') \in E[y] : \quad p(y|x') = e^{-\varepsilon} \cdot p(y|x).$$

Unrolling this recursion, we get

$$p(y_0|x_0) = \frac{\exp(-\varepsilon \cdot d(x_0, y_0))}{\sum_{x \in \mathcal{K}} \exp(-\varepsilon \cdot d(x, y_0))}.$$

The transitivity of the automorphism group of G implies that

$$\forall x_0, y_0 \in \mathcal{K} : \quad \sum_{x \in \mathcal{K}} \exp(-\varepsilon \cdot d(x, y_0)) = \sum_{y \in \mathcal{K}} \exp(-\varepsilon \cdot d(y, x_0)).$$

It follows that $p(y_0|x_0) = p(x_0|y_0)$. As a feasible solution of $\mathrm{LP}^G[3]$, p satisfies (7). Since $p(y_0|x_0) = p(x_0|y_0)$, it must also satisfy (2). Thus p is a feasible solution for $\mathrm{LP}^G[1]$. Since it is even optimal among the feasible solutions of the relaxation $\mathrm{LP}^G[2]$, it must be optimal for $\mathrm{LP}^G[1]$.   □

4   Proof of Claim 1 and Additional Remarks on LP[5]

Recall that $G = (\mathcal{K}, E)$ denotes a graph with $K = |\mathcal{K}|$ nodes and diameter D. Fix some $y \in \mathcal{K}$ and call it the "start node". Recall that $S_d = S_d(y)$ is the set of all nodes in $\mathcal{K}$ with distance d to y (the d-th layer in G). The cardinality of $S_d(y)$ is denoted by $s_d(y)$, or simply by s_d. For instance, $S_0 = \{y\}$ and S_1 is the set of all neighbors of y in G. An edge $e \in E$ either connects two nodes in subsequent layers or it connects two nodes in the same layer. Let again $E[y] \subseteq E$ be the set of edges of the former kind and let $G[y] = (\mathcal{K}, E[y])$. In other words, $G[y]$ is the layered graph that contains all shortest paths to the start node y. We consider an edge in $E[y]$ as being directed away from y, i.e., $(x, x') \in E[y]$ implies that $x \in S_d$ and $x' \in S_{d+1}$ for some $d \in [0 : D-1]$. Note that $E[y]$ naturally partitions into the (disjoint) union of $E_0, E_1, \ldots, E_{D-1}$ where $E_d = E[y] \cap (S_d \times S_{d+1})$. Let $0 < \gamma < 1$ denote a constant scaling factor. In this section, we consider the following two linear optimization problems:

Linear Program 4 (LP[4])	Linear Program 5 (LP[5])
$\mathbf{min}_{p=(p_d)} \; f_4(p) = \sum_{d=0}^{D} s_d p_d d$	$\mathbf{min}_{p=(p_x)} \; f_5(p) = \sum_{d=0}^{D} \left(\sum_{x \in S_d} p_x \right) d$
s.t. $\quad p \geq \mathbf{0} \,,\; \sum_{d=0}^{D} s_d p_d = 1 \,,$	**s.t.** $\quad p \geq \mathbf{0} \,,\; \sum_{x \in \mathcal{K}} p_x = 1 \,,$
(C4) $\quad \forall d \in [0 : d-1] : p_{d+1} \geq \gamma \cdot p_d.$	**(C5)** $\quad \forall (x, x') \in E[y] : p_{x'} \geq \gamma \cdot p_x$

In other words, we would like to find a probability distribution on $\mathcal{K}$ that minimizes the average distance to the start node y subject to (C4) resp. (C5). In Problem LP[5], we can assign individual probabilities to all nodes whereas, in Problem LP[4], we have to assign the same probability p_d to all nodes in the d-th layer S_d (so that the total probability mass assigned to S_d equals $s_d p_d$). Note that LP[5] yields the problem that occurs under the same name in the proof of Lemma 3 provided that we set $\gamma = e^{-\varepsilon}$ and $p_x = p(y|x)$.

As for LP[4], it is intuitively clear that we should move as much probability mass as possible to layers close to the start node y. Thus the following result (whose proof is omitted) does not come as surprise:

Lemma 4. LP[4] *is bounded and feasible. Moreover, there is a unique optimal solution that satisfies all constraints in (C4) with equality.*

Recall that G with start node y is said to have a regular layer sequence if nodes in the same layer of $G[y]$ have the same in-degree and the same out-degree. The next result is essentially a reformulation of Claim 1 from Sect. 3.

Lemma 5. LP[5] *is bounded and feasible. Moreover, if $G[y] = (\mathcal{K}, E[y])$ has a regular layer sequence, then* LP[5] *has an optimal solution that, first, assigns the same probability mass to all nodes in the same layer, and, second, satisfies all constraints in (C5) with equality.*

Proof. Clearly LP[5] is bounded. Showing the existence of a feasible solution is straightforward and hence omitted. Thus we have only to show that LP[5] has an optimal solution that satisfies all constraints in (C5) with equality. Call a feasible solution $p = (p_x)$ of LP[5] *normalized* if p assigns the same probability mass to all nodes in the same layer, say $p_x = \bar{p}_d$ for every node x in layer d. As for normalized feasible solutions, LP[5] collapses to LP[4]. According to Lemma 4, there is a unique optimal solution among all normalized feasible solutions of LP[5] that satisfies all constraints in (C5) with equality.[1] Thus, we now have to show that every feasible solution can be normalized without increasing its cost. To this end, let $p = (p_x)$ denote a fixed but arbitrary feasible solution for LP[5]. For $d = 0, 1, \ldots, D$, we set $\bar{p}_d = \frac{1}{s_d} \sum_{x \in S_d} p_x$, i.e., $\bar{p}_d$ is the probability mass assigned by p to nodes in S_d on the average. We claim that setting $p'_x = \bar{p}_d$ for every node $x \in S_d$ yields a normalized feasible solution of the same cost as p. Clearly $p' \geq 0$. Moreover $\sum_{x \in \mathcal{K}} p'_x = \sum_{x \in \mathcal{K}} p_x = 1$ because $p \mapsto p'$ leaves the total probability mass assigned to any layer S_d unchanged. For the same reason the cost of p' coincides with the cost of p, i.e., $f_5(p') = f_5(p)$. It remains to show that p' satisfies (C5). To this end, pick any $d \in [0 : D-1]$ and any $(x, x') \in E_d$. Let $\overrightarrow{t_d}$ denote the out-degree of x (or of any other node from S_d) and let $\overleftarrow{t_{d+1}}$ denote the in-degree of x' (or of any other node from S_{d+1}). A simple double counting argument shows that

$$s_d \overrightarrow{t_d} = |E_d| = s_{d+1} \overleftarrow{t_{d+1}}. \tag{9}$$

The following calculation shows that $p'_{x'} \geq \gamma p'_x$:

$$p'_{x'} = \frac{1}{s_{d+1}} \cdot \sum_{v \in S_{d+1}} p_v$$

$$\overset{*}{=} \frac{1}{s_{d+1} \overleftarrow{t_{d+1}}} \cdot \sum_{v \in S_{d+1}} \sum_{u:(u,v) \in E_d} p_v$$

$$\overset{(9)}{=} \frac{1}{s_d \overrightarrow{t_d}} \cdot \sum_{u \in S_d} \sum_{v:(u,v) \in E_d} p_v$$

$$\geq \gamma \cdot \frac{1}{s_d \overrightarrow{t_d}} \cdot \sum_{u \in S_d} \sum_{v:(u,v) \in E_d} p_u$$

$$\overset{*}{=} \gamma \cdot \frac{1}{s_d} \cdot \sum_{u \in S_d} p_u = \gamma \cdot p'_x$$

The equations marked "$*$" make use of our assumption that $G[y]$ has a regular layer sequence. The whole discussion can be summarized by saying that p' is a normalized feasible solution for LP[5] and its cost equals the cost of the feasible solution p that we started with. This concludes the proof. □

Let LP[4]$^\infty$ and LP[5]$^\infty$ denote the optimization problems that result from LP[4] and LP[5], respectively, when the underlying graph $G = (\mathcal{K}, E)$ has

---

[1] (C5) collapses to (C4) for normalized feasible solutions.

infinitely many nodes so that the layered graph $G[y] = (\mathcal{K}, E[y])$ might have infinitely many layers $S_0, S_1, S_2, \ldots$. In the formal definition of LP[4] and LP[5], we only have to substitute ∞ for D. An inspection of the proofs of Lemmas 4 and 5 reveals that they hold, mutatis mutandis, for the problems $\text{LP}[4]^\infty$ and $\text{LP}[5]^\infty$ as well:

Corollary 2. $\text{LP}[4]^\infty$ and $\text{LP}[5]^\infty$ are bounded and feasible. Moreover, there is a unique optimal solution for $\text{LP}[4]^\infty$ that satisfies all constraints in (C4) with equality and, if $G[y] = (\mathcal{K}, E[y])$ has a regular layer sequence, then $\text{LP}[5]^\infty$ has an optimal solution that satisfies all constraints in (C5) with equality.

Example 3. Let G_1 be an infinite path $y_0, y_1, y_2, \ldots$ with start node y_0. It follows from Corollary 2 that LP[5] has an optimal solution that satisfies all constraints in (C5) with equality. This leads to the following average distance from y_0:

$$\frac{\sum_{d \geq 1} \gamma^d d}{\sum_{d \geq 0} \gamma^d} = \frac{\frac{\gamma}{(1-\gamma)^2}}{\frac{1}{1-\gamma}} = \frac{\gamma}{1-\gamma}.$$

Let G_2 be the graph consisting of two infinite paths, $y_0, y_{-1}, \ldots$ and $y_0, y_1, \ldots$ both of which are starting from the start node y_0. Again Corollary 2 applies and the optimal average distance from y_0 is calculated as follows:

$$\frac{2 \cdot \sum_{d \geq 1} \gamma^d d}{1 + 2 \cdot \sum_{d \geq 1} \gamma^d} = \frac{\frac{2\gamma}{(1-\gamma)^2}}{1 + \frac{2\gamma}{1-\gamma}} = \frac{2\gamma}{1 - \gamma^2}. \tag{10}$$

As for finite paths, we have the following result:

Lemma 6. *Let P_ℓ be a path of length 2ℓ and let y_0 be the start node located in the middle of P_ℓ. Let $f(\ell)$ denote the optimal value that can be achieved in the linear program* LP[5] *w.r.t. to $G = P_\ell$. Then the following holds:*

1. LP[5] *has an optimal solution that satisfies all constraints in (C5) with equality so that*

$$f(\ell) = \frac{2 \cdot \sum_{d=1}^\ell \gamma^d d}{1 + 2 \cdot \sum_{d=1}^\ell \gamma^d}. \tag{11}$$

2. *The function $f(\ell)$ is strictly increasing with ℓ.*
3. *We have*

$$f(\ell) > \frac{2\gamma}{1 - \gamma^2} \cdot \left(1 - \gamma^\ell - \ell \gamma^\ell (1 - \gamma)\right). \tag{12}$$

Moreover, if $\ell \geq \frac{s}{1-\gamma}$, then

$$f(\ell) > \frac{2\gamma}{1 - \gamma^2} \cdot \left(1 - (s+1)e^{-s}\right). \tag{13}$$

4. $\lim_{\ell \to \infty} f(\ell) = \frac{2\gamma}{1-\gamma^2}$.

Proof. Let $P_\ell = y_{-\ell}, \ldots, y_{-1}, y_0, y_1, \ldots, y_\ell$.

1. Lemma 5 applies because $P_\ell[y_0]$ has a regular layer sequence.
2. An optimal solution for $P_{\ell+1}$ can be transformed into a feasible solution for P_ℓ by transferring the probability mass of the nodes $y_{-(\ell+1)}, y_{\ell+1}$ to the nodes $y_{-\ell}, y_\ell$, respectively. This transfer strictly reduces the cost. The optimal cost $f(\ell)$ that can be achieved on P_ℓ is, in turn, smaller than the cost of this feasible solution.
3. We start with the following calculation:

$$\sum_{d=1}^{\ell} \gamma^{d-1} d = \sum_{d \geq 1} \gamma^{d-1} d - \sum_{d \geq \ell+1} \gamma^{d-1} d$$

$$= \frac{1}{(1-\gamma)^2} - \gamma^\ell \cdot \sum_{d \geq 1} \gamma^{d-1}(d+\ell)$$

$$= \frac{1}{(1-\gamma)^2} - \gamma^\ell \cdot \left(\frac{1}{(1-\gamma)^2} + \frac{\ell}{1-\gamma} \right)$$

$$= \frac{1}{(1-\gamma)^2} \cdot \left(1 - \gamma^\ell - \ell\gamma^\ell(1-\gamma) \right)$$

Setting $F = 1 - \gamma^\ell - \ell\gamma^\ell(1-\gamma)$, it follows that

$$f(\ell) = \frac{\frac{2\gamma}{(1-\gamma)^2}}{1 + 2 \cdot \sum_{d=1}^{\ell} \gamma^d} \cdot F > \frac{\frac{2\gamma}{(1-\gamma)^2}}{1 + 2 \cdot \sum_{d \geq 1} \gamma^d} \cdot F.$$

Since the latter expression differs from (10) by the factor F only, we obtain (12).

The function $s \mapsto (s+1)e^{-s}$ is strictly monotonically decreasing for all $s \geq 0$. It suffices therefore to verify the bound (13) for $s = (1-\gamma)\ell$ so that

$$\gamma^\ell + \ell\gamma^\ell(1-\gamma) = \gamma^\ell(s+1).$$

Noting that

$$\gamma^\ell = \gamma^{s/(1-\gamma)} = (1 - (1-\gamma))^{s/(1-\gamma)} < e^{-s},$$

we may conclude that $\gamma^\ell + \ell\gamma^\ell(1-\gamma) < (s+1)e^{-s}$. From this, in combination with (12), the bound (13) is immediate.

4. The fourth assertion of Lemma 6 is immediate from the third one. □

Even though the regularity conditions for G (transitive automorphism group and regular layer sequence) are satisfied in simple settings (for instance, when each node in the graph corresponds to a binary database of size N), we do not expect this to be the case in most applications. For example, the regularity conditions are not fully satisfied by the graph representing sorted (N,T)-histograms introduced in Example 2. However, we conjecture, first, that these conditions are approximately satisfied for very large databases and, second, that the exponential mechanism is still approximately optimal when these conditions hold approximately. At the time being, we are not able to verify this conjecture for graphs G of practical interest. In the next section, we will illustrate the kind of arguments that we plan to bring into play by presenting a very precise analysis for the simple case where the graph G actually is a long but finite path. Developing these arguments further so as to analyze more reasonable classes of graphs (e.g., graphs representing the neighborhood relation for sorted histograms) remains a subject of future research.

5 A Toy Example: The Path Graph

Throughout this section, we consider the graph $G = (\mathcal{K}, E)$ whose nodes $y_1, \ldots, y_K$ form a path of length $K - 1$. Note that G does not satisfy the regularity condition: neither has G a transitive automorphism group nor has $G[y]$ a regular layer sequence (except for y being chosen as one of the endpoints and, if K is odd, for y being chosen as the point in the middle of the path). Let $\mathrm{OPT}^G[1]$ denote the smallest cost of a feasible solution for $\mathrm{LP}^G[1]$. We will show in this section that, despite the violation of the regularity condition, the exponential mechanism comes close to optimality provided that K is "sufficiently large". The main idea for proving this is as follows. We will split the set of nodes into a "central part" (nodes separated away from the endpoints of the path) and a "peripheral part" (nodes located close to the endpoints). Then we make use of the fact that all ε-differentially private mechanisms are on the horns of the following dilemma:

- If a feasible solution $p = (p(y|x))_{x,y \in \mathcal{K}}$ puts much probability mass on peripheral nodes y, then the cost contribution of the terms $p(y|x)$ with y "peripheral" and x "central" will be large.
- If not, then the cost contribution of the terms $p(y|x)$ with y "central" will be large. The proof of this statement will exploit the fact that, if y has distance at least ℓ to both endpoints of the path, then $G[y]$ contains the path P_ℓ from Lemma 6 (with y located in the middle of P_ℓ) as a subgraph. It is then easy to argue that the term $f(\ell)$ from Lemma 6 serves as a lower bound on the cost achieved by p.

We will now formalize these ideas. Let $\ell \geq 1$ be arbitrary but fixed. We assume that $K \geq 4\ell$. We define the following sets of "peripheral" nodes:

$$\mathcal{K}_1 = \{y_1, \ldots, y_\ell\} \cup \{y_{K-\ell+1}, \ldots, y_K\} \quad \text{and}$$
$$\mathcal{K}_2 = \{y_1, \ldots, y_{2\ell}\} \cup \{y_{K-2\ell+1}, \ldots, y_K\}.$$

In other words, $\mathcal{K}_1$ (resp. $\mathcal{K}_2$) contains all nodes that have a distance of at most $\ell - 1$ (resp. $2\ell - 1$) to one of the endpoints y_1 and y_K. The complements of these sets are denoted $\bar{\mathcal{K}}_1$ and $\bar{\mathcal{K}}_2$, respectively. Note that each node in $\bar{\mathcal{K}}_1$ (resp. $\bar{\mathcal{K}}_2$) has a distance of at least ℓ (resp. 2ℓ) to both of the endpoints. Moreover, any point in $\mathcal{K}_1$ has distance of at least ℓ to any node in $\bar{\mathcal{K}}_2$. For every set $M \subseteq \mathcal{K} \times \mathcal{K}$, we define

$$P(M) = \sum_{(x,y) \in M} p(x,y) = \frac{1}{K} \cdot \sum_{(x,y) \in M} p(y|x),$$

i.e., $P(M)$ is the total probability mass assigned to pairs $(x,y) \in M$ if $x \in \mathcal{K}$ is uniformly distributed and y has probability $p(y|x)$ conditioned to x. Then $P(\mathcal{K} \times \mathcal{K}_1)$ denotes the total probability assigned to pairs (x,y) with $y \in \mathcal{K}_1$. The total mass of pairs from $\bar{\mathcal{K}}_2 \times \mathcal{K}_1$ can then be bounded from below as follows:

$$P(\bar{\mathcal{K}}_2 \times \mathcal{K}_1) = P(\bar{\mathcal{K}}_2 \times \mathcal{K}) - P(\bar{\mathcal{K}}_2 \times \bar{\mathcal{K}}_1) \geq P(\bar{\mathcal{K}}_2 \times \mathcal{K}) - P(\mathcal{K} \times \bar{\mathcal{K}}_1)$$
$$= \left(1 - \frac{4\ell}{K}\right) - (1 - P(\mathcal{K} \times \mathcal{K}_1)) = P(\mathcal{K} \times \mathcal{K}_1) - \frac{4\ell}{K}.$$

Since $p(x,y) = p(y|x)/K$, we may rewrite the cost function $f^G(p)$ from (4) as follows:

$$f^G(p) = \sum_{(x,y) \in \mathcal{K} \times \mathcal{K}} p(x,y) d(x,y).$$

Since, as mentioned above already, $d(x,y) \geq \ell$ holds for all pairs $(x,y) \in \bar{\mathcal{K}}_2 \times \mathcal{K}_1$, we obtain a first lower bound on $f^G(p)$:

$$f^G(p) \geq P(\bar{\mathcal{K}}_2 \times \mathcal{K}_1) \cdot \ell \geq \left(P(\mathcal{K} \times \mathcal{K}_1) - \frac{4\ell}{K}\right) \cdot \ell. \tag{14}$$

The lower bound (14) is induced by the elements y taken from the "peripheral region" $\mathcal{K}_1$. In the next step, we derive a lower bound that is induced by the elements y taken from the "central region" $\bar{\mathcal{K}}_1$. We remind the reader of the short notation

$$K_y(p) = \sum_{x \in \mathcal{K}} p(y|x) \quad \text{and} \quad \bar{f}_y(p) = \sum_{x \in \mathcal{K}} \frac{p(y|x)}{K_y(p)} \cdot d(x,y)$$

and mention just another way of expressing the cost function:

$$f^G(p) = \sum_{y \in \mathcal{K}} \frac{K_y(p)}{K} \bar{f}_y(p). \tag{15}$$

We set $\bar{p}_y(x) = p(y|x)/K_y(p)$ and observe that $\sum_{x \in \mathcal{K}} \bar{p}_y(x) = 1$. In the sequel, we set $\gamma = e^{-\varepsilon}$. Let $f(\ell)$ be the function given by (11).

Claim 2. *If $y \in \bar{\mathcal{K}}_1$, then $\bar{f}_y(p) \geq f(\ell)$.*

The proof of Claim 2 is quite simple and hence omitted. In view of (15) and in view of the obvious identity

$$\sum_{y \in \bar{\mathcal{K}}_1} \frac{K_y(p)}{K} = P(\mathcal{K} \times \bar{\mathcal{K}}_1) = 1 - P(\mathcal{K} \times \mathcal{K}_1),$$

the above claim, in combination with Lemma 6, immediately implies the following second lower bound on the cost function:

$$f^G(p) \geq (1 - P(\mathcal{K} \times \mathcal{K}_1)) \cdot f(\ell) \tag{16}$$

$$> (1 - P(\mathcal{K} \times \mathcal{K}_1)) \cdot \frac{2\gamma}{1 - \gamma^2} \cdot \left(1 - \gamma^\ell - \ell\gamma^\ell(1 - \gamma)\right) \tag{17}$$

$$\geq (1 - P(\mathcal{K} \times \mathcal{K}_1)) \cdot \frac{2\gamma}{1 - \gamma^2} \cdot \left(1 - (s + 1)e^{-s}\right), \tag{18}$$

where the final inequality is valid provided that $\ell \geq \frac{s}{1-\gamma}$. If $P(\mathcal{K} \times \mathcal{K}_1) \geq 1/s$, we may invoke (14) and conclude that

$$f^G(p) \geq \left(\frac{1}{s} - \frac{4\ell}{K}\right) \cdot \frac{s}{1 - \gamma} = \left(1 - \frac{4s\ell}{K}\right) \cdot \frac{1}{1 - \gamma}.$$

Otherwise, if $P(\mathcal{K} \times \mathcal{K}_1) < 1/s$, we may invoke (18) and conclude that

$$f^G(p) > \frac{2\gamma}{1 - \gamma^2} \cdot \left(1 - \frac{1}{s}\right) \cdot \left(1 - (s + 1)e^{-s}\right).$$

We can summarize this discussion as follows.

Theorem 2. *Let $G = (\mathcal{K}, E)$ be a path of length $K - 1$. Suppose that $s \geq 1$, $0 < \gamma < 1$, $\ell \geq \frac{s}{1-\gamma}$ and $K \geq 4\ell$. Then,*

$$\mathrm{OPT}[1] \geq \frac{1}{1 - \gamma} \cdot \min\left\{1 - \frac{4s\ell}{K}, \ \frac{2\gamma}{1 + \gamma} \cdot \left(1 - \frac{1}{s}\right) \cdot \left(1 - (s + 1)e^{-s}\right)\right\}.$$

Corollary 3. *With the same notations and assumptions as in Theorem 2, the following holds. If $s \geq 2$ and $K \geq \frac{s^2\ell(1+\gamma)}{\gamma}$, then*

$$\mathrm{OPT}[1] \geq \frac{2\gamma}{1 - \gamma^2} \left(\left(1 - \frac{1}{s}\right) \cdot 1 - (s + 1)e^{-s}\right).$$

Proof. For $s \geq 2$ and $K \geq \frac{s^2\ell(1+\gamma)}{\gamma}$ the minimum in Theorem 2 is taken by the second of the two possible terms. □

We would like to show that the parameter vector $(p(y|x))$ which represents the exponential mechanism comes close to optimality. To this end, we need an upper bound on $f^G(p)$. In a first step, we determine an upper bound on the cost induced by the exponential mechanism which makes $p(y|x)$ proportional to $\gamma^{d(x,y)} = \exp(-\varepsilon d(x,y))$. This mechanism might achieve 2ε-differential privacy only. In a second step, we determine an upper bound on the cost induced by the ε-differentially private exponential mechanism which makes $p(y|x)$ proportional to $\gamma^{d(x,y)/2} = \exp(-\varepsilon d(x,y)/2)$. But let's start with the first step.

Lemma 7. *Suppose that the graph $G = (\mathcal{K}, E)$ forms a path of length $K - 1$. If p is determined by the 2ε-differentially private exponential mechanism which makes $p(y|x)$ proportional to $\gamma^{d(x,y)}$, then*

$$f^G(p) < \frac{2\gamma}{1 - \gamma^2}.$$

Note that this is optimal asymptotically, i.e., when K and the slack parameters ℓ, s in Corollary 3 approach infinity. The proof of Lemma 7 is quite simple and hence omitted. An application of Corollary 3 and of Lemma 7 (with $\gamma^{1/2} = \sqrt{\gamma} = e^{-\varepsilon/2}$ at the place of $\gamma = e^{-\varepsilon}$) immediately leads to the following result:

Corollary 4. *Suppose that the graph $G = (\mathcal{K}, E)$ forms a path of length $K - 1$, $s \geq 2$ and $K \geq \frac{s^2\ell(1+\gamma)}{\gamma}$. If p is determined by the ε-differentially private exponential mechanism which makes $p(y|x)$ proportional to $\gamma^{d(x,y)/2} = \exp(-\varepsilon d(x,y)/2)$, then*

$$\frac{\mathrm{OPT}^G[1]}{f^G(p)} \geq \frac{\gamma(1 - \sqrt{\gamma^2})}{\sqrt{\gamma}(1 - \gamma^2)} \cdot \left(1 - \frac{1}{s}\right) \cdot \left(1 - (s+1)e^{-s}\right)$$

$$\geq \frac{\sqrt{\gamma}}{1 + \gamma} \cdot \left(1 - \frac{1}{s}\right) \cdot \left(1 - (s+1)e^{-s}\right).$$

Note that $\frac{\sqrt{\gamma}}{1+\gamma}$ is close to $1/2$ if γ is close to 1.

6 Worst-Case Optimality: Sorting Function

In this section we briefly discuss a scenario where the exponential mechanism is optimal in terms of the worst-case error. More specifically, we consider the problem of publishing the output of the sorting function under differential privacy. Similarly to Example 1 in Sect. 3, we assume that a database $\mathcal{D}$ is associated with a vector $v \in \mathbb{R}^T$, and that neighboring databases lead to values $v, v' \in \mathbb{R}^T$ such that $\|v - v'\|_1 \leq 2$. For $r \leq T$, the *sorting function* $\pi \colon \mathbb{R}^T \to \mathbb{R}^r$ is defined as follows. For every $v \in \mathbb{R}^T$, take a permutation σ of $1, \ldots, T$ such that $v_{\sigma(1)} \geq \ldots \geq v_{\sigma(T)}$ and define $\pi(v) = (v_{\sigma(1)}, \ldots, v_{\sigma(r)})$. For instance, v may be a frequency list from a password dataset $\mathcal{D}$ and the sorting function applied to v would then return the frequency of the r most chosen passwords in the dataset. In this case, the sorting function π is actually defined over $\mathbb{N}^T$, and inputs can be thought of as histograms. Recent works [2,3] focus on the problem of releasing the whole list of password frequencies under differential privacy, i.e., for $r = T$. Here, we extend the analysis to a more general framework, where r can be arbitrary and the sorting functions are not restricted to histograms.

We first present a lower bound on the minimax risk (under the L_1-norm) that any differentially private mechanism must incur when releasing the output of the sorting function. The omitted proof is based on an application of Assouad's lemma. We underline that Theorem 3 is not entirely original, but carries over and extends a result that appears in a paper currently under review [1].

Theorem 3. *Let $\varepsilon \leq 1/8$. Then, any ε-differentially private mechanism for the sorting function $\pi\colon \mathbb{R}^T \to \mathbb{R}^r$, applied to values with L_1-norm upper-bounded by $N \leq T$, must incur the following minimax risks:*

1. *If $N \leq 1 + 1/(4\varepsilon)$, then $R^\star = \Omega(N)$;*
2. *If $N \geq 1/(2\varepsilon)$, then $R^\star = \Omega(\sqrt{N/\varepsilon})$; or*
3. *If $N \geq r(r+1)/(4\varepsilon) + r$, then $R^\star = \Omega(r/\varepsilon)$.*

We are now ready to prove the optimality of the exponential mechanism when it is used to release the output of the sorting function. For $s \in \mathbb{R}^r$, define $u(v,s) = -\|\pi(v) - s\|_1$. Note that the exponential mechanism instantiated with this utility function corresponds to the Laplace mechanism which adds Laplace noise with parameter $2/\varepsilon$ to the components of $\pi(v)$. It then is straightforward to show that the error introduced is $O(r/\varepsilon)$. Therefore, for sufficiently large values of N, this upper bound matches the corresponding lower bound in Theorem 3, concluding the analysis.

Acknowledgments. The research was supported by the DFG Research Training Group GRK 1817/1.

References

1. Aldà, F., Simon, H.U.: A lower bound on the release of differentially private integer partitions. Inf. Process. Lett. (2017, submitted)
2. Blocki, J.: Differentially private integer partitions and their applications (2016). tpdp.16.cse.buffalo.edu/abstracts/TPDP_2016_4.pdf. Accessed 08 Aug 2016
3. Blocki, J., Datta, A., Bonneau, J.: Differentially private password frequency lists. In: Proceedings of the 23rd Annual Network and Distributed System Security Symposium (2016)
4. Blum, A., Ligett, K., Roth, A.: A learning theory approach to non-interactive database privacy. J. ACM **60**(2), 12 (2013)
5. Brenner, H., Nissim, K.: Impossibility of differentially private universally optimal mechanisms. In: Proceedings of the 51st Annual IEEE Symposium on Foundations of Computer Science, pp. 71–80 (2010)
6. De, A.: Lower bounds in differential privacy. In: Cramer, R. (ed.) TCC 2012. LNCS, vol. 7194, pp. 321–338. Springer, Heidelberg (2012). doi:10.1007/978-3-642-28914-9_18
7. Duchi, J.C., Jordan, M.I., Wainwright, M.J.: Local privacy and statistical minimax rates. In: Proceedings of the 54th Annual IEEE Symposium on Foundations of Computer Science, pp. 429–438 (2013)
8. Dwork, C., McSherry, F., Nissim, K., Smith, A.: Calibrating noise to sensitivity in private data analysis. In: Halevi, S., Rabin, T. (eds.) TCC 2006. LNCS, vol. 3876, pp. 265–284. Springer, Heidelberg (2006). doi:10.1007/11681878_14
9. Dwork, C., McSherry, F., Talwar, K.: The price of privacy and the limits of LP decoding. In: Proceedings of the 39th Annual ACM Symposium on Theory of Computing, pp. 85–94 (2007)
10. Geng, Q., Kairouz, P., Oh, S., Viswanath, P.: The staircase mechanism in differential privacy. IEEE J. Sel. Top. Sign. Process. **9**(7), 1176–1184 (2015)

11. Ghosh, A., Roughgarden, T., Sundararajan, M.: Universally utility-maximizing privacy mechanisms. In: Proceedings of the 41st Annual ACM Symposium on Theory of Computing, pp. 351–360 (2009)
12. Hall, R., Rinaldo, A., Wasserman, L.: Random differential privacy. J. Priv. Confidentiality 4(2), 43–59 (2012)
13. Hardt, M., Talwar, K.: On the geometry of differential privacy. In: Proceedings of the 42nd ACM Symposium on Theory of Computing, pp. 705–714 (2010)
14. Hsu, J., Roth, A., Roughgarden, T., Ullman, J.: Privately solving linear programs. In: Esparza, J., Fraigniaud, P., Husfeldt, T., Koutsoupias, E. (eds.) ICALP 2014. LNCS, vol. 8572, pp. 612–624. Springer, Heidelberg (2014). doi:10.1007/978-3-662-43948-7_51
15. Kairouz, P., Oh, S., Viswanath, P.: Extremal mechanisms for local differential privacy. In: Advances in Neural Information Processing Systems, pp. 2879–2887 (2014)
16. Kasiviswanathan, S.P., Lee, H.K., Nissim, K., Raskhodnikova, S., Smith, A.: What can we learn privately. SIAM J. Comput. 40(3), 793–826 (2011)
17. Koufogiannis, F., Han, S., Pappas, G.J.: Optimality of the Laplace mechanism in differential privacy. arXiv preprint arXiv:1504.00065 (2015)
18. McGregor, A., Mironov, I., Pitassi, T., Reingold, O., Talwar, K., Vadhan, S.: The limits of two-party differential privacy. In: Proceedings of the 51st Annual IEEE Symposium on Foundations of Computer Science, pp. 81–90 (2010)
19. McSherry, F., Talwar, K.: Mechanism design via differential privacy. In: Proceedings of the 48th Annual IEEE Symposium on Foundations of Computer Science, pp. 94–103 (2007)

On Pairing Inversion of the Self-bilinear Map on Unknown Order Groups

Hyang-Sook Lee[1], Seongan Lim[2]([✉]), and Ikkwon Yie[3]

[1] Department of Mathematics, Ewha Womans University, Seoul, Korea
hsl@ewha.ac.kr
[2] Institute of Mathematical Sciences, Ewha Womans University, Seoul, Korea
seongannym@ewha.ac.kr
[3] Department of Mathematics, Inha University, Incheon, Korea
ikyie@inha.ac.kr

Abstract. A secure self-bilinear map is attractive since it can be naturally extended to a secure multi-linear map which has versatile applications in cryptography. However, it was known that a self-bilinear map on a cyclic group of a known order cannot be cryptographically secure. In 2014, Yamakawa et al. presented a self-bilinear map, the YYHK pairing, on unknown order groups by using an indistinguishability obfuscator as a building block. In this paper, we prove that the Pairing Inversion (PI) of the YYHK pairing is equivalently hard to the factorization of RSA modulus N as long as iO in the scheme is an indistinguishability obfuscator. First, we prove that the General Pairing Inversion (GPI) of the YYHK pairing $e : G \times G \to G$ is always solvable. By using the solvability of GPI, we prove that PI and BDHP for the YYHK-pairing e are equivalently hard to CDHP in the cyclic group G. This equivalence concludes that PI for the YYHK-pairing is equivalently hard to the factorization of N.

Keywords: Self-bilinear map · Pairing Inversion · General Pairing Inversion

1 Introduction

A secure self-bilinear map is attractive since it can be naturally extended to a secure multi-linear map which has versatile applications in cryptography. For a self-bilinear map $e : G \times G \to G$, the cyclic group G has its own well-defined computational problems such as Discrete Logarithm Problem (DLP) and Computational Diffie-Hellman Problem (CDHP). The bilinear map introduces several new computational problems such as Bilinear Diffie-Hellman Problem (BDHP), Pairing Inversion (PI) and General Pairing Inversion (GPI). In most cases, the hardness of PI is directly related to the security of cryptographic schemes using the pairing. Therefore, proving the exact hardness of PI is important for the cryptographic usage of the pairing. Results on the comparison of the hardness of PI and other classical computational problems are partially presented [3,6]. In [3], Galbraith et al. proved that PI for a pairing $e : G \times G \to G$ is at least as

© Springer International Publishing AG 2017
S. Dolev and S. Lodha (Eds.): CSCML 2017, LNCS 10332, pp. 86–95, 2017.
DOI: 10.1007/978-3-319-60080-2_6

hard as the CDHP in G. In fact, assessing the exact strength of PI is one of the interesting challenges in the research on the cryptographic pairings [1,4,5].

For a self-bilinear map $e : G \times G \to G$, if the order of the cyclic group G is known then the CDHP in G is known to be solvable by [2]. Therefore, the self-bilinear map on a group of known order is not useful in cryptography. Recently, a self-bilinear map with auxiliary input on a cyclic group G of unknown order was presented by T. Yamakawa, S. Yamada, G. Hanaoka and N. Kunihiro [7] (from hereon it referred to simply as "the YYHK-pairing"). The YYHK-pairing requires indistinguishability obfuscation iO as one of the building block algorithms, which has no known practical construction as of today. However, we think it is worth to assess the strength of PI of the YYHK-pairing independently of practical construction of iO. In [7], they proved that BDHP on the cyclic group G in the YYHK-pairing is equivalently hard to the integer factorization if the underlying iO is an indistinguishability obfuscator. This implies that PI of the YYHK-pairing is at least as hard as the integer factorization under the same assumption. In this paper, we prove that PI of the YYHK-pairing is, in fact, equivalently hard to the integer factorization under the same assumption. In order to assess the strength of PI, we prove the solvability of GPI for the YYHK-pairing. We note that GPI itself has no known direct security impact on the pairing based cryptography. However, we found that the GPI is useful to analyze the strength of computational problems related to a pairing $e : G \times G \to G$. By using the solvability of GPI, we prove that PI and BDHP for the YYHK-pairing is equivalently hard to CDHP in G. Therefore, PI is equivalently hard to the factorization of N as long as iO in the scheme is an indistinguishability obfuscator.

The rest of the paper is organized as follows. In Sect. 2, we review the definitions of computational problems for a self-bilinear map $e : G \times G \to G$ and known relations among these problems. We also present a brief description of the YYHK pairing. In Sect. 3, we present our results. We add some comments on the parameters and prove GPI is solvable for the YYHK pairing under a reasonable assumption on its parameters. By using the solvability of GPI, we prove that all the problems PI, BDHP for the YYHK-pairing and CDHP in G are equivalently hard and their equivalence concludes that PI for the YYHK-pairing is equivalently hard to the factorization of N. In Sect. 4, we conclude our paper.

2   Preliminaries

In this section, we review some definitions and known results related to the main parts of this paper.

2.1   Self-bilinear Maps

First, we recall the definition of a self-bilinear map [7].

Definition 1 (Self-bilinear Map). *For a cyclic group G, a self-bilinear map $e : G \times G \to G$ has the following properties.*

– *The map $e : G \times G \to G$ is efficiently computable.*
– *For all $g_1, g_2 \in G$ and $\alpha \in \mathbb{Z}$, it holds that*

$$e(g_1^\alpha, g_2) = e(g_1, g_2^\alpha) = e(g_1, g_2)^\alpha.$$

– *The map e is non-degenerate, that is, if $g_1, g_2 \in G$ are generators of G, then $e(g_1, g_2)$ is a generator of G.*

We consider a self-bilinear map $e : G \times G \to G$ with unknown order $|G|$ in this paper. We express the group G as a multiplicative group.

2.2 Computational Problems Related to Bilinear Maps

There are several computational problems related to a bilinear map $e : G \times G \to G$ which is required to be hard for cryptographic application of the bilinear map.

Discrete Logarithm Problem in G (DLP): the problem of computing $a \in \mathbb{Z}_{|G|}$ for a given pair $(g, y = g^a)$ with randomly chosen $g \in G$ and $a \in \mathbb{Z}_{|G|}$. In this case, we denote $\mathsf{DLP}(g, y) = a$.

Computational Diffie-Hellman Problem in G (CDHP): the problem of computing $g^{ab} \in G$ for a given triple (g, g^a, g^b) with randomly chosen $g \in G$ and $a, b \in \mathbb{Z}_{|G|}$. In this case, we denote $\mathsf{CDHP}(g, g^a, g^b) = g^{ab}$.

Pairing Inversion Problem (PI): the problem of computing $g_1 \in G$ such that $e(g, g_1) = z$ for any given $g, z \in G$. In this case, we denote $\mathsf{PI}(g, z) = g_1$.

Bilinear Diffie-Hellman Problem (BDHP): the problem of computing $e(g, g)^{abc} \in G$ for a given (g, g^a, g^b, g^c) with randomly chosen $g \in G$ and $a, b, c \in \mathbb{Z}_{|G|}$). In this case, we denote $\mathsf{BDHP}(g, g^a, g^b, g^c) = e(g, g)^{abc}$.

There is another computational problem, GPI, which is related to a bilinear map $e : G \times G \to G$ where its security influence is not known.

General Pairing Inversion Problem (GPI): the problem of finding a pair $(g_1, g_2) \in G \times G$ such that $e(g_1, g_2) = z$ for a randomly given $z \in G$. In this case, we denote $\mathsf{GPI}(z) = (g_1, g_2)$.

A computational problem is said to be solvable, or easy to solve, if there is a polynomial time algorithm that output a correct solution for any instance of the problem. The following relations are known for a bilinear map $e : G \times G \to G$ [3, 6]:

$$\mathsf{DLP} \longleftarrow \mathsf{PI} \longleftarrow \mathsf{CDHP} \longleftarrow \mathsf{BDHP}$$

Here, $\mathsf{Prob}_1 \to \mathsf{Prob}_2$ means that the problem Prob_1 is solvable by using a polynomial time algorithm that solves the problem Prob_2.

For a self-bilinear map $e : G \times G \to G$, if the order of the cyclic group G is known, then the problem CDHP is known to be solvable by [2], which clearly implies the solvability of BDHP for the bilinear map. Therefore, the self-bilinear map on a group of known order is not useful in cryptography and this is why the YYHK self-bilinear map in [7] is defined on a group of unknown order.

2.3 Group of Signed Quadratic Residues

We recall the definition of a group of signed quadratic residues $\mathbb{QR}_N^+$. Let N be a RSA modulus of ℓ_N-bit which is a Blum integer, that is, $N = pq$ where p and q are distinct primes with the same length and $p \equiv q \equiv 3 \bmod 4$ and $\gcd(p - 1, q - 1) = 2$ hold. We represent the modular ring $\mathbb{Z}_N$ as follows:

$$\mathbb{Z}_N = \{-\frac{N-1}{2}, ..., -1, 0, 1, ..., \frac{N-1}{2}\}.$$

We define the set of signed quadratic residues as

$$\mathbb{QR}_N^+ = \{|u^2 \bmod N| \| u \in \mathbb{Z}_N^*\},$$

where $|u^2 \bmod N|$ is the absolute value of u^2 as an element of $\mathbb{Z}_N$. We define the group of signed quadratic residues $(\mathbb{QR}_N^+, \circ)$ with the binary operation '$\circ$' defined as

$$g \circ h = |gh \bmod N| \text{ for } g, h \in \mathbb{QR}_N^+.$$

Then the group of signed quadratic residues $(\mathbb{QR}_N^+, \circ)$ is cyclic of order $\frac{(p-1)(q-1)}{4}$ which is unknown if the factorization of N is unknown.

2.4 YYHK Pairing

T. Yamakawa, S. Yamada, G. Hanaoka and N. Kunihiro presented a self-bilinear map (YYHK pairing) on a cyclic group of unknown order [7]. We briefly review the YYHK pairing. We refer [7] for details of the YYHK pairing. The YYHK pairing use an Indistinguishability Obfuscator(iO) as a building block. The definition of iO is given as follows [7].

Definition 2 (Indistinguishability Obfuscator(iO)). *Let $\mathcal{C}_\lambda$ be the class of circuits of size at most λ. An efficient randomized algorithm iO is called an indistinguishability obfuscator if the following conditions are satisfied:*

– *For all security parameter $\lambda \in \mathbb{N}$, for all $C \in \mathcal{C}_\lambda$, we have that*

$$Pr\left[\forall x C'(x) = C(x) | C' \leftarrow iO(\lambda, C)\right] = 1.$$

– *For any efficient algorithm $\mathcal{A} = (\mathcal{A}_1, \mathcal{A}_2)$, there exists a negligible function negli such that the following holds: if $\mathcal{A}_1(1^\lambda)$ always outputs (C_0, C_1, σ) with $C_0, C_1 \in \mathcal{C}_\lambda$ and $C_0(x) = C_1(x)$ for all x, then we have*

$$|Pr\left[\mathcal{A}_2(\sigma, iO(\lambda, C_0) = 1 : (C_0, C_1, \sigma) \leftarrow \mathcal{A}_1(1^\lambda)\right]$$
$$- Pr\left[\mathcal{A}_2(\sigma, iO(\lambda, C_1) = 1 : (C_0, C_1, \sigma) \leftarrow \mathcal{A}_1(1^\lambda)\right]| \leq negli(\lambda)$$

The iO makes circuits C_0 and C_1 computationally indistinguishable if they have exactly the same functionality.

The YYHK pairing considers the following set $\mathcal{C}_{N,x}$ of circuits for the RSA modulus N of ℓ_N-bits and an integer x. For input $y \in \{0, 1\}^{\ell_N}$, the circuit

$C_{N,x} \in \mathcal{C}_{N,x}$ interprets y as an element of $\mathbb{Z}_N$. If $y \in \mathbb{QR}_N^+$, $C_{N,x}$ returns $y^x \in \mathbb{QR}_N^+$. Otherwise it returns 0^{ℓ_N}. The canonical circuit $\tilde{C}_{N,x} \in \mathcal{C}_{N,x}$ is defined in a natural way of exponentiation.

The YYHK pairing consists of (InstGen, Sample, AlGen, Map, AlMult) which are described as follows:

InstGen(1^λ) $\to$ params: Run RSAGen(1^λ) to obtain (N, p, q) and set

$$\text{params} = N.$$

The params defines the followings:

- the cyclic group $G = \mathbb{QR}_N^+$ while representing $\mathbb{Z}_N = \{-\frac{N-1}{2}, ..., 0, ..., \frac{N-1}{2}\}$.
- $Approx(G) = \frac{N-1}{4}$,
- the set

$$T_x^\ell = \{iO(M_\ell, C_{N,2x}; r) : C_{N,2x} \in \mathcal{C}_{N,2x} \text{ such that } |C_{N,2x}| \le M_\ell, r \in \{0,1\}^*\}$$

for M_ℓ is chosen appropriately.

Sample(params) $\to g$: Choose a random element $h \in \dot{\mathbb{Z}}_N^*$, compute $h^2 \in \mathbb{Z}_N^*$ outputs $g = |h^2 \bmod N| \in \mathbb{QR}_N^+$. When params $= N$ and a generator $g \in G$ are fixed, the self-bilinear map $e_g : G \times G \to G$ is defined as $e_g(g^x, g^y) = g^{2xy}$.
AlGen(params, ℓ, x) $\to \tau_x$: Define the range of x as $R := [\frac{N-1}{2}]$. Take the canonical circuit $\tilde{C}_{N,2x} \in \mathcal{C}_{N,2x}$, set $\tau_x \leftarrow iO(M_\ell, \tilde{C}_{N,2x})$ and output τ_x.
Map(params, g^x, g^y, τ_y) $\to e_g(g^x, g^y)$: Compute $\tau_y(g^x)$ and output it.
AlMult(params, ℓ, τ_x, τ_y) $\to \tau_{x+y}$: Compute $\tau_{x+y} \leftarrow iO(M_\ell, Mult(\tau_x, \tau_y)$ and output it.

The hardness of BDHP for the YYHK-pairing e_g is proven in [7] in terms of multi-linear map. Here we recall it in terms of a bilinear map.

Theorem 1 (Theorem 1 in [7]). *The BDH assumption holds for the YYHK-pairing e_g if the factoring assumption holds with respect to RSAGen and iO is an indistinguishability obfuscator.*

3   Main Results

In this section, we show that PI for the YYHK-pairing is equivalently hard to the factorization of N as long as iO in the scheme is an indistinguishability obfuscator. Our proof consists of several parts. First we show that GPI for the YYHK-pairing is solvable. And then we prove the equivalence of the solvability of PI for the YYHK-pairing and the solvability of CDHP in G by using the solvability of GPI. We also prove the equivalence of the solvability of CDHP in G and the solvability of BDHP for the YYHK-pairing by using the solvability of GPI. Therefore, we see that BDHP and PI are equivalently solvable for the YYHK-pairing. Now one conclude the desired result by Theorem 1. It is noteworthy that our reduction proof is under the assumption that the oracles for hard problems solve every instance of the problem as we have defined for solvability of problems in the previous section.

3.1   Solvability of GPI for the YYHK-pairing

First, we note that the YYHK pairing is defined with respect to the selected generator g of $G = \mathbb{QR}_N^+$. For this moment, we denote the YYHK pairing using a generator g as $e_g : G \times G \to G$. We prove that GPI for the YYHK-pairing is solvable.

Lemma 1. *For any generator $g \in G = \mathbb{QR}_N^+$ with $g^{1/2} \in G$, $e_g(g^{1/2}, g^k) = g^k$ for any integer k.*

Proof. Since $g^{1/2} \in G$, $X = e_g(g^{1/2}, g^k) \in G$ is well-defined. We see that $e_g(g, g^k) = g^{2k} = (g^k)^2$ and $e_g(g, g^k) = e_g(g^{1/2} \cdot g^{1/2}, g^k) = X^2$. Therefore, we have $|X^2 \bmod N| = |(g^k)^2 \bmod N|$, which implies that $X = \pm g^k \bmod N$. Since $-1 \notin \mathbb{QR}_N$ and $g^k \in G$, we have $X = g^k$.   □

Lemma 1 implies that the problem GPI is solvable for the YYHK-pairing $e_g : G \times G \to G$ if $g^{1/2} \in G$ is given.

There is an important difference in the description of the YYHK pairing compare with the pairing using elliptic curve groups. The computation of the YYHK pairing requires a trusted system manger. The procedures of algorithms InstGen and AlGen contains information which should be kept secret by the system manager, that is, the prime numbers p and q for InstGen and the information on x for AlGen. Considering the YYHK pairing as a publicly computable map, it is desirable that all algorithms other than InstGen and AlGen do not contain secret information. In particular, we can assume that AlGen do not contain any secret information which means that it is reasonable to assume that the element $h = g^{1/2} \in G$ is public as well as g for $e_g : G \times G \to G$. Therefore, we can assume that the GPI for e_g is always solvable for the YYHK-pairing $e_g : G \times G \to G$.

Remark 1. We have followings remarks on the YYHK pairing:

– As in the definition of the YYHK pairing, the map $e : G \times G \to G$ is defined with respect to the selected generator g of G. It is important to note that different generators can define different pairings. For example, suppose that all of g, g^2, h, h^2 are generators of G, then we see that the pairings defined by g^2 and h^2 are two different pairings from Lemma 1:

$$e_{g^2}(g, h) = h, e_{h^2}(g, h) = g$$

Therefore, the map $e : G \times G \to G$ should include the considered generator g explicitly since it is not well-defined otherwise.
– The RSA modulus $N = pq$ for the YYHK pairing should be of the form $p = 2p' + 1, q = 2q' + 1$ where p' and q' are odd primes. In this way we have $e_g : G \times G \to G$ which is surjective.

Now we show that for a fixed N, it is desirable to fix one generator g, too.

Theorem 2. *One can solve CDHP in the cyclic group G by using pairing computations for some generators as an oracle.*

Proof. This is clear from the equality: CDHP $(g, g_1, g_2)) = e_{g^2}(g_1, g_2)$. $\square$

Theorem 2 says that if there are more generators which define pairings on the group $G = \mathbb{QR}_N^+$ then there are more instances of CDHP which are solvable. Therefore, it is best to fix one generator, that is, one pairing for one group $G = \mathbb{QR}_N^+$. In this sense, one can omit the index g of the YYHK pairing even though the definition the YYHK pairing is determined by g.

3.2 PI $\leftrightarrow$ CDHP for the YYHK-pairing

We prove a more general result for the hardness of problems PI and CDHP.

Theorem 3. *If GPI is solvable for a pairing* $e : G \times G \to G$, *then we have*

$$CDHP \ in \ G \ \leftrightarrow PI \ for \ the \ pairing \ e.$$

Proof. We note that PI $\leftarrow$ CDHP is clear from the following equality,

$$\text{CDHP}(g_0, g_0^a, g_0^b) = \text{PI}(g_0, e(g_0^a, g_0^b)) \text{ for any } g_0 \in G \text{ and } a, b \in \mathbb{Z}_{|G|}.$$

Now we show that PI $\to$ CDHP. Assume that the problem CDHP is solvable in the cyclic group G. We recall that GPI is solvable for the pairing e by the hypothesis. Therefore, we can assume that the computations CDHP($*$) and GPI($*$) can be done in polynomial time. The followings explain how to solve PI using CDHP and GPI solvers:

Input: an instance (g_0, w) of the problem PI for the pairing e
(i) Solve GPI and get GPI$(w) = (g_1, g_2)$, that is, $e(g_1, g_2) = w$.
(ii) Solve CDHP and get CDHP$(g_0, g_1, g_2) = g_3$.
(iii) Output PI$(g_0, w) \leftarrow g_3$

Clearly, the computation above is completed in polynomial time. Now we show that PI$(g_0, w) = g_3$, that is, $e(g_0, g_3) = w$. We see that $g_1 = g_0^a$ and $g_2 = g_0^b$ for some $a, b \in \mathbb{Z}_{|G|}$ which means that $g_3 = g_0^{ab}$. Therefore, we have

$$e(g_0, g_3) = e(g_0, g_0^{ab}) = e(g_0^a, g_0^b) = e(g_1, g_2) = w.$$ $\square$

3.3 CDHP $\leftrightarrow$ BDHP for the YYHK-pairing

We prove a more general result for the relations on the problems CDHP and BDHP.

Theorem 4. *Suppose that GPI is solvable for a pairing* $e : G \times G \to G$. *Then BDHP for the pairing e is solvable if and only if CDHP in the cyclic group G is solvable.*

Proof. We note that CDHP $\leftarrow$ BDHP is clear from the following equality,

$$\mathsf{BDHP}(g_0, g_0^a, g_0^b, g_0^c) = e(\mathsf{CDHP}(g_0, g_0^a, g_0^b), g_0^c) \text{ for any } g_0 \in G \text{ and } a, b, c \in \mathbb{Z}_{|G|}.$$

Now we show that CDHP $\rightarrow$ BDHP. Assume that the problem BDHP is solvable for the pairing e. We recall that GPI is solvable for the pairing by the hypothesis. Therefore, we can assume that the computations BDHP($*$) and GPI($*$) can be done in polynomial time. The following describes how to solve CDHP using BDHP and GPI solvers:

Input: an instance $(\hat{g}, \hat{g}^a, \hat{g}^b)$ of the problem CDHP
  (i) Solve the problem GPI and get $(g_i, h_i)_{i=0,1,2}$:

$$\mathsf{GPI}(\hat{g}) = (g_0, h_0), \mathsf{GPI}(\hat{g}^a) = (g_1, h_1), \mathsf{GPI}(\hat{g}^b) = (g_2, h_2)$$

$$(\text{that is }, e(g_0, h_0) = \hat{g}, e(g_1, h_1) = \hat{g}^a, e(g_2, h_2) = \hat{g}^b)$$

  (ii) Solve the problem BDHP and get w_3, w_4:

$$\mathsf{BDHP}(g_0, h_0, g_1, g_2) = w_3, \mathsf{BDHP}(h_0, g_0, h_1, h_2) = w_4$$

  (iii) Solve the problem GPI and get $(g_3, h_3), (g_4, h_4)$:

$$\mathsf{GPI}(w_3) = (g_3, h_3), \mathsf{GPI}(w_4) = (g_4, h_4)$$

  (iv) Solve BDHP and get $\mathsf{BDHP}(g_0, g_3, h_3, g_4) = w_5$.
  (v) Solve GPI and get $\mathsf{GPI}(w_5) = (g_5, h_5)$
  (vi) Solve the problem BDHP and get $\mathsf{BDHP}(h_0, h_4, g_5, h_5) = w_6$
  (vii) Output $\mathsf{CDHP}(\hat{g}, \hat{g}^a, \hat{g}^b) \leftarrow w_6$.

Clearly, the computation above is completed in polynomial time. Now we show that $\mathsf{CDHP}(\hat{g}, \hat{g}^a, \hat{g}^b) = w_6$, that is, $w_6 = \hat{g}^{ab}$. For $1 \leq i \leq 5$, we see that $g_i = g_0^{\alpha_i}$ and $h_i = h_0^{\beta_i}$ for some $\alpha_i, \beta_i \in \mathbb{Z}_{|G|}$. We note that the followings are true for the unknowns $a, b, \alpha_i, \beta_i, |G|$,

$$ab \equiv \alpha_1 \alpha_2 \beta_1 \beta_2 \bmod |G|$$
$$\alpha_3 \beta_3 \equiv \alpha_1 \alpha_2 \bmod |G|$$
$$\alpha_4 \beta_4 \equiv \beta_1 \beta_2 \bmod |G|$$
$$\alpha_5 \beta_5 \equiv \alpha_3 \beta_3 \alpha_4 \bmod |G|.$$

Combining altogether, we get

$$w_6 = e(g_0, h_0)^{\beta_4 \alpha_5 \beta_5} = \hat{g}^{\beta_4 \alpha_5 \beta_5} \text{ and } \beta_4 \alpha_5 \beta_5 \equiv \beta_4 \alpha_3 \beta_3 \alpha_4 \equiv \alpha_1 \alpha_2 \beta_1 \beta_2 \equiv ab \bmod |G|.$$

Therefore, we have $w_6 = \hat{g}^{ab}$.   □

From Theorems 3 and 4, we now conclude the following equivalences if GPI is solvable for a pairing $e : G \times G \to G$:

$$\mathsf{PI} \leftrightarrow \mathsf{CDHP} \leftrightarrow \mathsf{BDHP}.$$

Therefore, we see that the following problems are equivalently hard for the YYHK-pairing $e : G \times G \to G_T$:

$$\mathsf{PI} \leftrightarrow \mathsf{CDHP} \leftrightarrow \mathsf{BDHP}.$$

From Theorem 1, we see that the PI assumption holds for the YYHK-pairing $e : G \times G \to G_T$ if the factorization of N infeasible as long as iO is an indistinguishability obfuscator. Moreover, if one can factor, the order $\frac{(p-1)(q-1)}{4}$ of the cyclic group G is known which implies that CDHP in G is solvable by [2] and so does PI from the equivalence PI $\leftrightarrow$ CDHP.

Therefore, we have

Theorem 5. *The PI assumption holds for the YYHK-pairing e if and only if the factoring assumption holds with respect to RSAGen as long as iO is an indistinguishability obfuscator.*

4   Conclusion

In this paper, prove the pairing inversion (PI) of the YYHK self-bilinear map $e : G \times G \to G$ for $G = \mathbb{QR}_N^+$ is equivalently hard to the factorization of RSA modulus N as long as iO in the scheme is an indistinguishability obfuscator. We review the YYHK pairing and add some remarks on its parameters. The YYHK pairing on the group $G = \mathbb{QR}_N^+$ is defined in terms of a generator of g. We see that only one generator g can be used, that is, one pairing can be defined for the cyclic group G. Moreover, we see that it is desirable to assume that $h = g^{1/2} \in G$ is known as well as the generator g which defines the pairing. From these observation, we show that GPI of the YYHK pairing $e : G \times G \to G$ is always solvable. By using the solvability of GPI for the YYHK pairing, we prove that PI and BDHP for the YYHK-pairing are equivalently hard. Therefore, we conclude that PI is equivalently hard to the factorization of N as long as iO in the scheme is an indistinguishability obfuscator.

Acknowledgments. We thank the anonymous reviewers for useful comments. Hyang-Sook Lee was supported by Basic Science Research Programs through the National Research Foundation of Korea (NRF) funded by the Ministry of Science, ICT and Future Planning (Grant Number: 2015R1A2A1A15054564). Seongan Lim was also supported by Basic Science Research Programs through the NRF (Grant Number: 2016R1D1A1B01008562).

References

1. Chang, S., Hong, H., Lee, E., Lee, H.-S.: Pairing inversion via non-degenerate auxiliary pairings. In: Cao, Z., Zhang, F. (eds.) Pairing 2013. LNCS, vol. 8365, pp. 77–96. Springer, Cham (2014). doi:10.1007/978-3-319-04873-4_5
2. Cheon, J.-H., Lee, D.-H.: A note on self-bilinear maps. Bull. KMS **46**(2), 303–309 (2009)
3. Galbraith, S., Hess, F., Vercauteren, F.: Aspects of pairing inversion. IEEE Trans. Inf. Theor. **54**(12), 5719–5728 (2008)
4. Hess, F.: Pairings, 3rd Bar-Ilan Winter School on Cryptography (2013). http:crypto.biu.ac.il
5. Kanayama, N., Okamoto, E.: Approach to pairing inversions without solving Miller inversion. IEEE Trans. Inf. Theor. **58**(2), 1248–1253 (2012)
6. Satoh, T.: On pairing inversion problems. In: Takagi, T., Okamoto, T., Okamoto, E., Okamoto, T. (eds.) Pairing 2007. LNCS, vol. 4575, pp. 317–328. Springer, Heidelberg (2007). doi:10.1007/978-3-540-73489-5_18
7. Yamakawa, T., Yamada, S., Hanaoka, G., Kunihiro, N.: Self-bilinear map on unknown order groups from indistinguishability obfuscation and its applications. In: Garay, J.A., Gennaro, R. (eds.) CRYPTO 2014. LNCS, vol. 8617, pp. 90–107. Springer, Heidelberg (2014). doi:10.1007/978-3-662-44381-1_6

Brief Announcement: Anonymous Credentials Secure to Ephemeral Leakage

Łukasz Krzywiecki$^{(\boxtimes)}$, Marta Wszoła, and Mirosław Kutyłowski

Faculty of Fundamental Problems of Technology,
Wrocław University of Science and Technology, Wrocław, Poland
lukasz.krzywiecki@pwr.edu.pl

Abstract. We present a version of Camenisch-Lysyanskaya's anonymous credential system immune to attacks based on leakage of ephemeral values used during protocol execution. While preserving "provable security" of the original design, our scheme improves its security in a realistic scenario of an imperfect implementation on a cryptographic device.

Keywords: Anonymous credential · CL signature · Leakage · Adversary

Anonymous Credentials. An anonymous credentials system enable a user to prove his attributes without revealing his identity. This is possible when a trusted entity – *the Issuer* – provides the user appropriate cryptographic *credentials*, presumably after checking the user's attributes.

In this paper we focus on Camenisch-Lysyanskaya anonymous credentials system, *CL system* for short, based on their signature scheme [1]. It consists of two main protocols: issuing a credential to the user (see Fig. 1) by the issuer holding the private keys $(x, y, \{z\}_1^l)$ and proving attributes against a verifier (see Fig. 2) holding the public keys $(X = g^x, Y = g^y, \{Z_i\}_1^l)$, where $Z_i = g^{z_i}$.

CL system is provably secure: it is infeasible to prove possession of a credential without prior receiving such a credential from a party holding the private keys $(x, y, \{z\}_1^l)$. The proof is a reduction to the LRSW assumption [2]. In also supports anonymity by cryptographic means - a verifier learns nothing about the prover but the attributes presented.

Implementation Related Threats. *Provable security* of a scheme does not mean that a particular implementation, even based on a cryptographic device, is secure. In practice, it is risky hard to assume that the cryptographic device is unconditionally tamper resistant and that it contains neither trapdoors nor implementation faults.

These problems concern in particular CL system: it becomes insecure once the ephemeral values are not well protected. Indeed, once the ephemeral random values r_i and r'' are leaked, then immediately the secret m_i can be computed from s_i presented by the prover. Leakage of the random values can be caused by, say, poor design of the (pseudo)random number generator, leaking the seed of the generator, or lack of tamper resistance.

© Springer International Publishing AG 2017
S. Dolev and S. Lodha (Eds.): CSCML 2017, LNCS 10332, pp. 96–98, 2017.
DOI: 10.1007/978-3-319-60080-2_7

User($\{m\}_0^l$)		Issuer($x,\ y,\ \{z\}_1^l$)
$M = g^{m_0} \Pi_{i=1}^l Z_i^{m_i}$		
$(r_0, \dots r_l) \leftarrow_\$ \mathbb{Z}_q^*$		
$T = g^{r_0}\ \Pi_{i=1}^l Z_i^{r_i}$	$\xrightarrow{M,\ T}$	$c \leftarrow_\$ \mathbb{Z}_q^*$
	$\xleftarrow{c}$	
$\forall_{i \in \{0,\dots,l\}}\ s_i = r_i - cm_i$	$\xrightarrow{\{s_i\}_0^l}$	$T \overset{?}{=} M^c\ g^{s_0}\ \Pi_{i=1}^l Z_i^{s_i}$
		$a_0 \leftarrow_\$ \mathbb{Z}_q^*,\ A_0 = g^{a_0}$
		$\forall_{i \in \{1,\dots,l\}}\ A_i = A_0^{z_i}$
		$\forall_{i \in \{0,\dots,l\}}\ B_i = A_i^y$
Store($\{A_i\}_0^l,\ \{B_i\}_0^l,\ C$)	$\xleftarrow{\{A_i\}_0^l,\ \{B_i\}_0^l,\ C}$	$C = A_0^x\ M^{a_0 x y}$

Fig. 1. CL system: issuing a credential.

User($\{m_i\}_0^l, \{A_i\}_0^l, \{B_i\}_0^l,\ C$)		Verifier($X, Y, \{Z_i\}_1^l$)
$(r', r'', r_a, r_0, \dots, r_l) \leftarrow_\$ \mathbb{Z}_q^*$		
$\forall_{i \in \{0,\dots,l\}}\ \tilde{A}_i = A_i^{r'},\ \tilde{B}_i = B_i^{r'}$		
$\tilde{C} = C^{r'r''}$		
$\hat{t} = \hat{e}(X, \tilde{A}_0)^{r_a}\ \Pi_{i=0}^l \hat{e}(X, \tilde{B}_i)^{r_i}$	$\xrightarrow{\{\tilde{A}_i\}_0^l, \{\tilde{B}_i\}_0^l, \tilde{C}, \hat{t}}$	$\forall_{i \in \{1,\dots,l\}}\ e(\tilde{A}_0, Z_i) \overset{?}{=} \hat{e}(g, \tilde{A}_i)$
		$\forall_{i \in \{0,\dots,l\}}\ \hat{e}(\tilde{A}_i, Y) \overset{?}{=} \hat{e}(g, \tilde{B}_i)$
$s_a = r_a - cr''$	$\xleftarrow{c}$	$c \leftarrow_\$ \mathbb{Z}_q^*$
$\forall_{i \in \{0,\dots,l\}}\ s_i = r_i - cm_i r''$	$\xrightarrow{s_a, \{s_i\}_0^l}$	$\hat{t} \overset{?}{=} \hat{e}(g, \tilde{C})^c \hat{e}(X, \tilde{A}_0)^{s_a} \Pi_{i=0}^l \hat{e}(X, \tilde{B}_i)^{s_i}$

Fig. 2. CL system: attribute verification.

In order to deal with this situation we concern the security model, for which an adversary may get access to all ephemeral values created during the protocol execution. On the other hand we assume that the secret keys are implemented in a very minimalistic way: there is a separate component storing the secret keys, say $\bar{k}$, which on request $\bar{y}$ computes the value $f(\bar{k}, \bar{y})$, for some fixed, deterministic function f. In case of the protocol proposed below, the function f computes means exponentiation, where the secret keys are used as exponents.

Modified Anonymous Credentials Scheme. The idea is to eliminate the Schnorr-like signatures s_i which are the source of the problem in case of leakage. Instead, we follow the strategy from [3] and these values are given in the exponent. The protocol is depicted in Figs. 3 and 4.

While the number of modifications in the CL system is very small, it is a priori unclear whether it preserves the original security properties. It is well known that even tiny changes might have catastrophic consequences. E.g., forging a proof might be easier, since the values S_i are presented and not their discrete logarithms. Fortunately, it turns out that using a slightly modified LRSW Assumption one can prove similar properties as before, but for the adversary with access to the ephemeral values.

User($\{m_i\}_0^l$)		Issuer($x, y, \{z_i\}_1^l$)
$M = g^{m_0} \Pi_{i=1}^l Z_i^{m_i}$		
$(r_0, \ldots, r_l) \leftarrow_\$ \mathbb{Z}_q^*$		
$T = g^{r_0} \Pi_{i=1}^l Z_i^{r_i}$	$\xrightarrow{M,\,T}$	$(c, \omega) \leftarrow_\$ \mathbb{Z}_q^*,\ \tilde{g} = g^\omega$
	$\xleftarrow{c,\tilde{g}}$	
$\forall_{i \in \{0,\ldots,l\}}\ S_i = \tilde{g}^{r_i - c m_i}$	$\xrightarrow{\{S_i\}_0^l}$	$\hat{e}(\tilde{g}, T/M^c) \stackrel{?}{=} \hat{e}(S_0, g)\, \Pi_{i=1}^l \hat{e}(S_i, Z_i)$
		$a_0 \leftarrow_\$ \mathbb{Z}_q^*,\ A_0 = g^{a_0}$
		$\forall_{i \in \{1,\ldots,l\}}\ A_i = A_0^{z_i}$
		$\forall_{i \in \{0,\ldots,l\}}\ B_i = A_i^y$
Store($\{A_i\}_0^l, \{B_i\}_0^l,\ C$)	$\xleftarrow{\{A_i\}_0^l,\ \{B_i\}_0^l,\ C}$	$C = A_0^x\, M^{a_0 x y}$

Fig. 3. Credential issuance protocol for the modified system.

User($\{m_i\}_0^l, \{A_i\}_0^l, \{B_i\}_0^l, C$)		Verifier($X, Y, \{Z_i\}_1^l$)
$(r', r'', r_a, r_0, \ldots, r_l) \leftarrow_\$ \mathbb{Z}_q^*$		
$\forall_{i \in \{0,\ldots,l\}}\ \tilde{A}_i = A_i^{r'}$		
$\forall_{i \in \{0,\ldots,l\}}\ \tilde{B}_i = B_i^{r'}$		
$\tilde{C} = C^{r' r''}$		
$\hat{t} = \hat{e}(X, \tilde{A}_0)^{r_a} \Pi_{i=0}^l \hat{e}(X, \tilde{B}_i)^{r_i}$	$\xrightarrow{\{\tilde{A}_i\}_0^l, \{\tilde{B}_i\}_0^l, \tilde{C}, \hat{t}}$	$\forall_{i \in \{1,\ldots,l\}}\ \hat{e}(\tilde{A}_0, Z_i) \stackrel{?}{=} \hat{e}(g, \tilde{A}_i)$
		$\forall_{i \in \{0,\ldots,l\}}\ \hat{e}(\tilde{A}_i, Y) \stackrel{?}{=} \hat{e}(g, \tilde{B}_i)$
		$(\omega, c) \leftarrow_\$ \mathbb{Z}_q^*$
$s_a = r_a - c r''$	$\xleftarrow{c,\ \overline{X}}$	$\overline{X} = X^\omega$
$\forall_{i \in \{0,\ldots,l\}}\ S_i = \overline{X}^{r_i - c m_i r''}$	$\xrightarrow{s_a, \{S_i\}_0^l}$	$\hat{t}^\omega \stackrel{?}{=} \hat{e}(g^{\omega c}, \tilde{C}) \hat{e}(\overline{X}, \tilde{A}_0)^{s_a} \Pi_{i=0}^l \hat{e}(S_i, \tilde{B}_i)$

Fig. 4. Credential verification protocol for the modified system.

Acknowledgments. The paper was partially supported by the Polish National Science Center, based on the decision DEC-2013/08/M/ST6/00928, project HARMONIA.

References

1. Camenisch, J., Lysyanskaya, A.: Signature schemes and anonymous credentials from bilinear maps. In: Franklin, M. (ed.) CRYPTO 2004. LNCS, vol. 3152, pp. 56–72. Springer, Heidelberg (2004). doi:10.1007/978-3-540-28628-8_4
2. Lysyanskaya, A., Rivest, R.L., Sahai, A., Wolf, S.: Pseudonym systems. In: Heys, H., Adams, C. (eds.) SAC 1999. LNCS, vol. 1758, pp. 184–199. Springer, Heidelberg (2000). doi:10.1007/3-540-46513-8_14
3. Krzywiecki, L.: Schnorr-like identification scheme resistant to malicious subliminal setting of ephemeral secret. In: Bica, I., Reyhanitabar, R. (eds.) SECITC 2016. LNCS, vol. 10006, pp. 137–148. Springer, Cham (2016). doi:10.1007/978-3-319-47238-6_10

The Combinatorics of Product Scanning Multiplication and Squaring

Adam L. Young[1]([✉]) and Moti Yung[2,3]

[1] Cryptovirology Labs, New York, USA
ayoung235@gmail.com
[2] Snap Inc., Los Angeles, USA
[3] Deptartment of Computer Science, Columbia University, New York, USA

Abstract. Multiprecision multiplication and squaring are fundamental operations used heavily in fielded public key cryptosystems. The method called *product scanning* for both multiplication and squaring requires fewer memory accesses than the competing approach called *operand scanning*. A correctness proof for product scanning loop logic will assure that the method works as intended (beyond engineering testing) and will improve understanding of it. However, no proofs of correctness for product scanning multiplication loop logic nor product scanning squaring loop logic has been provided before, to our knowledge. To this end, in this note we provide exact combinatorial characterizations of the loop structure for both product scanning multiplication and product scanning squaring and then use these characterizations to present the first proofs of correctness for the iterative loops of these methods. Specifically, we identify the two combinatorial families that are inherently present in the loop structures. We give closed form expressions that count the size of these families and show successor algorithms for them. The combinatorial families we present may help shed light on the structure of similar methods. We also present loop control code that leverages these two successor algorithms. This has applications to implementations of cryptography and multiprecision libraries.

Keywords: Product scanning multiplication · Operand scanning multiplication · Multiple-precision arithmetic · Algorithmic combinatorics · Successor algorithm

1 Introduction

Multiple precision multiplication and squaring is at the heart of cryptographic libraries that implement public-key cryptography. Fast algorithms for multiplication and squaring have long been studied in the academic literature. Two main categories of algorithms are asymptotically fast algorithms and scanning algorithms. Asymptotically fast algorithms include Karatsuba [6], Toom-Cook [2,10], and Schönhage-Strassen [9]. These are advantageous for large multiplicands. Scanning algorithms for multiplication and squaring vary in performance

© Springer International Publishing AG 2017
S. Dolev and S. Lodha (Eds.): CSCML 2017, LNCS 10332, pp. 99–114, 2017.
DOI: 10.1007/978-3-319-60080-2_8

based on such factors as the number of single-precision multiplications and the number of memory accesses. Scanning multiplication and squaring algorithms are highly relevant in modern cryptography. They are typically used to instantiate the base-case of Karatsuba and are good alternatives in embedded applications where the overhead of asymptotically fast multipliers is prohibitive. In this paper we focus on product scanning multiplication and squaring.

From the perspective of algorithmic understanding and proving correctness we take a close look at product scanning. We found that the prior work on these algorithms for multiplication and squaring (developed mainly by practitioners) overlooked the problem of proving the correctness of the loop logic. We summarize our contributions as follows.

1. We define the two combinatorial families that are inherently present in product scanning multiplication and product scanning squaring.
2. We give closed-form solutions to counting the size of these two families.
3. We give "successor algorithms" for these two families and prove their correctness. These are the first proofs of correctness for product scanning multiplication and product scanning squaring loop logic. They are proper proofs by induction, thus holding for loops of all sizes.
4. We present product scanning multiplication and product scanning squaring loop code that leverages the two successor algorithms we develop. The correctness of product scanning is therefore further assured by our proof that the loop logic holds for all loop sizes.

Organization: Background and related work is presented in Sect. 2. Our results on product scanning multiplication and product scanning squaring are covered in Sects. 3 and 4, respectively. We conclude in Sect. 5.

2   Background and Related Work

Two forms of multiprecision multiplication are operand scanning and product scanning. Operand scanning has the outer loop move through the words of one of the operands. Product scanning has the outer loop move through the words of the final product. The product scanning method is described by Kaliski and concretely expressed as `BigMult` [5]. An integer x is expressed as an n-digit array $x[0], ..., x[n-1]$. BIG_MULT_A, BIG_MULT_B, BIG_MULT_C, and BIG_MULT_N correspond to variables a, b, c, and n. Observe that four **for** loops are used instead of two. Kaliski unrolls the outermost loop by partitioning the computation along the half-way point. In so doing, the control code of the unrolled portions is simple. In [5] a proof of correctness of the control code is not provided and a product scanning implementation of squaring is not covered.

The **for** loop structure in Algorithm 2 of [3] is identical to Kaliski's `BigMult` with the following exception. The final single-precision value of the product is assigned outside of the last loop in Algorithm 2 whereas Kaliski assigns it within his **for** loop.

Algorithm 2.10 [4] is the product scanning form of integer multiplication. The issue is how to implement the **for** loop control code in step 2.1 efficiently.

```
/* Computes a = b*c. Lengths: a[2*n], b[n], c[n]. */
void BigMult (void)
{
  unsigned long x;
  unsigned int i, k;
  x = 0;
  for (k = 0; k < BIG_MULT_N; k++) {
    for (i = 0; i <= k; i++)
      x += ((unsigned long)BIG_MULT_B[i])*BIG_MULT_C[k-i];
    BIG_MULT_A[k] = (unsigned char)x;
    x >>= 8;
  }
  for (; k < (unsigned int)2*BIG_MULT_N; k++) {
    for (i = k-BIG_MULT_N+1; i < BIG_MULT_N; i++)
      x += ((unsigned long)BIG_MULT_B[i])*BIG_MULT_C[k-i];
    BIG_MULT_A[k] = (unsigned char)x;
    x >>= 8;
  }
}
```

This is set notation and leaves a significant amount of implementation choices up to the programmer.

Hankerson et al. point out that, generally speaking, operand scanning multiplication has more memory accesses whereas Algorithm 2.10 has more complex control code unless the loops are unrolled (p. 70, [4]). Comba also compared operand scanning with product scanning. He did so in detail for 16-bit Intel processors. Comba stated, "The required control code is then rather complicated and time-consuming. The problem can be avoided by a radical solution: unraveling not only the inner loop, but the outer loop as well" [1]. The unrolled product scanning method is present in OpenSSL.

When squaring a multiprecision number about half of the single-precision multiplications can be avoided due to the fact that $a[i] * a[j] = a[j] * a[i]$. This is in contrast to multiplication that computes $a[i] * b[j]$ and $a[j] * b[i]$ since it may be the case that $a[i] * b[j] \neq a[j] * b[i]$. Tuckerman presented a summation formula for fast multiprecision squaring that leverages this concept [11]. However, he did not address the complexity of the control code that was used. This technique is also used in Algorithm 2.13 from [4]. Algorithm 2.13 is the product scanning form of integer squaring. An open problem is how to implement the **for** loop control code in step 2.1 of Algorithm 2.13 efficiently.

The prior work leaves open the problem of proving the correctness of the loop logic for product scanning multiplication and product scanning squaring. We characterize the combinatorial families present in product scanning multiplication and product scanning squaring. We then use these characterizations to provide control code in which loops are not unrolled.

We now recall the basic combinatorial family known as a *composition of n into k parts*.

Algorithm 2.10: Integer Multiplication (product scanning form)

INPUT : Integer $a, b \in [0, p-1]$.
OUTPUT : $c = a \cdot b$.
1. $R_0 \leftarrow 0, R_1 \leftarrow 0, R_2 \leftarrow 0$.
2. For k from 0 to $2t-2$ do
    2.1 For each element of $\{(i,j) \mid i+j = k, 0 \le i, j \le t-1\}$
        $(UV) \leftarrow A[i] \cdot B[j]$.
        $(\varepsilon, R_0) \leftarrow R_0 + V$.
        $(\varepsilon, R_1) \leftarrow R_1 + U + \varepsilon$.
        $R_2 \leftarrow R_2 + \varepsilon$.
    2.2 $C[k] \leftarrow R_0, R_0 \leftarrow R_1, R_1 \leftarrow R_2, R_2 \leftarrow 0$.
3. $C[2t-1] \leftarrow R_0$.
4. Return(c).

Algorithm 2.13: Integer Squaring (product scanning form)

INPUT : Integer $a \in [0, p-1]$.
OUTPUT : $c = a^2$.
1. $R_0 \leftarrow 0, R_1 \leftarrow 0, R_2 \leftarrow 0$.
2. For k from 0 to $2t-2$ do
    2.1 For each element of $\{(i,j) \mid i+j = k, 0 \le i \le j \le t-1\}$
        $(UV) \leftarrow A[i] \cdot A[j]$.
        if $(i < j)$ then do: $(\varepsilon, UV) \leftarrow (UV) \cdot 2, R_2 \leftarrow R_2 + \varepsilon$.
        $(\varepsilon, R_0) \leftarrow R_0 + V$.
        $(\varepsilon, R_1) \leftarrow R_1 + U + \varepsilon$.
        $R_2 \leftarrow R_2 + \varepsilon$.
    2.2 $C[k] \leftarrow R_0, R_0 \leftarrow R_1, R_1 \leftarrow R_2, R_2 \leftarrow 0$.
3. $C[2t-1] \leftarrow R_0$.
4. Return(c).

Definition 1. *Nijenhuis-Wilf: Let n and k be fixed positive integers. By a composition of n into k parts, we mean a representation of the form:*

$$n = r_1 + r_2 + ... + r_k$$

in which $r_i \ge 0$ for $i = 1, 2, ..., k$ and the order of the summands is important.

The number of compositions of n into k parts is $J(n,k) = \binom{n+k-1}{n}$ [8]. For example, there are exactly 28 compositions of 6 into 3 parts:

$$
\begin{aligned}
6 + 0 + 0 &= 0 + 6 + 0 = 0 + 0 + 6 = 1 + 2 + 3 = \\
5 + 1 + 0 &= 5 + 0 + 1 = 1 + 5 + 0 = 2 + 1 + 3 = \\
1 + 0 + 5 &= 0 + 1 + 5 = 0 + 5 + 1 = 2 + 2 + 2 = \\
4 + 2 + 0 &= 4 + 0 + 2 = 0 + 4 + 2 = 2 + 4 + 0 = \\
2 + 0 + 4 &= 0 + 2 + 4 = 4 + 1 + 1 = 1 + 4 + 1 = \\
1 + 1 + 4 &= 3 + 3 + 0 = 3 + 0 + 3 = 0 + 3 + 3 = \\
3 + 2 + 1 &= 3 + 1 + 2 = 1 + 3 + 2 = 2 + 3 + 1 = 6
\end{aligned}
$$

3 Product Scanning Multiplication

3.1 A Characterization of Multiplication Control Code

Definition 2. $R_{k,t} = \{(i,j) \mid i + j = k, 0 \le i, j \le t - 1\}$.

$R_{k,t}$ is the set used in step 2.1 of Algorithm 2.10. The key to the control code we develop is to come up with an alternate characterization of $R_{k,t}$. The following are examples of this set.

$$
\begin{aligned}
R_{3,5} &= \{(0,3), (1,2), (2,1), (3,0)\} \\
R_{4,5} &= \{(0,4), (1,3), (2,2), (3,1), (4,0)\} \\
R_{8,7} &= \{(2,6), (3,5), (4,4), (5,3), (6,2)\}
\end{aligned}
$$

This resembles a composition of n into k parts. The compositions of 3 into 2 parts are $0 + 3, 1 + 2, 2 + 1, 3 + 0$. The compositions of 8 into 2 parts are $0 + 8, 1 + 7, 2 + 6, 3 + 5, 4 + 4, 5 + 3, 6 + 2, 7 + 1, 8 + 0$. We characterize $R_{k,t}$ from the perspective of combinatorics as follows.

$R_{k,t}$ is all compositions (i,j) of k into 2 parts satisfying $0 \le i, j \le t - 1$

A successor algorithm imposes a total ordering over the elements of a combinatorial family (for an introduction to algorithmic combinatorics see [7]). So, a combinatorial family has a "smallest" element. To simplify the control code we seek an ANSI C style **for** loop for step 2.1 that cycles through the needed (i,j) pairs in the correct order. From a combinatorics perspective this **for** loop will be in the form of "**for** (A ; B ; C)" where A sets (i,j) to be the smallest element in the family, B is a test that the current (i,j) of the loop has a successor, and C is the successor algorithm.

3.2 Successor Algorithm for Multiplication

We develop the successor algorithm by analyzing the case for $k = 4$ and $t = 5$ (see Table 1). Pick a k, say, $k = 4$. Observe that all pairs for $k = 4$ are all possible sums of two numbers that equal 4 where $0 \le i, j \le t - 1$: $0 + 4, 1 + 3, 2 + 2, 3 + 1, 4 + 0$. So, we see that the sequence generated by step 2.1 of Algorithm 2.10 is the compositions of k into 2 parts where $0 \le i, j \le t - 1$. Given this observation we can derive a function $X(k,t)$ that computes the cardinality of $R_{k,t}$.

We start with the number of compositions of k into 2 parts. Consider the case when k is even, e.g., $k = 4$. This set is $0 + 4, 4 + 0, 1 + 3, 3 + 1, 2 + 2$ which is 5

Table 1. Case of t=5 for multiplication

r	(i, j)	k	fmin	X(k, t)
0	(0, 0)	0	0	1
1	(0, 1)	1	1	2
2	(1, 0)	1	1	2
3	(0, 2)	2	2	3
4	(1, 1)	2	2	3
5	(2, 0)	2	2	3
6	(0, 3)	3	3	4
7	(1, 2)	3	3	4
8	(2, 1)	3	3	4
9	(3, 0)	3	3	4
10	(0, 4)	4	4	5
11	(1, 3)	4	4	5
12	(2, 2)	4	4	5
13	(3, 1)	4	4	5
14	(4, 0)	4	4	5
15	(1, 4)	5	3	4
16	(2, 3)	5	3	4
17	(3, 2)	5	3	4
18	(4, 1)	5	3	4
19	(2, 4)	6	2	3
20	(3, 3)	6	2	3
21	(4, 2)	6	2	3
22	(3, 4)	7	1	2
23	(4, 3)	7	1	2
24	(4, 4)	8	0	1

in number. These are exactly the pairs for $k = 4$ in Table 1. So, $J(n, k)$ seems to work so far. Now consider the case that k is odd. For example, $k = 3$. The compositions of 3 into 2 parts are $0 + 3, 1 + 2, 2 + 1, 3 + 0$. These are exactly the pairs for $k = 3$ in Table 1. So, $J(n, k)$ still seems to work.

But there is a problem. $J(n, k)$ does not give the correct size for all k. For $t = 5$ and $k = 6$ we have $J(6, 2) = 7$ which is not the cardinality of $R_{6,5}$. We need the answer to be 3 corresponding to the size of $\{(2, 4), (3, 3), (4, 2)\}$. Looking more closely at Table 1 it can be seen that the upper and lower halves of the table are symmetric in the following sense. In the column for k note that there is one 8 for one 0, two 7s for two 1s, three 6s for three 2s, four 5s for four 3s, and five 4s right in the middle. We can reflect k in the upper half down into the lower half using the below function. This gives the penultimate column in Table 1.

We are now ready to fix $X(k,t)$. Observe that by using fmin to compute the first argument to J, we correctly count the compositions of k for the bottom half of the table (fmin is written in Rust). $X(k,t)$ is therefore defined as follows. $X(k,t) = J(\text{fmin}(k,t),2)$. This simplifies to $X(k,t) = \text{fmin}(k,t) + 1$. This gives the rightmost column in Table 1.

```
fn fmin(k:usize,t:usize) -> usize
{
let min : usize = if k < (2*t-2-k) {k} else {2*t-2-k};
min
}
```

Let r denote the row number. We make the following observations.

1. Note that, (i): for rows $0, 1, 3, 6$, and 10, $j = k$ (these rows are the first pair for compositions of k into 2 parts for the stated value of k), and (ii): for rows $15, 19, 22$, and 24, $j = 4$. It follows from (i) and (ii) that j, in the first pair in a given composition of k into 2 parts family, is the minimum of k and $t - 1$.
2. The successor of a pair in a composition of k into 2 parts family is found by adding 1 to i and subtracting 1 from j. Take $k = 4$ for example. The family is $(0, 4), (1, 3), (2, 2), (3, 1), (4, 0)$.

These observations lead to the successor algorithm multsuccessor. We generate the pairs (i, j) in the order needed for product scanning multiplication in genmultcompositions.

```
fn multsuccessor(obj : (usize,usize),t : usize) -> (usize,usize)
{ // obj is the combinatorial object (i,j)
let (i,j) = obj;
if i == t-1 || j == 0
  {panic!("no more compositions");}
let iprime : usize = i+1;
let jprime : usize = j-1;
(iprime,jprime)
}
```

We now prove the correctness of the successor algorithm multsuccessor.

Theorem 1. *Let $x_0 = (i,j) \in R_{k,t}$ satisfy $j = \min(k, t - 1)$ and $i = k - j$ and let Φ be the outputs of* multsuccessor *applied recursively to (x_0, t). Then $R_{k,t} = \{x_0\} \cup \Phi$.*

Proof. Clearly $x_0 \in R_{k,t}$. If $k = 0$ or $k = 2t - 2$ the claim holds since there are no successors. Consider the case that $0 < k < 2t - 2$. We prove correctness of the outputs by strong induction on the output sequence. Hypothesis: The 1st, 2nd, etc. up to and including the uth pair (i, j) that is output are all contained in $R_{k,t}$ and there are no duplicates in this sequence. Base case: It clearly holds for the first output $(k - \min(k, t - 1) + 1, \min(k, t - 1) - 1)$. Induction step: Let the uth pair be (i, j). (i, j) must fall within one of the following mutually exclusive cases:

```
fn genmultcompositions(t : usize) -> ()
{
for k in 0..(t<<1)-1 // range is 0..2t-2 inclusive
  {
  let mut j : usize = if k < t-1 {k} else {t-1};
  let mut i : usize = k-j;
  loop
    {
    println!("({},{})",i,j);
    if i == t-1 || j == 0
      {break;}
    let (iprime,jprime) = multsuccessor((i,j),t);
    i = iprime;
    j = jprime;
    }
  }
}
```

Case 1: $j = 0$. There are no more compositions.

Case 2: $j > 0$ and $i = t - 1$. There are no more compositions.

Case 3: $j > 0$ and $i < t - 1$. Then the successor is (i', j') where $i' = i + 1$ and $j' = j - 1$. It follows that $i' + j'$ is a composition of k into 2 parts since $i' + j' = i + j = k$. We also have that $0 \leq i', j' \leq t - 1$. So, $(i', j') \in S_{k,t}$. Furthermore, (i', j') cannot have been output previously since j decreases by 1 in every iteration.

We now show that the length of this set matches $|R_{k,t}|$. Either $k < t - 1$ or not. Consider the case that $k < t - 1$. Then $x_0 = (0, k)$. The outputs stop when i reaches k since at that point $j = 0$. So, the length of the set is $k + 1$. This matches exactly $X(k,t)$. Now consider the case that $k \geq t - 1$. Then $x_0 = (k - t + 1, t - 1)$. The outputs stop when i reaches $t - 1$. So, the length of the set is $1 + t - 1 - (k - t + 1) = 2t - 2 - k + 1$. This matches exactly $X(k,t)$. $\square$

Theorem 1 proves that `multsuccessor` correctly implements step 2.1 of Algorithm 2.10. We have therefore shown a drop-in-replacement **for** loop for step 2.1 in Algorithm 2.10.

3.3   Integer Multiplication Algorithm

We can implement the **for** loop in step 2.1 using the successor algorithm in ANSI C as follows.

```
for (int j=(k<t-1)?k:t-1,i=k-j;i<=t-1 && j>=0;i++,j--)
```

This simplifies further by replacing $i \leq t - 1$ by $i < t$. We then arrive at genmultcompositions2 that generates all the (i, j) for multiplication using merely two **for** loops. From this we revise Algorithm 2.10 (see Fig. 1). Keep in mind that j is an integer type and can assume a negative value (once negative control leaves the loop).[1] We show a **for** loop for step 2.1 based on the combinatorial family inherently present in product scanning.

```
void genmultcompositions2(int t)
{
for (int k=0;k<(t<<1)-1;k++)
  {
  for (int j=(k<t-1)?k:t-1,i=k-j;i<t && j>=0;i++,j--)
    {
    printf("got one for k=%d: (i,j) = (%d,%d)\n",k,i,j);
    }
  }
}
```

Algorithm: Integer Multiplication (product scanning form)

INPUT : Integer $a, b \in [0, p-1]$.
OUTPUT : $c = a \cdot b$.
1. $R_0 \leftarrow 0, R_1 \leftarrow 0, R_2 \leftarrow 0$.
2. for $(k = 0 \; ; \; k < (t << 1) - 1 \; ; \; k{+}{+})$
    2.1 for $(j = (k < t - 1) \; ? \; k \; : \; t - 1, i = k - j \; ; \; i < t \;\&\&\; j \geq 0 \; ; \; i{+}{+}, j{-}{-})$
        $(UV) \leftarrow A[i] \cdot B[j]$.
        $(\varepsilon, R_0) \leftarrow R_0 + V$.
        $(\varepsilon, R_1) \leftarrow R_1 + U + \varepsilon$.
        $R_2 \leftarrow R_2 + \varepsilon$.
    2.2 $C[k] \leftarrow R_0, R_0 \leftarrow R_1, R_1 \leftarrow R_2, R_2 \leftarrow 0$.
3. $C[2t - 1] \leftarrow R_0$.
4. Return(c).

Fig. 1. Integer multiplication with the successor algorithm

Having developed control code for product scanning multiplication, it is instructive to review Kaliski's `BigMult` algorithm. A couple things become apparent. First, Kaliski develops his inner loops around i whereas we developed ours around j. Second, the inner loop conditions can be combined into one. This means we can roll-up the outer loop that Kaliski unrolled (see Fig. 2).

This observation places the development of product scanning control code for the case of multiplication into perspective. Technically, the Hankerson et al. conclusion that product scanning multiplication will have complex control code unless loops are unrolled is consistent with what was developed in [5]. However, Kaliski was incredibly close to having a single line of code for the control code in step 2.1 of Algorithm 2.10.

---

[1] Programmer's note: if j is erroneously implemented using an unsigned type then the check $j \geq 0$ would be flagged by a good compiler as superfluous and even worse the output would not be correct.

Algorithm: Integer Multiplication

INPUT : Integer $a, b \in [0, p-1]$.
OUTPUT : $c = a \cdot b$.
1. $R_0 \leftarrow 0, R_1 \leftarrow 0, R_2 \leftarrow 0$.
2. for $(k = 0 ; k < (t << 1) - 1 ; k{+}{+})$
$\quad$ 2.1 for $(i = (k < t)\ ?\ 0\ :\ k - t + 1, j = k - i\ ;\ i \leq k\ \&\&\ i < t\ ;\ i{+}{+}, j{-}{-})$
$\qquad (UV) \leftarrow A[i] \cdot B[j]$.
$\qquad (\varepsilon, R_0) \leftarrow R_0 + V$.
$\qquad (\varepsilon, R_1) \leftarrow R_1 + U + \varepsilon$.
$\qquad R_2 \leftarrow R_2 + \varepsilon$.
$\quad$ 2.2 $C[k] \leftarrow R_0, R_0 \leftarrow R_1, R_1 \leftarrow R_2, R_2 \leftarrow 0$.
3. $C[2t - 1] \leftarrow R_0$.
4. Return(c).

Fig. 2. Outer loop on k rolled up

4 Product Scanning Squaring

4.1 A Characterization of Squaring Control Code

Definition 3. $S_{k,t} = \{(i, j) \mid i + j = k, 0 \leq i \leq j \leq t - 1\}$.

$S_{k,t}$ is the set used in step 2.1 of Algorithm 2.13. The key to the control code we develop is to come up with an alternate characterization of $S_{k,t}$. The following are examples of this set. They resemble a composition of n into k parts.

$$S_{3,5} = \{(1, 2), (0, 3)\}$$
$$S_{4,5} = \{(2, 2), (1, 3), (0, 4)\}$$
$$S_{6,7} = \{(3, 3), (2, 4), (1, 5), (0, 6)\}$$
$$S_{8,7} = \{(4, 4), (3, 5), (2, 6)\}$$

We characterize $S_{k,t}$ from the perspective of combinatorics as follows.

$S_{k,t}$ is all compositions$(i, j)$of$k$into 2 parts satisfying$0 \leq i \leq j \leq t - 1$

In the sections that follow we develop a closed form solution to counting the elements in this family and we develop a successor algorithm for the family.

4.2 Successor Algorithm for Squaring

We develop the successor algorithm by analyzing the case for $k = 4$ and $t = 5$ (see Table 2). Pick a k, say, $k = 4$. Observe that all pairs for $k = 4$ are all possible sums of two numbers that equal 4 where $0 \leq i \leq j \leq t - 1$: $2 + 2, 1 + 3, 0 + 4$. So, we see that the sequence generated by step 2.1 is the compositions of k into 2 parts where $0 \leq i \leq j \leq t - 1$. Given this observation we can derive a function $Y(k, t)$ that computes the cardinality of $S_{k,t}$.

Table 2. Case of t = 5 for squaring

r	(i, j)	k	fmin	Y(k, t)
0	(0, 0)	0	0	1
1	(0, 1)	1	1	1
2	(1, 1)	2	2	2
3	(0, 2)	2	2	2
4	(1, 2)	3	3	2
5	(0, 3)	3	3	2
6	(2, 2)	4	4	3
7	(1, 3)	4	4	3
8	(0, 4)	4	4	3
9	(2, 3)	5	3	2
10	(1, 4)	5	3	2
11	(3, 3)	6	2	2
12	(2, 4)	6	2	2
13	(3, 4)	7	1	1
14	(4, 4)	8	0	1

We start with the number of compositions of k into 2 parts. Consider the case when k is even, e.g., $k = 4$. This set is $0 + 4, 4 + 0, 1 + 3, 3 + 1, 2 + 2$ which is 5 in number. By computing $(J(k, 2) + 1)/2$ we arrive at 3, the number we are after, the cardinality of $\{(2, 2), (1, 3), (0, 4)\}$. We can think of this as calculating the cardinality of the set $\{(0, 4), (4, 0), (1, 3), (3, 1), (2, 2), (2, 2)\}$ and dividing it by two, matching the cardinality of $\{(2, 2), (1, 3), (0, 4)\}$. The additional $(2, 2)$ in this multiset is accounted for by the $+1$ in the numerator of $(J(k, 2) + 1)/2$.

Now consider the case that k is odd. For example, $k = 3$. The compositions of 3 into 2 parts are $0 + 3, 3 + 0, 1 + 2, 2 + 1$. Dividing the cardinality of this set by 2 yields 2. So, by computing $J(k, 2)/2$ we get the cardinality we are after, in this case corresponding to $\{(1, 2), (0, 3)\}$.

So, we have so far $Y(k, t) = (J(k, 2) + 1)/2$ when k is even and $Y(k, t) = J(k, 2)/2$ when k is odd. But there is a problem. This is not correct for all k. For $t = 5$ and $k = 6$ we have $Y(6, 5) = 4$ which is wrong. We need $Y(6, 5) = 2$ corresponding to $\{(3, 3), (2, 4)\}$. Looking more closely at Table 2 it can be seen that the upper and lower halves of the table are symmetric in the following sense. In the column for k note that there is one 8 for one 0, one 7 for one 1, two 6s for two 2s, two 5s for two 3s, and three 4s right in the middle. We can reflect k in the upper half down into the lower half using the `fmin` function. This gives the penultimate column in Table 2.

We are now ready to fix $Y(k, t)$. Observe that by using `fmin` to compute the first argument to J, we correctly count the compositions of k for the bottom half of the table. $Y(k, t)$ is therefore defined as follows. $Y(k, t) = (J(\texttt{fmin}(k, t), 2) + 1)/2$ when k is even and $Y(k, t) = J(\texttt{fmin}(k, t), 2)/2$ when k is odd. It can be shown that this simplifies to the following.

$$Y(k, t) = \begin{cases} (\texttt{fmin}(k, t) + 2)/2 & \text{if } k \text{ is even} \\ (\texttt{fmin}(k, t) + 1)/2 & \text{if } k \text{ is odd} \end{cases}$$

The above definition gives the rightmost column in Table 2.

```
fn sqrsuccessor(obj : (usize,usize),t : usize) -> (usize,usize)
{ // obj is the combinatorial object (i,j)
let (i,j) = obj;
if i == 0 || j == t-1
  {
  panic!("no more compositions");
  }
let iprime : usize = i-1;
let jprime : usize = j+1;
(iprime,jprime)
}
```

Lemma 1. *Let k be even, let $x_0 = (\frac{k}{2}, \frac{k}{2})$, and let Φ be the $Y(k, t) - 1$ outputs of* `sqrsuccessor`(x_0, t) *applied recursively. Then $S_{k,t} = \{x_0\} \cup \Phi$.*

Proof. Clearly $(\frac{k}{2}, \frac{k}{2}) \in S_{k,t}$. If $k = 0$ or $k = 2t - 2$ the claim holds since there are no successors. Consider the case that $0 < k < 2t - 2$. We prove correctness of the outputs by strong induction on the output sequence. Hypothesis: The 1st, 2nd, etc. up to and including the uth pair (i, j) that is output are contained in $S_{k,t}$ and there are no duplicates in this sequence. Base case: It clearly holds for the first output $(\frac{k}{2} - 1, \frac{k}{2} + 1)$. Induction step: Let the uth pair be (i, j). The pair (i, j) must fall within one of the following mutually exclusive cases:

Case 1: $i = 0$. There are no more compositions.

Case 2: $i > 0$ and $j = t - 1$. There are no more compositions.

Case 3: $i > 0$ and $j < t - 1$. Then the successor is (i', j') where $i' = i - 1$ and $j' = j + 1$. It follows that $i' + j'$ is a composition of k into 2 parts since $i' + j' = i + j = k$. We also have that $0 \leq i' \leq j' \leq t - 1$. Furthermore, (i', j') cannot have been output previously since j increases by 1 in every iteration.

We now show that the length of the resulting set matches $|S_{k,t}|$. Since i is decremented by 1 and j is incremented by 1 in each invocation the number of pairs in this set is $1 + \min(\frac{k}{2}, t - 1 - \frac{k}{2})$. This is $Y(k, t)$ for even k. The lemma therefore holds.   □

Lemma 2. *Let k be odd, let $x_0 = (\lfloor\frac{k}{2}\rfloor, 1 + \lfloor\frac{k}{2}\rfloor)$, and let Φ be the $Y(k,t) -$
1 outputs of* sqrsuccessor$((\lfloor\frac{k}{2}\rfloor, 1 + \lfloor\frac{k}{2}\rfloor), t)$ *applied recursively. Then $S_{k,t} =$
$\{x_0\} \cup \Phi$.*

Proof. Clearly $(\lfloor\frac{k}{2}\rfloor, 1 + \lfloor\frac{k}{2}\rfloor) \in S_{k,t}$. If $k = 1$ or $k = 2t - 3$ the claim holds
since there are no successors. Consider the case that $1 < k < 2t - 3$. We prove
correctness of the outputs by strong induction on the output sequence. Hypothesis: The 1st, 2nd, etc. up to and including the uth pair (i, j) that is output are
all contained in $S_{k,t}$ and there are no duplicates in this sequence. Base case: It
clearly holds for the first output $(-1 + \lfloor\frac{k}{2}\rfloor, 2 + \lfloor\frac{k}{2}\rfloor)$. Induction step: Let the
u^{th} pair be (i, j). The pair (i, j) must fall within one of the following mutually
exclusive cases:

Case 1: $i = 0$. There are no more compositions.

Case 2: $i > 0$ and $j = t - 1$. There are no more compositions.

Case 3: $i > 0$ and $j < t - 1$. Then the successor is (i', j') where $i' = i - 1$
and $j' = j + 1$. It follows that $i' + j'$ is a composition of k into 2 parts since
$i' + j' = i + j = k$. We also have that $0 \le i' \le j' \le t - 1$. Furthermore, (i', j')
cannot have been output previously since j increases by 1 in every iteration.

We now show that the length of the resulting set matches $|S_{k,t}|$. Since i is
decremented by 1 and j is incremented by 1 in each invocation the number of
pairs in the set is $1 + \min(\lfloor\frac{k}{2}\rfloor, t - 1 - (1 + \lfloor\frac{k}{2}\rfloor))$. This is $Y(k,t)$ for odd k. The
lemma therefore holds. □

Theorem 2. *Let x_0 be $(\frac{k}{2}, \frac{k}{2})$ for even k and let x_0 be $(\lfloor\frac{k}{2}\rfloor, 1 + \lfloor\frac{k}{2}\rfloor)$ for odd k.
Define Φ to be the $Y(k,t) - 1$ outputs of* sqrsuccessor(x_0, t) *applied recursively.
Then $S_{k,t} = \{x_0\} \cup \Phi$.*

Proof. Follows directly from Lemmas 1 and 2. □

Theorem 2 proves that sqrsuccessor correctly implements step 2.1 of Algorithm 2.13. We have therefore shown a drop-in-replacement **for** loop for step 2.1
in Algorithm 2.13.

4.3   Integer Squaring Algorithm

The code for generating (i, j) fits into two lines in ANSI C, one line for each
loop. gensqrcompositions2 is the generator in ANSI C. We use this to revise
Algorithm 2.13 (see Fig. 3 in Appendix A). Keep in mind that i *is an integer
type* and can assume a negative value.[2]

---

[2] Programmer's note: if i is erroneously implemented using an unsigned type then the
check $i \ge 0$ would be flagged by a good compiler as superfluous and even worse the
output would not be correct.

Below we give another variant of the **for** loop for step 2.1.

```rust
fn gensqrcompositions(t : usize) -> ()
{
for k in 0..(t<<1)-1 // range is 0..2t-2 inclusive
  {
  let mut i : usize = k>>1;  // this is \lfloor k/2 \rfloor
  let mut j : usize = i+(k&1);
  loop
    {
    println!("({},{})",i,j);
    if i == 0 || j == t-1
      {break;}
    let (iprime,jprime) = sqrsuccessor((i,j),t);
    i = iprime;
    j = jprime;
    }
  }
}
```

```c
void gensqrcompositionsunrolled(int t)
{
int k=0;
for (;k<t;k++)
  {
  for (int i=k>>1,j=i+(k&1);i>=0;i--,j++)
    {printf("got one for k=%d: (i,j) = (%d,%d)\n",k,i,j);}
  }
for (;k<(t<<1)-1;k++)
  {
  for (int i=k>>1,j=i+(k&1);j<t;i--,j++)
    {printf("got one for k=%d: (i,j) = (%d,%d)\n",k,i,j);}
  }
}
```

```c
for (unsigned j=(k>>1)+(k&1);k>=j && j<t;j++)
```

This leverages the fact that $i = k - j$. Note that this increases slightly the complexity of the body of the **for** loop since i appears twice in the body (and needs to be replaced by $k - j$).

Finally, we adapt Kaliski's technique of unrolling once about the midway point to the case of product scanning squaring. `gensqrcompositionsunrolled` shows how to unroll our outermost loop about the midway point such that the inner loop conditions are each a single inequality test. This algorithm is given in Appendix A.

5   Conclusion

In this note we provided exact characterizations of the combinatorial families inherently present in product scanning multiplication and product scanning squaring. We leveraged the two successor algorithms we developed to provide loop logic for product scanning multiplication and product scanning squaring, and we provided the first proofs of correctness of the loop logic for product scanning multiplication and product scanning squaring.

A   Integer Squaring

```
void gensqrcompositionsunrolled(int t)
{
int k=0;
for (;k<t;k++)
  {
  for (int i=k>>1,j=i+(k&1);i>=0;i--,j++)
    {printf("got one for k=%d: (i,j) = (%d,%d)\n",k,i,j);}
  }
for (;k<(t<<1)-1;k++)
  {
  for (int i=k>>1,j=i+(k&1);j<t;i--,j++)
    {printf("got one for k=%d: (i,j) = (%d,%d)\n",k,i,j);}
  }
}
```

Algorithm: Integer Squaring (product scanning form)

INPUT : Integer $a \in [0, p-1]$.
OUTPUT : $c = a^2$.
1. $R_0 \leftarrow 0, R_1 \leftarrow 0, R_2 \leftarrow 0$.
2. for $(k = 0 \ ; \ k < (t << 1) - 1 \ ; \ k{+}{+})$
$\quad$ 2.1 for $(i = k >> 1, j = i + (k\&1) \ ; \ i \geq 0 \ \&\& \ j < t \ ; \ i{-}{-}, j{+}{+})$
$\quad\quad (UV) \leftarrow A[i] \cdot A[j]$.
$\quad\quad$ if $(i < j)$ then do: $(\varepsilon, UV) \leftarrow (UV) \cdot 2, R_2 \leftarrow R_2 + \varepsilon$.
$\quad\quad (\varepsilon, R_0) \leftarrow R_0 + V$.
$\quad\quad (\varepsilon, R_1) \leftarrow R_1 + U + \varepsilon$.
$\quad\quad R_2 \leftarrow R_2 + \varepsilon$.
$\quad$ 2.2 $C[k] \leftarrow R_0, R_0 \leftarrow R_1, R_1 \leftarrow R_2, R_2 \leftarrow 0$.
3. $C[2t - 1] \leftarrow R_0$.
4. Return(c).

Fig. 3. Integer squaring with the successor algorithm

References

1. Comba, P.: Exponentiation cryptosystems on the IBM PC. IBM Syst. J. **29**, 526–538 (1990)
2. Cook, S.A.: On the minimum computation time of functions. Ph.D. thesis, Harvard University (1966)
3. Großschädl, J., Avanzi, R.M., Savaş, E., Tillich, S.: Energy-efficient software implementation of long integer modular arithmetic. In: Rao, J.R., Sunar, B. (eds.) CHES 2005. LNCS, vol. 3659, pp. 75–90. Springer, Heidelberg (2005). doi:10.1007/11545262_6
4. Hankerson, D., Menezes, A., Vanstone, S.: Guide to Elliptic Curve Cryptography. Springer, New York (2004)
5. Kaliski, Jr., B.S.: The Z80180 and big-number arithmetic. Dr. Dobb's J. 50–58, September 1993. https://www.linkedin.com/in/burtkaliskijr
6. Karatsuba, A., Ofman, Y.: Multiplication of multidigit numbers on automata. Dokl. Akad. Nauk SSSR **145**, 293–294 (1962)
7. Kreher, D.L., Stinson, D.R.: Combinatorial Algorithms-Generation, Enumeration, and Search. CRC Press, Boca Raton (1999)
8. Nijenhuis, A., Wilf, H.S.: Combinatorial Algorithms for Computers and Calculators, 2nd edn. Academic Press, Inc., New York (1978)
9. Schönhage, A., Strassen, V.: Schnelle Multiplikation großer Zahlen. Computing **7**, 281–292 (1971)
10. Toom, A.L.: The complexity of a scheme of functional elements realizing the multiplication of integers. Dokl. Akad. Nauk SSSR **150**(3), 496–498 (1963)
11. Tuckerman, B.: The 24th Mersenne prime. Proc. Nat. Acad. Sci. **68**(10), 2319–2320 (1971)

Stylometric Authorship Attribution
of Collaborative Documents

Edwin Dauber[✉], Rebekah Overdorf, and Rachel Greenstadt

Drexel University, Philadelphia, PA 19104, USA
egd34@drexel.edu

Abstract. Stylometry is the study of writing style based on linguistic features and is typically applied to authorship attribution problems. In this work, we apply stylometry to a novel dataset of multi-authored documents collected from Wikia using both relaxed classification with a support vector machine (SVM) and multi-label classification techniques. We define five possible scenarios and show that one, the case where labeled and unlabeled collaborative documents by the same authors are available, yields high accuracy on our dataset while the other, more restrictive cases yield lower accuracies. Based on the results of these experiments and knowledge of the multi-label classifiers used, we propose a hypothesis to explain this overall poor performance. Additionally, we perform authorship attribution of pre-segmented text from the Wikia dataset, and show that while this performs better than multi-label learning it requires large amounts of data to be successful.

Keywords: Stylometry · Authorship attribution · Machine learning · Multi-label learning

1 Introduction

Authorship attribution methods have been used successfully to uncover the author of documents in many different domains and areas. These methods can be used to compromise privacy and uncover the author of any anonymous text on the web. There is an important caveat, however, to the use of current state-of-the-art authorship attribution techniques. While they are very effective with documents written by a single person, they are not designed to handle collaboratively written documents. With the rise of Internet collaborative writing platforms such as Wikipedia[1] and GoogleDrive[2], the development of new techniques to handle multi-authored text is necessary.

Multi-label machine learning models are designed to assign multiple labels to an unlabeled sample in a classification task. These methods are well studied and have been used to great success in different real world learning problems in many distinct areas of research, such as image recognition and text categorization.

---

[1] http://en.wikipedia.org.
[2] https://drive.google.com.

© Springer International Publishing AG 2017
S. Dolev and S. Lodha (Eds.): CSCML 2017, LNCS 10332, pp. 115–135, 2017.
DOI: 10.1007/978-3-319-60080-2_9

In this work, we study the multi-label problem in the context of authorship attribution.

Collaboration has also been considered as a stylometric defense [3]. By either having another author rewrite text to obfuscate it or writing collaboratively, standard stylometric methods fail to identify the correct author. We present an analysis of new stylometric methods specifically designed for multi-label classification that address this type of obfuscation.

Our contributions are as follows. We define five variations of the multi-label stylometry problem based on the availability of training data, test both traditional single-label stylometric techniques and multi-label classification techniques as methods to solve our variations on authentic collaborative documents collected from the Internet, and identify successes and limitations of these techniques. Specifically, we identify one of these variations, which we call *consistent collaboration*, for which these techniques are promising, at least for small closed-world scenarios, and we demonstrate that these techniques are insufficient as-is to solve the other four variations for even small closed-world scenarios. We also present a hypothesis to explain the performance on these different variations. We then show that account attribution using pre-segmented texts is possible given sufficient data and present an analysis of the level of separation in collaboration on these real-world documents, as a way of predicting the viability of supervised segmentation as an alternative to multi-label stylometry.

We formally define the multi-author stylometry problem in Sect. 2. We examine previous work related to multi-authored documents and Wikipedia in Sect. 3. We discuss our dataset in Sect. 5 and our methodology in Sect. 4. We demonstrate the results of our experiments in Sect. 6, discuss our results in Sect. 7, and discuss future work in Sect. 8.

2   Problem Statement

We consider two problems in which the authors of a collaborative document are in question. In the first problem, the only documents of known authorship are non-collaborative, single-authored documents. In the second problem, multi-authored documents of known authorship are available. While in general we acknowledge that we are not likely to know the number of authors per document, in order to provide best case results, for our experiments we impose the assumption that we do have that knowledge. We note, however, that while this assumption is reflected in our training and evaluation datasets, it only affects our treatment of single-label classifiers, not multi-label classifiers.

2.1   Non-collaborative Training Documents

We define two variations in which the available training documents are non-collaborative.

Complete suspect set: Non-collaborative documents of known authorship are available for each suspect. More formally: given a set of n authors $\mathcal{A} = \{A_1, A_2, \ldots, A_n\}$, and a set of documents $\mathcal{D}_i$ for each $\mathcal{A}_i$ which we know to be written by only that author; we want to identify the k authors of a document of unknown authorship d.

Partial suspect set: Non-collaborative documents of known authorship are available for some of the suspects. More formally: given a set of n authors $\mathcal{A} = \{A_1, A_2, \ldots, A_n\}$, and a set of documents $\mathcal{D}_i$ for each $\mathcal{A}_i$ which we know to be written by only that author, and a document of unknown authorship d written by k authors, of which c authors are in our suspect set, we want to identify those c authors.

2.2 Collaborative Training Documents

In the case where suspect authors have collaborative writings, we consider a subproblem in which all documents have the same number of authors. This problem has three variations.

Consistent collaboration: The suspect set consists of pairings or groups of authors who are suspected of collaboratively writing the document in question together. Formally: given a set of n author groups $\mathcal{G} = \{G_1, G_2, \ldots, G_n\}$, where $G_i = \{A_1, A_2, \ldots, A_m\}$ and each G_i has a set of documents $\mathcal{D}_i$ which we know to be written collaboratively by $\{A_1, A_2, \ldots, A_m\}$, identify the true group of authors $G_t \in \mathcal{G}$ of a document of unknown authorship d. This provides us with a best-case scenario in which we know all of the possible combinations of authors of d and have sufficient training data.

Mixed collaboration: Collaborative documents written by some of the suspect groups are unavailable, but other collaborative works by suspect authors are available. Formally: given a set of n author groups $\mathcal{G} = \{G_1, G_2, \ldots, G_n\}$ where $G_i = \{A_1, A_2, \ldots, A_m\}$ and each G_i has a set of documents $\mathcal{D}_i$ which we know to be written collaboratively by $\{A_1, A_2, \ldots, A_m\}$, identify the true group of authors G_t of a document of unknown authorship d, such that G_t may or may not be an element of $\mathcal{G}$, but all authors A_{t_i} in G_t are covered by groups which are in $\mathcal{G}$. This provides us with an average-case scenario for which we know some of the possible combinations of authors of d and have sufficient training data for some of them while having limited training data for others.

Inconsistent collaboration: Collaborative documents written by the suspect groups are unavailable, but other collaborative works by suspect authors are available. Formally: given a set of n author groups $\mathcal{G} = \{G_1, G_2, \ldots, G_n\}$ where $G_i = \{A_1, A_2, \ldots, A_m\}$ and each G_i has a set of documents $\mathcal{D}_i$ which we know to be written collaboratively by $\{A_1, A_2, \ldots, A_m\}$, identify the true group of authors $G_t \notin \mathcal{G}$ of a document of unknown authorship d. Although G_t is not part of $\mathcal{G}$, all authors A_{t_i} in G_t are covered by groups which are in $\mathcal{G}$. This provides us with the worst-case scenario that the authors of d have not collaborated in the past or such data is unavailable.

2.3   Pre-segmented Text

We consider one more problem in this paper, in which we have text which has already been segmented by anonymized author. Specifically, we use the revision history to segment the wiki articles by user account at the sentence level. In this case, we want to attribute the author's account. More formally: given a set of n authors $\mathcal{A} = \{A_1, A_2, \ldots, A_n\}$, each of whom has a set of documents $\mathcal{D}_i$ which we know to be written by only that author; we want to identify the author of an account a containing a set of k document segments $a = \{s_1, s_2, \ldots, s_k\}$.

3   Background and Related Work

3.1   Multi-label Learning

There have been a number of proposed techniques for multi-label learning that we consider in this work. All of these methods have been tested on various multi-label problems, but to the best of our knowledge, none of them have been proposed for solving the collaborative authorship attribution problem.

Multi-Label k-Nearest Neighbors (MLkNN) [22] is a lazy learning approach derived from the popular k-nearest neighbors classifier that utilizes MAP estimation based on the count of each label in the nearest neighbor set. Because of the likelihood estimation, this method performs well at ranking authors by likelihood of being one of the collaborators. It is also cheap, computationally, which is especially beneficial in authorship attribution when linking identities on a large scale, for example, Wikipedia.

While MLkNN is an adaptation of an algorithm to assign multiple labels to a sample, the following methods transform the problem to achieve multi-label classification. That is, they transform a multi-label classification problem into a single-label classification problem.

The most straightforward of these methods is binary relevance (BR) [20]. Binary relevance trains a binary yes or no classifier for each label. While this method is straightforward, it serves as a baseline since many methods easily outperform it. Label powerset (LP) [20] for example, instead creates a single-label classifier with each possible combination of the labels as one of the new labels, which captures label dependencies. We take advantage of this method especially, because authorship attribution of collaborative writings does not only include authors appending their writing together, but also editing or co-writing each other's work.

Another problem transform method is Hierarchy Of Multi-label classifiERs (HOMER) [19]. HOMER is a multi-label learning method that recursively breaks down the label set into smaller label sets creating a balanced tree structure where the leaves are single labels and all other nodes represent a single multi-label classification problem. Another method is RAndom k-labELsets (RAkEL) [21]. The RAkEL algorithm randomly selects a k-sized subset of labels m times and trains an LP classifier on each. Each iteration yields a binary solution for each label in the k-sized subset of labels. The average decision for each label is calculated and the labels with averages above a certain threshold are considered positive. We attempt to use these methods, but they offer no noticable accuracy improvement over the basic methods.

Madjarov et al. wrote an experiments paper with various multi-label learning algorithms and datasets [13] including 6 datasets for text classification. While some of these datasets proved difficult to classify, others were less so. One dataset involving classifying airplane problems from aviation safety reports yielded exact match accuracy of 81.6% and example-based accuracy, which measures the percentage of correctly predicted labels, of 91.4%. From this, we can see that, depending on the specific problem, multi-label learning can be very applicable to text.

Prior work in multi-label authorship attribution is limited to de-anonymizing academic submissions. Payer et al. proposed a framework called *deAnon* to break the anonymity of academic submissions [16]. Along with common features used in stylometry (e.g. bag-of-words, letter frequencies), they included information about which papers were cited. They use an ensemble of linear SVMs, a common classifier used in authorship attribution; MLkNN, a multi-label classifier; and ranking by average cosine similarity. From 1,405 possible authors, the ensemble classifier obtained a 39.7% accuracy that one of the authors was the first guess and 65.6% accuracy than an author is within the first ten guesses.

Our work differs from this for a few reasons. First, we leverage the clear ground truth of Wikia's revision history to set up controlled experiments. We also compare other proposed multi-label techniques described previously against ranking techniques. We extend our evaluation to include a sample of multi-label metrics. These differences lead us to obtain better results and demonstrate by comparison the results we would obtain not only against ranking techniques but also against results on single-authored documents in our domain of interest.

Macke and Hirshman attempted to use a recursive neural network to perform sentence-level attribution of Wikipedia articles, however they showed that even at 10 suspect authors naive bayes outperformed the neural network, and due to the most common words identified, it is likely that they were actually detecting topic, rather than style [12].

It is not always the case that, when given a multi-authored document, we want to know the set of contributing authors. In some cases, we want to know which authors wrote which pieces. In this case, methods that break apart the document in question can be very useful. This has been achieved through a sliding window approach [8] and sentence level classification [2,11]. However, both of these techniques were developed for large texts, as opposed to the short texts typically found on the internet. So, while they may be applicable for collaboratively written books, they are poorly suited as-is for use on wiki-scale text.

3.2   Single-Author Stylometry

In the case in which we know all documents in our dataset have only a single-author, we formally define the problem of authorship attribution as follows: given a set of n authors $\mathcal{A} = \{A_1, A_2, \ldots A_n\}$, for each of whom we have a set of documents $\mathcal{D}_i$ which we know to be written by that author, we want to identify the author of a document of unknown authorship d. This problem has been studied extensively [1,4,7,15] and we borrow feature sets and methods from prior work. Juola wrote an extensive review of authorship attribution literature [10]. Because of the high accuracies reported by many of these works, we would consider that multi-authored stylometry might be an application for which multi-label learning could be applied.

The Writeprints feature set [1] is a set of popular features used in authorship attribution. It includes lexical, syntactic, content, structural, and idiosyncratic features. Because these features have been repeatedly shown to be effective at performing authorship attribution of English text, we use this feature set, limited in size to the most common features for computational purposes.

Linear support vector machines (SVM) are often used in stylometry for classification and produce a high precision and high recall [7] for this problem. Later studies, including [1], similarly found that linear SVMs were a good classifier for stylometry. For our single-label technique, we also use a linear SVM.

In the specific domain of Wikipedia, authorship identification is studied as a way to combat sockpuppets [17] and vandalism [9]. Sockpuppet detection, however, has been studied through the text and metadata on talk pages and not on the text of articles or text written collaboratively. While vandalism detection does study the style of specific edits in the article text, the goal is not to determine authorship, collaborative or otherwise.

4   Methodology

For all evaluations for the multi-authored text, we use the Writeprints Limited feature set, extracted using the tool *JStylo* [14]. We experimented with many different multi-label classifiers, and will only be presenting the best results. In addition, for all experiments with multi-authored testing documents we use a best-case scenario evaluation of a linear SVM which takes the top m predicted authors for a testing document written by m actual authors out of the set of n suspects. For real application, this would prove optimistic, since techniques would be needed to compensate for not knowing the exact number of authors.

For the evaluations of the pre-segmented data, we use a partial normalized version of the Writeprints feature set, also extracted through JStylo. We also re-extract features for the first revision dataset using this set to directly compare to the pre-segmented samples. Table 1 shows the number and type of features used for both feature sets.

4.1   Experimental Design

We begin by establishing the effectiveness of stylometry techniques in the Wikia domain on documents by single-authors. We do this by performing 5-fold cross-validation on our single-authored dataset. The purpose of this experiment is

Table 1. This table demonstrates the types and amounts of various features used in the two feature sets we use in this paper. Bigrams refer to sequential pairs, while trigrams are sequential triples.

Feature type	Count (single-authored and multi-authored)	Count (pre-segmented)
Basic counts	1 (characters)	2 (characters, words)
Average characters per word	1	1
Character percentage	3 (digits, total, uppercase)	3 (digits, lowercase, uppercase)
Letter frequency	26	26
Letter bigram frequency	≤ 50	≤ 676
Letter trigram frequency	≤ 50	≤ 1000
Digit frequency	10	10
Digit bigram frequency	≤ 100	≤ 100
Digit trigram frequency	≤ 1000	≤ 1000
Word length frequency	Variable	Variable
Special character, punctuation frequency	Variable	Variable
Function word frequency	≤ 50	≤ 512
Part of speech tag frequency	≤ 50	≤ 1000
Part of speech bigram frequency	≤ 50	≤ 1000
Part of speech trigram frequency	≤ 50	≤ 1000
Word frequency	≤ 50	≤ 1000
Word bigram frequency	≤ 50	≤ 1000
Word trigram frequency	≤ 50	≤ 1000
Misspelling frequency	≤ 50	≤ 1000
Special word counts	0	3 (unique, large, used twice)

to establish a baseline of the performance of our techniques in this domain for solving the traditional authorship attribution problem.

For each variation we defined, we test both the single-label linear SVM and a wide range of multi-label classifiers. We evaluate *complete suspect set* and *partial suspect set* using a train-test technique. We had 60 authors for each experiment, with 9 single-authored training files each. For both of these experiments, the best multi-label classifier was a label powerset classifier with a linear SVM as the base classifier and a threshold of 0.5, so for all result analysis of these experiments we will examine this classifier as well as the standard linear SVM. We also experimented with decision trees, naive bayes, and random forests, but none of these outperformed the linear SVM.

We evaluate *consistent collaboration* through 5-fold cross-validation. We evaluate *inconsistent collaboration* and *mixed collaboration* through a train-test technique, which we also use for *complete suspect set* and *partial suspect set*. For the training data for the *inconsistent collaboration* and *mixed collaboration* cases, we use the *copy transformation* in which each document is counted once for each label (author) to whom it belongs [18]. For *consistent collaboration* and *mixed collaboration*, the same label powerset classifier with linear SVM base and threshold of 0.5 was the best multi-label classifier. However, for *inconsistent collaboration* a binary relevance classifier with naive bayes (NB) base classifier was the best multi-label classifier.

For *mixed collaboration*, we have on average 3.7 training collaborative groups per author, each with on average 3.4 training documents. On average, 5% of the test documents have author groups distinct from those in the training.

Additionally, we have experiments on pre-segmented data. Here, we use a linear SVM as our classifier, and perform cross-validation experiments. We adapt a technique proposed by Overdorf and Greenstadt for tweets and reddit comments and by Dauber et al. for pre-segmented source code to perform account attribution, as well as performing simple attribution of individual samples [6, 15]. For account attribution, we average the SVM output probabilities for the samples belonging to the account in order to attribute the samples as a group. We experiment with account sizes of 2, 5, and 10 samples. We perform experiments with 10 training samples per author, ranging from 10 to 50 authors, for each. We also experiment with the effect of adding more training samples, and perform experiments using an account size of 10 with both 20 and 30 training samples.

4.2   Evaluation Metrics

In the multi-label classification case, simple accuracy as a metric does not give sufficient information to understand the performance of the classifier. Traditional accuracy corresponds to an "exact match" of guessed and correct authors. Indeed, this metric has been proposed and tested in the case of academic papers under the name *guess-all*. In multi-label machine learning literature, *guess-all* is referred to as *subset accuracy* [20]. A broader metric, *guess-one*, measures the frequency with which we correctly predict any author of the document in question. However, *guess-one* does not exactly match to any multi-label learning metric, so while we consider subset accuracy, we do not use *guess-one*, except for our experiments training on single authored documents. In that scenario, we treat *guess-one* as accuracy if we were performing a single-label attribution task, for which any author of the document counts as a correct attribution.

Subset accuracy is considered ineffective at portraying the actual success of the classifier due to ignoring the complexities of how a classification can be partially correct in multi-label learning [16]. Therefore, we also consider *example-based accuracy* (EBA), which describes the average correctness of the label assignments per example. It is calculated by taking the average of the number of correctly predicted labels divided by the total number of actual and predicted labels per example. This shows how many authors we have correctly predicted on average per example. In real-world applications, both subset accuracy and EBA have value in determining the believability of the predictions of our classifiers.

Finally, in order to compare directly between our linear SVM and multi-label techniques, we calculate a version of EBA for our linear SVM which considers the top m ranked authors as predicted labels. As a result, for the SVM each two-authored document will have an accuracy contribution of 0, $\frac{1}{3}$, or 1. In the more general case, the accuracy contribution for partially correct attributions ranges from $\frac{1}{2m-1}$ when only one of our selected labels is correct to $\frac{m-1}{m+1}$ when we only select one incorrect label. For a multi-label classifier with n labels, the accuracy contribution of each document for which we were partially correct can range from $\frac{1}{n}$ when we choose all incorrect labels and one of the correct labels to $\frac{m}{m+1}$ when we select all of the correct labels as well as an additional label.

5 Data

Our dataset was collected from the Star Wars Wiki, Wookieepedia, a Wikia site focused on Star Wars related topics[3]. The dataset was collected by crawling the wiki through the Special:AllPages page, going through the revision history of each article. Our dataset includes 359,685 revisions by over 20,000 authors distributed over 29,046 articles. However, many of those authors had fewer than 5000 words from first revisions, allowing us no more than 75 authors for single-authored only experiments, and fewer for experiments training on single-authored documents and testing on multi-authored documents. While this suspect set is too small to make claims of scalability, it does allow us to showcase the overall difficulty of the problem and overall ineffectiveness of the existing techniques.

We chose to use this dataset because it is the largest English language Wikia and has enough data to run controllable experiments with authentic collaborative documents. Additionally, it has the property that text is naturally organized into topics so we can control for the topic vocabulary, ensuring that we are classifying authorship and not just topic or subject of the writing. This dataset also contains articles of a range of sizes, from under 100 words to a few over 3,000 words. Most importantly, this dataset has clear ground truth in the form of revision histories. However, some of the potential problems from Wikipedia persist in Wikia, including the possibility of sockpuppets and the various writing guidelines, including rules of style.

For the *mixed collaboration* and *consistent collaboration* cases, we note that the number of potential suspects is actually much larger. This is because most collaborative groupings are rare, occurring only once or twice in the entire dataset, and therefore in order to have sufficient training data for any given author, many other authors need to be introduced into the dataset. As such, we do not have firm control over the number of suspect authors, but will make note of the number of suspects when presenting results. We also have limited ability to control for the number of training samples per author and so will also present the total number of training samples in the dataset. It is important to note that while the total number of suspects may be large, the number of actually significant suspects is closer to the 75 authors for which we had single-authored training data, and in some cases may be even less. This is because most authors only contribute to a few documents at the truncated level which we observe. These documents are used to boost the amount of training text and range of collaborative training groups available for the other authors. Due to lack of data and collaborative groupings for these rare authors, the chances of any given

---

[3] http://starwars.wikia.com/wiki/Main_Page.

sample being attributed to them is unlikely, unless in combination with their collaborators.

We attribute the overall small number of principal authors to the wiki environment. In general, wikis have a few very active members and include many people who make occasional edits and corrections. Therefore, it is not surprising that most authors have very little data available.

5.1   Training and Testing Data

For experiments with single-authored documents, we collected data only from first revisions of articles to guarantee that documents have only a single-author. We gathered 5,000 words of text for each author, chunked into 500 word documents, appending articles as necessary. If text from an article would extend beyond 500 words, we truncated the article and discarded the remaining text so that cross-validation would not train and test on parts of the same article. We used chunks of 500 words because this is a value which has been found to work well in the past to balance number of documents and presence of style [5].

For multi-authored data, we chunked in the same manner as above, with the caveat that we controlled for the split of authorship in the multi-authored documents. We truncated the revision history as soon as we had sufficient authors for the experiment. We set thresholds for authorship based on the number of authors, and if the threshold was not met we took only the initial version as part of our single-authored dataset.

We also performed some experiments attributing pre-segmented samples of text. For this dataset, we determined authorship on the sentence level by locating the first revision in which the sentence appeared. We then took consecutive sentences by the same author as a sample, and restricted the dataset to samples between 100 and 400 words.

5.2   Collaborative Examples

In Fig. 1, we demonstrate the collaborative process on two short documents to increase understanding of the possible forms collaboration can take in this setting. We use colors to denote text originating in different revisions or revision sets. In the interest of conserving space, for each set of consecutive revisions by the same author we take only the last such revision.

In the Alpha Charge page edits, we can see that sometimes collaboration takes the form of editing and expanding. Notice that the first author wrote most of the text, but the second author changed the first word and expanded the end of the first sentence. Segmentation methods would be forced to lose information on the first sentence, because it is the work of two authors but can only be assigned to one.

In the Bark Mite page edits, we can observe a very different kind of collaboration. Here, notice that the first author wrote two sentences. The second author added some front matter, which would be placed in a table on the wiki

---

Example Revisions

Alpha Charge
This is an article stub with little special content.[a]

This is the first revision set by the first author.

> The alpha charge was a discreet type of explosive, often used by Trever Flume. Alpha charges came in full, half, and quarter charge varieties, offering different blast strengths.

This is the final revision by a second author.

> ~~The~~An alpha charge was a discreet type of explosive, often used by Trever Flume due to the explosives lack of noise and smoke. Alpha charges came in full, half, and quarter charge varieties, offering different blast strengths.

Bark Mite
This is an article stub with a table as well as text.[b]

This is the first revision set by the first author.

> Bark mites were arthropods on Rori and Endor. They ate bark, and made large hives in trees and caves.

This is the second revision set by a second author.

> Arthropod   Trees   Bark   Bark mites were arthropods on Rori and Endor. They ate bark, and made large hives in trees and caves.

This is the final revision by a third author.

> Arthropod   Trees   Bark   Bark mites were arthropods on Rori and Endor. They ate bark, and made large hives in trees and caves. Bark mites appeared in the video game Star Wars Galaxies, a massively multiplayer online-role playing game developed by Sony and published by LucasArts, prior to its closure on December 15, 2011.

---

[a] http://starwars.wikia.com/wiki/Alpha_charge
[b] http://starwars.wikia.com/wiki/Bark_mite

Fig. 1. Above are examples of the changes in two small articles as they are revised by multiple authors.

to better define the subject of the page. The third author then adds a single long sentence to the end of the article, which makes up over half of the words in the article. This kind of collaboration is more receptive to segmentation, and a suitably powerful segmentation algorithm with sufficient data would lose little to no information.

6   Results

6.1   Single-Authored Baseline

In order to set a baseline and to form a context for multi-author stylometry, single-authored documents in the same domain must be analyzed. With traditional methods used in other single-authored stylometry problems, we analyze *first edits* of a Wikia page, guaranteeing a single author wrote all of the text. With a SVM classifier and Writeprints Limited feature set, described in Sect. 4, 5-fold cross validation achieved an accuracy of 51.3% with 10 authors and 14.2% with 75 authors. Note that accuracy here is number of correct classifications over the total number of classifications, so it is most similar to subset accuracy in that a correctly classified instance is completely, and not partially, correct.

We notice that even in these purely single-author results, our accuracies are lower than those reported in other literature [1,4]. We believe that this is in part due to the rules of style adhered to by Wikia editors. To some extent, Wikia authors attempt to mutually imitate each other in order to have an encyclopedic tone and unified style.

6.2   Non-collaborative Training Documents

For *complete suspect set* and *partial suspect set*, we ran experiments using 60 authors with 9 single-authored first edit training documents per author. We used the same 60 authors for both problems, with different test instances, and experimented ranging from 2-authored documents to 4-authored documents. For *complete suspect set*, all test instances only had authors from within the suspect set, and for *partial suspect set* all test instances had at least one author in the suspect set and at least one author outside the suspect set. That means that for the 2-authored documents EBA and subset accuracy are identical for *partial suspect set*.

Figure 2 shows the results of our experiments for these problems. We do not show the subset accuracy results for *complete suspect set*. This is because we only have non-zero subset accuracy for 2-authored documents for this case. The linear SVM taking the top two authors had subset accuracy of 4.3% and the binary relevance classifier with naive bayes as the base had subset accuracy of 1.5%. Along with the low EBA results, which cap at 23.2% for label powerset with a linear SVM base and 21.7% for a linear SVM taking the top two authors and get worse as the number of authors increase, this shows that predicting the authors of a collaboratively written document from singularly written documents is not practical.

The fact that the EBA results for *partial suspect set* are similar to the results for *complete suspect set* suggests that in the general case these problems aren't very different. The notable difference comes from subset accuracy, due to the fact that for *partial suspect set* some samples reduce to identifying if a suspect author is one of the authors of the document. We show that while this still is a hard problem, it is easier than identifying the set of authors of a collaboratively written document from singularly written documents. The other notable trend in the results is that as we add more authors to the testing document, accuracy decreases. This suggests that single authored training documents are less effective the further the testing document gets from being single authored.

As expected, *guess-one* is easier than the other metrics for both problems. However, counter to the intuition that increasing the number of collaborators should make it easier to successfully identify one of them, *guess-one* accuracy decreases as the number of authors per document increases. We offer a possible explanation in Sect. 7, by observing these results in combination with our other results.

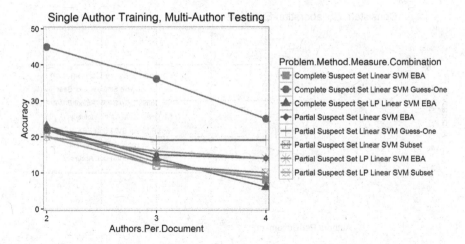

Fig. 2. This graph shows the results of training on single authored documents and testing on multi-authored documents, with the number of collaborators per test document on the x-axis and accuracy on the y-axis. There are a constant 60 suspect authors with 9 training documents each. Linear SVM attributions were performed by taking the top x authors for a document by x authors.

6.3 Consistent Collaboration

In Fig. 3, we examine the results of the *consistent collaboration* experiments. We note that the number of suspects is not held constant here, and neither is the number of overlapping suspect groups. However, we can make some observations by comparing to the results from purely single authored results. We can note we have far better accuracy on consistent collaboration pairs than purely single authored documents with comparable numbers of suspects, and that the magnitude of the difference increases as we have more authors collaborating on the document. The two primary factors which could account for this are the number of collaborators and the amount of overlap between collaborative groups, which decreases in our dataset as we increase the number of collaborators. While these results are not conclusive due to lack of data, they suggest that *consistent collaboration* is one subproblem of multi-authored authorship attribution which current tools can deal with.

One likely explanation for these observations is that collaborators' styles blend into the overall style of the document. As a result, collaboration groups would have a more distinct style than individuals, and as the groups grow they become more distinctive. Another is that as collaboration groups grow, the percentage contribution by any one member decreases, reducing the influence of overlapping members and of more difficult to attribute members. While it would take more evaluation on more datasets to confirm these hypotheses, they would explain these observations, and if true would mean that this particular subproblem is generally easy among authorship attribution tasks, which presents a

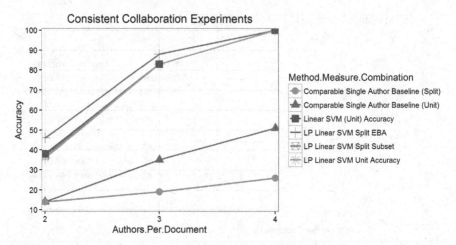

Fig. 3. This graph shows the results of the consistent collaboration experiments. The difference between Split and Unit is that Unit investigates the group of authors exclusively as a set, while Split investigates the group of authors as separate entities. For 2-authored documents, we had 400 documents from 116 authors in 134 pairs. For 3-authored documents, we had 58 documents from 49 authors in 22 triples. For 4-authored documents, we had 20 documents by 28 authors in 8 groups. Beyond that, we had too little data to continue. Additionally, we show the accuracies for the closest size suspect set to both Split and Unit cases from the single authored experiments for comparison purposes. For 2-authored documents, both of those are 75 suspects. For 3-authored documents, this is 50 suspects for Split and 20 suspects for Unit. For 4-authored documents, this is 30 suspects for Split and 10 suspects for Unit.

significant privacy risk to any people who have frequent collaborators in both the public and anonymous spaces.

6.4 Mixed and Inconsistent Collaboration

Figure 4 shows the results of both the *mixed collaboration* and *inconsistent collaboration* cases. We note that the number of suspects and amount of training data are not held constant here. However, we can still make some important observations. The primary observation is that, regardless of the changes in the number of suspects or the number of collaborators per document, EBA for *mixed collaboration* is higher or approximately equal to EBA for *inconsistent collaboration*, which is greater than or approximately equal to subset accuracy for *mixed collaboration*. Subset accuracy for *inconsistent collaboration* is not shown because it is only non-zero for the linear SVM at 2-authors per document and 3-authors per document, and for each of those it is 1.5%.

This trend is not surprising, given two basic facts. First, EBA is a much easier metric than subset accuracy, as discussed in Sect. 4. Secondly, *inconsistent collaboration* is a strictly harder special case of *mixed collaboration*. More interesting is the fact that the best performing multi-label classifier for *inconsistent collaboration* was a binary relevance classifier based on naive bayes, while for all other experiments it was the label powerset classifier based on the linear SVM. Combined with the results from *consistent collaboration*, this suggests a reason why multi-label classification does not work well in the general case for authorship attribution.

Label powerset is a classifier which attempts to treat combinations of labels as single labels in order to make the multi-label learning problem into a single-label learning problem. In contrast, binary relevance transforms the multi-label problem into a binary classification problem for each label. The fact that normally label powerset works better, and that *consistent collaboration* seems to work well, suggests that for stylometric authorship attribution the combination of authors causes a shift in features distinctive to the combination, which can no longer be easily linked back to the styles of the original authors individually by traditional techniques. Therefore, when training data is lacking for combinations of authors, as occurs somewhat for *mixed collaboration* and completely for *inconsistent collaboration*, we are either left with a less well-trained label powerset classifier or forced to fall back on an ineffective binary relevance classifier. This also shows why training on single-authored documents and testing on multi-authored documents works poorly, since that is a similar process to that

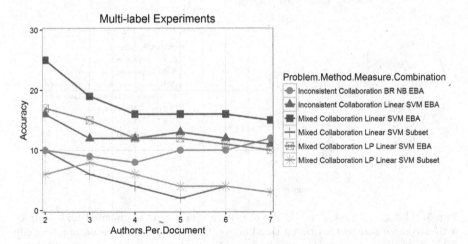

Fig. 4. This graph shows the results of the *mixed collaboration* and *inconsistent collaboration* experiments. For all experiments, there are many suspect authors serving as distractors with only a couple of training instances, due to the small number of occurrences for most collaborative groupings. For 2-authored documents, we had over 360 training instances and over 360 suspect authors. For 3-authored documents we had over 320 training instances and over 470 suspect authors. For 4-authored documents, we had about 360 training instances and over 630 suspect authors. For 5-authored documents, we had over 420 training instances and over 840 suspect authors. For 6-authored documents, we had over 470 training instances and over 1030 suspect authors. For 7-authored documents, we had over 500 training instances and over 1200 suspect authors. Due to lack of training data, most of these suspects have little impact.

of binary relevance, without the benefit of having training documents with input from other authors.

6.5   Authorship Attribution of Pre-segmented Text Samples

Figure 5 shows the results of the experiments with pre-segmented text samples. Not shown in the graph is the result of a single experiment with accounts of 10 samples and 10 suspect authors with 90 training samples each, which had accuracy of 63.6%. Along with the results in the graph, we can conclude that, like shown in [6] with source code, both the number of training samples and the number of samples in the account to be attributed are important to increasing accuracy. Unlike the work with source code, which showed a relatively modest number of samples needed to reach high accuracy, in this work we show that we would need more samples than are present in our dataset to reach high accuracy. However, we do show that we can surpass the base accuracy for standard stylometric chunks with at least 20 training samples and 10 account samples to attribute.

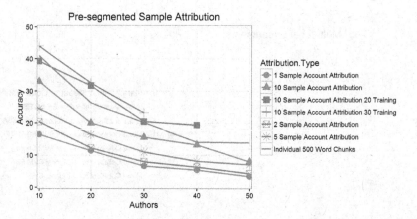

Fig. 5. This graph shows the results of experiments on pre-segmented text samples, with a comparison to traditional chunking performed on our single author first edit dataset. The samples were identified on a per-sentence basis, with a sample consisting of a set of consecutive sentences originating from the same author. Samples used for experimentation were between 100 and 400 words with normalized features, and for consistency we used the same feature extraction process for our comparison chunks. The line labeled Individual 500 Word Chunks is this comparison experiment, and uses 10-fold cross-validation with 10 chunks per author. The experiments labeled with 20 and 30 training samples were performed with 3-fold and 4-fold cross-validation respectively, and end early due to lack of authors with sufficient data. The remaining experiments were performed with 2-fold cross-validation and 10 training samples.

7   Discussion

For the *consistent collaboration* case, we notice that the subset accuracy and example-based accuracy of the multi-label techniques are similar. The fact that example-based accuracy is somewhat higher than subset accuracy here suggests that some, but not many, of the mis-classifications are between overlapping groups. We note that while this case is multi-authored, the authors occur in repeating groups, making it closer to a single-authored case overall.

However, in the other cases, example-based accuracy is clearly better than subset accuracy for all approaches. This indicates that once we lose the similarity to single-author stylometry, it becomes noticeably harder to make the exact correct predictions than to make partially correct predictions, just as in other applications of multi-label learning.

Applications which are single-author, or are multi-author but reducible to single-label problems, are best handled with an SVM. Applications which are purely multi-author are best handled with a multi-label classifier. However, for any multi-author problem, it is essential to have multi-authored training data. While we can obtain some correct predictions from multi-authored documents in which the combinations of authors in the training documents does not reflect the combinations of the documents we want to evaluate, if we do not know that the authors we are interested in only collaborate with certain people it is best to have as wide a range of collaborative groups as possible.

In our experiments, label powerset was the best multi-label classifier. Combined with our results between subproblems, we hypothesize that, stylometrically speaking, collaboration causes a new style to emerge distinct to each set of collaborators. Our observations with decreasing *guess-one* accuracy in the experiments with non-collaborative training documents as we add collaborators and with the increasing accuracy for *consistent collaboration* as we add collaborators suggests that this new style becomes less like the individual styles of the authors as more authors collaborate on a document. This would mean that in order to achieve high accuracy comparable to the accuracy typically observed for single authored documents using these multi-label classifiers we would need sufficient training data for all combinations. In other words, this hypothesis would mean that of the five problems we have defined, only *consistent collaboration* can yield high accuracy in typical use. However, our own experience shows that it can be difficult to gather sufficient data to enable this special case.

Based on our experiments, we believe a defense like the one proposed in [3] to be an effective tool against the current limitations of stylometry. Because their defense relies on crowdsourcing, rather than using contacts, they avoid both the *complete suspect set* and *consistent collaboration* cases. If the crowdsourcing participants rarely rewrite or their rewrites are difficult to identify for training data, then this defense forces the *partial suspect set* case. If the crowdsourcing participants rewrite often, and their rewrites can be identified, then this defense allows no better than the *mixed collaboration* case, and if used sparingly either forces linkability as explored in the original paper or the *inconsistent collaboration* case. Because each of those cases yield poor accuracy, it is unlikely that an analyst would be able to achieve a confident attribution against this defense. However, we stress that this is based on current limitations, and future breakthoughs may still show that this defense is insufficient.

8   Future Work

We are interested in combining single-authored and multi-authored documents in both the training and test sets. In doing this, we hope to determine if we can lessen the burden of acquiring training data while expanding the range of document authorship sets which can be correctly classified.

Our dataset is small, so we also would like to evaluate on a larger dataset, potentially gathered from Wikipedia. Our current results suggest that scalability might be a greater problem for stylometry on collaborative documents than for conventional stylometry. More importantly, we wish to determine if there is a point at which training an SVM on combinations of authors becomes computationally impractical, even if the training data was present.

While wiki articles are one type of collaborative document, they are not the only one. We would like to extend our evaluation to other collaborative domains, including more traditional text and source code. While source code has an easy collection from GitHub[4], it is difficult to find a data set for traditional text with clear ground truth outside of the wiki setting. This is especially important since our single-authored baseline results are so poor, so we hope to find a collaborative data source which is easier to attribute in the base case.

We are also interested in investigating other multi-label learning techniques and different parameterizations. It is likely that optimizing the parameters and finding the optimal algorithm will greatly improve the results in the multi-label case. It is also possible that doing so will improve the single-label results and be able to better compensate for non-representative training data, such as only having single-author training documents or only having collaborative groups which do not occur in the data of interest.

Additionally, we are interested in investigating the effects of changing our feature set, both by admitting more n-gram features and by pruning based on heuristics such as information gain. We would also like to experiment with different chunk sizes and amounts of training text, to determine if it is necessary to include more information to find authors' styles in the multi-author case.

Because we have identified a potential cause for the difficulty of multi-label stylometric attribution, we would like to further investigate to see if we can find a method which works around the issues we have identified. Alternately, we would like to find a way to perform supervised segmentation on documents of this small scale.

9   Conclusion

Collaborative documents require a different way of thinking about stylometry. With single-authored documents, the privacy concern comes from analysts

---

[4] https://github.com/.

collecting a corpus of the author's text and comparing a sensitive document to them. With collaboratively written documents, current techniques require the analyst to collect a corpus of documents written by collaborative groups.

We show that with sufficient training data, the *consistent collaboration* case is the only case in which multi-label stylometry is viable using currently available techniques. We also show that even in other cases, the multi-label learning algorithm which attempts to perform the same transformation, label powerset, performs the best as long as there is some data for the combination of authors. Because of this, we hypothesize that the feature values of collaborations are distinguishable by collaborative group, rather than by member of the group. From a theoretical standpoint, that would mean that label powerset is the correct type of multi-label classifier for the problem. However, in practice it is rare that sufficient data exists for training on all possible collaborative groups of interest. We conclude that this is the greatest difficulty for the application of conventional multi-label learning to stylometry.

Prior work has suggested that collaboration may provide an effective defense against stylometry. While we are not ready to conclude that stylometric attribution of small collaborative documents without training data written by the same collaborative group is impossible, it is clearly a much harder problem than conventional stylometry and requires the development of new techniques. Therefore, those seeking temporary anonymity may find it safer to have people with whom they have never written another publicly available document collaborate with them.

We also investigate the viability of performing segmentation in these situations. We show from the structure of the collaboration that while in some cases authors work on distinct sections of the document, in others authors work not only in the same section but on the same sentences. Therefore, while segmentation may work well in some cases, there are others for which it is difficult to fully capture the collaborative nature of the document with segmentation techniques. We also present results from attempts to attribute pre-segmented text. We demonstrate that, while it is harder to attribute individual segments than it is to perform traditional document attribution, once there are sufficient training and evaluation samples it is possible to attribute an account of such samples. Between these, we believe that supervised segmentation methods, especially with overlapping segments, may allow for a reasonable attribution in some cases, with the caveat that some information may be lost and that they need to be tailored to smaller text sizes than current unsupervised methods require.

Acknowledgements. This work was supported by the National Science Foundation under grant #1253418.

References

1. Abbasi, A., Chen, H.: Writeprints: a stylometric approach to identity-level identification and similarity detection in cyberspace. ACM Trans. Inf. Syst. (TOIS) **26**(2), 7 (2008)
2. Akiva, N., Koppel, M.: A generic unsupervised method for decomposing multiauthor documents. J. Am. Soc. Inf. Sci. Technol. **64**(11), 2256–2264 (2013)
3. Almishari, M., Oguz, E., Tsudik, G.: Fighting authorship linkability with crowdsourcing. In: Proceedings of the 2nd of the ACM Conference on Online Social Networks, pp. 69–82. ACM (2014)
4. Brennan, M., Afroz, S., Greenstadt, R.: Adversarial stylometry: circumventing authorship recognition to preserve privacy and anonymity. ACM Trans. Inf. Syst. Secur. (TISSEC) **15**(3), 12 (2012)
5. Corney, M.W., Anderson, A.M., Mohay, G.M., de Vel, O.: Identifying the authors of suspect email. Comput. Secur. (2001)
6. Dauber, E., Caliskan, A., Harang, R., Greenstadt, R.: Git blame who?: Stylistic authorship attribution of small, incomplete source code fragments. arXiv preprint arXiv:1701.05681 (2017)
7. Diederich, J., Kindermann, J., Leopold, E., Paass, G.: Authorship attribution with support vector machines. Appl. Intell. **19**(1–2), 109–123 (2003)
8. Fifield, D., Follan, T., Lunde, E.: Unsupervised authorship attribution. arXiv preprint arXiv:1503.07613 (2015)
9. Harpalani, M., Hart, M., Singh, S., Johnson, R., Choi, Y.: Language of vandalism: improving wikipedia vandalism detection via stylometric analysis. In: Proceedings of the 49th Annual Meeting of the Association for Computational Linguistics: Human Language Technologies: Short Papers, vol. 2, pp. 83–88. Association for Computational Linguistics (2011)
10. Juola, P., et al.: Authorship attribution. Found. Trends® Inf. Retrieval **1**(3), 233–334 (2008)
11. Koppel, M., Akiva, N., Dershowitz, I., Dershowitz, N.: Unsupervised decomposition of a document into authorial components. In: Proceedings of the 49th Annual Meeting of the Association for Computational Linguistics: Human Language Technologies, vol. 1, pp. 1356–1364. Association for Computational Linguistics (2011)
12. Macke, S., Hirshman, J.: Deep sentence-level authorship attribution (2015)
13. Madjarov, G., Kocev, D., Gjorgjevikj, D., Džeroski, S.: An extensive experimental comparison of methods for multi-label learning. Pattern Recogn. **45**(9), 3084–3104 (2012)
14. McDonald, A.W.E., Afroz, S., Caliskan, A., Stolerman, A., Greenstadt, R.: Use fewer instances of the letter "i": toward writing style anonymization. In: Fischer-Hübner, S., Wright, M. (eds.) PETS 2012. LNCS, vol. 7384, pp. 299–318. Springer, Heidelberg (2012). doi:10.1007/978-3-642-31680-7_16
15. Overdorf, R., Greenstadt, R.: Blogs, twitter feeds, and reddit comments: cross-domain authorship attribution. PoPETs **2016**(3), 155–171 (2016)
16. Payer, M., Huang, L., Gong, N.Z., Borgolte, K., Frank, M.: What you submit is who you are: a multi-modal approach for deanonymizing scientific publications. IEEE Trans. Inf. Forensics Secur. **10**, 200–212 (2015)
17. Solorio, T., Hasan, R., Mizan, M.: Sockpuppet detection in wikipedia: a corpus of real-world deceptive writing for linking identities. arXiv preprint arXiv:1310.6772 (2013)

18. Tsoumakas, G., Katakis, I.: Multi-label classification: an overview. Int. J. Data Warehouse. Min. **3**(3), 13 (2007)
19. Tsoumakas, G., Katakis, I., Vlahavas, I.: Effective and efficient multilabel classification in domains with large number of labels. In: Proceedings of ECML/PKDD 2008 Workshop on Mining Multidimensional Data (MMD 2008), pp. 30–44 (2008)
20. Tsoumakas, G., Katakis, I., Vlahavas, I.: Mining multi-label data. In: Maimon, O., Rokach, L. (eds.) Data Mining and Knowledge Discovery Handbook, pp. 667–685. Springer, New York (2010)
21. Tsoumakas, G., Katakis, I., Vlahavas, I.: Random k-labelsets for multilabel classification. IEEE Trans. Knowl. Data Eng. **23**(7), 1079–1089 (2011)
22. Zhang, M.L., Zhou, Z.H.: ML-KNN: a lazy learning approach to multi-label learning. Pattern Recogn. **40**(7), 2038–2048 (2007)

A Distributed Investment Encryption Scheme: Investcoin

Filipp Valovich[✉]

Faculty of Mathematics, Horst Görtz Institute for IT Security,
Ruhr-Universität Bochum, Universitätsstraße 150, 44801 Bochum, Germany
filipp.valovich@rub.de

Abstract. This work presents a new framework for Privacy-Preserving Investment systems in a distributed model. In this model, independent investors can transfer funds to independent projects, in the same way as it works on crowdfunding platforms. The framework protects the investors' single payments from being detected (by any other party), only the sums of each investor's payments are revealed (e.g. to the system). Likewise, the projects' single incoming payments are concealed and only the final sums of the incoming payments for every project are revealed. In this way, no other party than the investor (not even the system administration) can detect how much she paid to any single project. Though it is still possible to confidentially exchange any part of an investment between any pair of investors, such that market liquidity is unaffected by the system. On top, our framework allows a privacy-preserving return of a multiple of all the held investments (e.g. interest payments or dividends) to the indivdual investors while still revealing nothing else than the sum of all returns for every investor. We introduce Investcoin as practicable instantiation for this framework. It is a proper combination of three cryptographic protocols, namely a Private Stream Aggregation scheme, a Commitment scheme and a Range test. The security of the three protocols is based on the Decisional Diffie-Hellman (DDH) assumption. Thus, by a composition theorem, the security of Investcoin is also based on the DDH assumption. Furthermore, we provide a simple decentralised key generation protocol for Investcoin that supports dynamic join, leave and fault-tolarance of investors and moreover achieves some security guarantees against malicious investors.

1 Introduction

The promise of performance benefit by using technologies like online-outsourcing and cloud-computing goes along with the loss of control over individual data. Therefore the public awareness of data protection increases. We use encryption and privacy technologies to protect our electronic messages, our consumer behaviour or patient records. In this work, we put the following question up for discussion: why is there only minor attention paid to the protection of sensitive

The research was supported by the DFG Research Training Group GRK 1817/1.

S. Dolev and S. Lodha (Eds.): CSCML 2017, LNCS 10332, pp. 136–154, 2017.
DOI: 10.1007/978-3-319-60080-2_10

financial data in the public? Indeed the requirement to trust in financial institutions may be an obstacle for the trade secrecy of companies. On the one hand, transactions on organised markets are registered by electronic systems, audited and eventually get under the control of the system administration (e.g. it can refuse a transaction). In some cases this is desired: e.g. it should be possible to detect a company financing criminal activities. On the other hand, we would like to protect the trade secrecy of the companies. In this sense, there is a transparency/confidentiality trade-off in organised markets, such as exchange or to some extent also crowdfunding platforms.

In this work we address the problem of providing adequate privacy guarantees to investors. As observed in [18], although there is no observable significant effect concerning "the impact of privacy violations on the investment amount, (...) one has to remember that trust influences behavior (...) and privacy issues influences trust (...) and therefore an indirect influence still exists". Conversely, this means that individuals would participate more in investments if their privacy is protected. As an effect, the reporting of investors concerning wins and losses (and therefore risks) becomes more reliable [3,9,14]. As further motivation of our work, the possibility to circumvent certain regulatories may be desired, e.g. financial sanctions by order of "repressive" countries. Investors may look for ways to invest in sanctioned companies without being traced by their home country.

Consequently, the objective of this work is to solve privacy issues by concealing particular investment decisions but offering transparency of "aggregated" investment decisions. In this regard we introduce a cryptographically secure distributed investment encryption (DIE) scheme for the aggregation of the investments of a number of different investors funding different projects on an electronic platform. A DIE scheme maintains market liquidity, i.e. the scheme does not affect the possibility to trade assets among investors. Informally, a DIE scheme conceals the single payments of investors from being traced but reveals to the system only the aggregates of the investors' payments. Similarly, the projects' single incoming payments are concealed and only the final sums of the incoming payments for every project are revealed. Moreover, a DIE scheme conceals the single returns (representing interest payments, coupons or dividends) from every single project to every single investor but reveals (to the system administration) the aggregated return of every single investor. Therefore, up to a certain extent, a DIE scheme simultaneously maintains transparency (e.g. taxes on the final return of every investor can be raised) and trade secrecy.

As a particular DIE scheme we present Investcoin, a combination of three cryptographic protocols: a Private Stream Aggregation (PSA) scheme, first introduced in [23], a (homomorphic) Commitment scheme and a Range test for committed values, all secure under the Decisional Diffie-Hellman assumption. Informally, the PSA scheme is used for the secure aggregation of funds for every particular project and the homomorphic Commitment scheme is used for the secure aggregation of all investments and returns of every particular investor. The Range test ensures that investments are not negative. We provide a simple secret sharing key generation protocol for Investcoin, that allows investors to dynamically join, leave or fail during the protocol execution and prevents investors from some malicious cheating.

Related work. The notion of Commitment schemes (first in [4,6]) is well-established in the literature. The notion of PSA was introduced in [23]. A PSA scheme is a cryptographic protocol which enables a number of users to individually and securely send encrypted time-series data to an untrusted aggregator requiring each user to send exactly one message per time-step. The aggregator is able to decrypt the aggregate of all data per time-step, but cannot retrieve any further information about the individual data. In [23] a security definition for PSA and a secure instantiation were provided. In [15] a scheme with a tighter security reduction was provided and the work [2] generalised the scheme in [15]. By lowering the security requirements established in [23], the work [24] provided general conditions for the existence of secure PSA schemes, based on key-homomorphic weak PRFs.

Investcoin is not a classical cryptocurrency. It can be thought of as a cryptographic layer on top of any currency used for *investments*, similar to what Zerocoin is intended to be for Bitcoin (or other *payment* systems). Bitcoin is the first cryptocurrency, introduced in [17]. Currently it is the cryptocurrency with the largest market capitalisation. Bitcoin is as a peer-to-peer payment system where transactions are executed directly between users without interaction of any intermediate party. The transactions are verified by the users of the network and publicly recorded in a blockchain, a distributed database. Zerocoin was proposed in [16] as an extension for Bitcoin (or any other cryptocurrency) providing cryptographic anonymity to recorded transactions in the blockchain (Bitcoin itself provides only pseudonymity). This is achieved by the use of a seperate mixing procedure based on Commitment schemes. Therefore particular transactions cannot be publicly traced back to particular Bitcoin adresses anymore. This is also the main principle of Investcoin: no investment in a particular project can be traced back to a particular investor. In this regard, Investcoin has similarities with Zerocoin.

Methods for market regulation through aggregated privacy-preserving risk reporting were studied in [1]. They constructed protocols allowing a number of users to securely compute aggregated risk measures based on summations and inner products. In [12], cryptographic tools for statistical data privacy were investigated to balance transparency and confidentiality for financial regulation.

2   Preliminaries

In this section we provide the description of our investment model, the basic protocols underlying Investcoin and their corresponding security definitions.

2.1   Model

As initial situation we consider a network consisting of n investors and λ projects to be funded by the investors. As an analogy from the real world one can think of a crowdfunding platform or an exchange system where projects or companies try to collect funds from various individual investors. Each investor N_i, $i = 1, \ldots, n$, is willing to invest the amount $x_{i,j} \geqslant 0$ into the project P_j, $j = 1, \ldots, \lambda$, thus the total amount invested by N_i is $\sum_{j=1}^{\lambda} x_{i,j}$ and the total amount received by project P_j is $\sum_{i=1}^{n} x_{i,j}$. Moreover, there exists an administration (which may be the administration of the crowdfunding platform). The investors and the project managements *are not required to trust the administration.*

We consider a series of investment rounds. An investment round denotes the moment when the payments of all participating investors are registered by the administration of the system. From round to round the values n and λ may change, i.e. investors and projects may join or leave the network before a round.

After an investment round is over and the time comes to give a return to the investors (i.e. at maturity), the management of each project P_j publishes some value α_j defining the return for each investor (i.e. an indicator of economic growth, interest yield, dividend yield or similar). The untrusted system administration (or simply system) serves as a pool for the distribution of investments to the projects and of returns to the investors: first, for all $i = 1, \ldots, n$ it collects the total amount $\sum_{j=1}^{\lambda} x_{i,j}$ invested by investor N_i and rearranges the union of the total amounts into the aggregated investment $\sum_{i=1}^{n} x_{i,j}$ for project P_j for all $j = 1, \ldots, \lambda$; at maturity date, for all $j = 1, \ldots, \lambda$ it collects the total returns $\alpha_j \sum_{i=1}^{n} x_{i,j}$ of the projects and rearranges the union of the total returns into the returns $\sum_{j=1}^{\lambda} \alpha_j x_{i,j}$ of the investors.

While the investors do not have to trust each other nor the system (i.e. an investor doesn't want the others to know her financial data), we aim at constructing a computationally secure protocol (in the random oracle model) for transferring the funds in an honest-but-curious model where the untrusted system administration tries to compromise investors in order to build a coalition (i.e. parties may collude). This coalition tries to infer additional information about uncompromised investors, but under the constraint of honestly following the investment protocol. On the other hand, we allow investors that are not part of the coalition, to execute some malicious behaviour, i.e. stealing funds, stealing returns or simply distorting the overall computations (formal details will be clear in Sect. 4.1). Thereby we have the following objectives in each investment round:

Security

- Hiding: For all $i = 1, \ldots, n$ the only financial data of investor N_i (if uncompromised) known to the system is $C_i = \sum_{j=1}^{\lambda} x_{i,j}$ and $E_i = \sum_{j=1}^{\lambda} \alpha_j x_{i,j}$. For all $j = 1, \ldots, \lambda$ the only financial data of project P_j known to the system is $X_j = \sum_{i=1}^{n} x_{i,j}$ and α_j. Particularly, no other party than N_i should get to know her respective investments to $P_1, \ldots, P_\lambda$.
- Binding: Investors may not announce an incorrect investment, i.e. if N_i has send $x_{i,j}$ to P_j, then P_j should also receive $x_{i,j}$ from N_i.
- For all i, j: $x_{i,j} \geqslant 0$. I.e. no investor can 'steal' money from a project.

Correctness

- For all i, if C_i is the real aggregate of N_i's investments, then $C_i = \sum_{j=1}^{\lambda} x_{i,j}$, the system knows C_i and can charge N_i's bank account with amount C_i.
- For all j, if X_j is the real aggregate of P_j's funds, then $X_j = \sum_{i=1}^{n} x_{i,j}$, the system knows X_j and transfers the amount X_j to the bank account of P_j.
- For all i, if E_i is the real aggregate of N_i's returns, then $E_i = \sum_{j=1}^{\lambda} \alpha_j x_{i,j}$, the system knows E_i and transfers it to the bank account of N_i.
- If one of these conditions is violated (e.g. on the purpose of stealing money), then the injured party should be able to detect this fact and to prove it to the network latest after the end of the corresponding investment round.

Now we provide the building blocks for a scheme satisfying these objectives.

2.2 Private Stream Aggregation

In this section, we define Private Stream Aggregation (PSA) and provide a security definition. This notion was introduced in [23].

The Definition of PSA. A PSA scheme is a protocol for safe distributed time-series data transfer which enables the receiver (here: the system administrator) to learn nothing else than the sums $\sum_{i=1}^{n} x_{i,j}$ for $j = 1, 2, \ldots$, where $x_{i,j}$ is the value of the ith participant in (time-)step j and n is the number of participants (here: investors). Such a scheme needs a key exchange protocol for all n investors together with the administrator as a precomputation, and requires each investor to send exactly one message (namely the amount to spend for a particular project) in each step $j = 1, 2, \ldots$.

Definition 1 (Private Stream Aggregation [23]). *Let κ be a security parameter, $\mathcal{D}$ a set and $n, \lambda \in \mathbb{N}$ with $n = poly(\kappa)$ and $\lambda = poly(\kappa)$. A Private Stream Aggregation (PSA) scheme $\Sigma = (\mathsf{Setup}, \mathsf{PSAEnc}, \mathsf{PSADec})$ is defined by three ppt algorithms:*

Setup: *$(\mathsf{pp}, T, s_0, s_1, \ldots, s_n) \leftarrow \mathsf{Setup}(1^\kappa)$ with public parameters pp, $T = \{t_1, \ldots, t_\lambda\}$ and secret keys s_i for all $i = 1, \ldots, n$.*
PSAEnc: *For $t_j \in T$ and all $i = 1, \ldots, n$: $c_{i,j} \leftarrow \mathsf{PSAEnc}_{s_i}(t_j, x_{i,j})$ for $x_{i,j} \in \mathcal{D}$.*
PSADec: *Compute $\sum_{i=1}^{n} x'_{i,j} = \mathsf{PSADec}_{s_0}(t_j, c_{1,j}, \ldots, c_{n,j})$ for $t_j \in T$ and ciphers $c_{1,j}, \ldots, c_{n,j}$. For all $t_j \in T$ and $x_{1,j}, \ldots, x_{n,j} \in \mathcal{D}$ the following holds:*

$$\mathsf{PSADec}_{s_0}(t_j, \mathsf{PSAEnc}_{s_1}(t_j, x_{1,j}), \ldots, \mathsf{PSAEnc}_{s_n}(t_j, x_{n,j})) = \sum_{i=1}^{n} x_{i,j}.$$

The system parameters pp are public and constant for all t_j with the implicit understanding that they are used in Σ. Every investor encrypts her amounts $x_{i,j}$ with her own secret key s_i and sends the ciphertext to the administrator.

If the administrator receives the ciphertexts of *all* investors for some t_j, it can compute the aggregate of the investors' data using the decryption key s_0.

While in [23], the $t_j \in T$ were considered to be time-steps within a time-series (e.g. for analysing time-series data of a smart meter), here the $t_j \in T$ are associated with projects P_j, $j = 1, \ldots, \lambda$, to be funded in an investment round.

Security of PSA. Our model allows an attacker to compromise investors. It can obtain auxiliary information about the values of investors or their secret keys. Even then a secure PSA scheme should release no more information than the aggregates of the uncompromised investors' values.

Definition 2 (Aggregator Obliviousness, informal). *A PSA scheme achieves adaptive Aggregator Obliviousness or* AO *if for all ppt adversaries with adaptive control over a coalition of compromised users the generated ciphers are indistinguishable under a Chosen Plaintext Attack.*

Feasibility of AO. In the *random oracle model* we can achieve AO for some constructions [2,23,24]. Because of its simplicity and efficient decryption, we use the PSA scheme proposed in [24] and present it in Fig. 1. It achieves AO in the random oracle model based on the DDH assumption (see [24] for the proof).

The original scheme proposed in [23] is similar to the one in Fig. 1, but its decryption is inefficient, if the plaintext space is super-polynomially large in the security parameter. These schemes also achieve the non-adaptive version of AO in the *standard model*. See [23,24] for the formal security definitions.

2.3   Commitment Schemes

A Commitment scheme allows a party to publicly commit to a value such that the value cannot be changed after it has been committed to (binding) and the value itself stays hidden to other parties until the owner reveals it (hiding). For the basic definitions we refer to [4,6]. Here we just recall the Pedersen Commitment introduced in [19] (Fig. 2), which is computationally binding under the dlog assumption and perfectly hiding. In Sect. 3 we will combine the Pedersen Commitment with the PSA scheme from Fig. 1 for the construction of Investcoin and thereby consider the input data $x = x_{i,j}$ to the Commitment scheme as investment amounts from investor N_i to project P_j. An essential property for the construction of Investcoin is that the Pedersen Commitment contains a homomorphic commitment algorithm, i.e. $\mathsf{Com}_{pk}(x, r) * \mathsf{Com}_{pk}(x', r') = \mathsf{Com}_{pk}(x + x', r + r')$.

2.4   Range Test

To allow the honest verifier to verify in advance, that the (possibly malicious) prover commits to an integer x in a certain range, a Range test must be applied. Range tests were studied in [5,7,20,21]. For Investcoin, an interactive procedure can be applied. It is a combination of the Pedersen Commitment to the binary representation of x and the extended Schnorr proof of knowledge [22] (Fig. 3) applied to proving knowledge of one out of two secrets as described in [10]. Its basic idea was described in [5].

By the interactive procedure from Fig. 4, the prover shows to the verifier, that the committed value x lies in the interval $[0, 2^l - 1]$ without revealing anything else about x. For the security of the construction in Fig. 3 we refer to [10], where a more general protocol was considered (particularly the special honest verifier zero-knowledge property is needed). We use the Fiat-Shamir heuristic [11] to make the Range test non-interactive.

2.5 Secure Computation

Our security definition for Investcoin will be twofold. In the first part, we consider an honest-but-curious coalition consisting of the untrusted system administration together with its compromised investors and the group of honest investors. Here we refer to notions from Secure Multi-Party Computation (SMPC). In the second part, we identify reasonable malicious behaviour that an investor could execute and show, how the system can be secured against such malicious investors. In this Section we focus on defining security against the honest-but-curious coalition.

Definition 3. *Let κ be a security parameter and $n, \lambda \in \mathbb{N}$ with $n = poly(\kappa)$. Let ρ be a protocol executed by a group of size $u \leqslant n$ and a coalition of honest-but-curious adversaries of size $n-u+1$ for computing the deterministic functionality f_ρ. The protocol ρ performs a* secure computation *(or securely computes f_ρ), if there exists a ppt algorithm $\mathcal{S}$, such that*

$$\{\mathcal{S}(1^\kappa, y, f_\rho(x,y))\}_{x,y,\kappa} \approx_c \{\mathsf{view}^\rho(x,y,\kappa)\}_{x,y,\kappa},$$

where $\mathsf{view}^\rho(x,y,\kappa) = (y,r,m)$ is the view of the coalition during the execution of the protocol ρ on input (x,y), x is the input of the group, y is the input of the coalition, r is its random tape and m is its vector of received messages.

This definition follows standard notions from SMPC (as in [13]) and is adapted to our environment: first, we consider two-party protocols where each party consists of multiple individuals (each individual in a party has the same goals) and second, we do not consider security of the coalition against the group, since the system administration has no input and thus its security against honest-but-curious investors is trivial. Rather we will later consider its security against malicious investors.

Investcoin is the combination of various protocols, so we will prove the security of these protocols separately and then use the composition theorem from [8].

Theorem 1 (Composition in the Honest-but-Curious Model [8]). *Let κ be a security parameter and let $m = poly(\kappa)$. Let π be a protocol that computes a functionality f_π by making calls to a trusted party computing the functionalities $f_1, \ldots, f_m$. Let $\rho_1, \ldots, \rho_m$ be protocols computing $f_1, \ldots, f_m$ respectively. Denote by $\pi^{\rho_1, \ldots, \rho_m}$ the protocol π, where the calls to a trusted party are replaced by executions of $\rho_1, \ldots, \rho_m$. If $\pi, \rho_1, \ldots, \rho_m$ non-adaptively perform secure computations, then also $\pi^{\rho_1, \ldots, \rho_m}$ non-adaptively performs a secure computation.*

3 Investcoin: The Scheme

In this section, we introduce Investcoin. This protocol is build from a combination of the PSA scheme from Fig. 1, the homomorphic Commitment scheme

---

PSA scheme

Setup: Public parameters are primes $q > m \cdot n, p = 2q + 1$ and a hash function $H : T \to \mathcal{QR}_{p^2}$ modelled as random oracle. Secret keys are $s_0, \ldots, s_n \leftarrow_R \mathbb{Z}_{pq}$ with $\sum_{i=0}^n s_i = 0 \bmod pq$.
PSAEnc: For $t_j \in T$, $i = 1, \ldots, n$, encrypt $x_{i,j} \in [0, m]$ by $c_{i,j} \leftarrow (1 + p \cdot x_{i,j}) \cdot H(t_j)^{s_i} \bmod p^2$.
PSADec: For $t_j \in T$ and ciphers $c_{1,j}, \ldots, c_{n,j}$ compute $V_j \in \{1 - p \cdot mn, \ldots, 1 + p \cdot mn\}$ with

$$V_j \equiv H(t_j)^{s_0} \cdot \prod_{i=1}^n c_{i,j} \equiv \prod_{i=1}^n (1 + p \cdot x_{i,j}) \equiv 1 + p \cdot \sum_{i=1}^n x_{i,j} \bmod p^2$$

and compute $\sum_{i=1}^n x_{i,j} = (V_j - 1)/p$ over the integers (holds if the $c_{i,j}$ encrypt the $x_{i,j}$).

Fig. 1. PSA scheme secure in the random oracle model.

Commitment scheme

GenCom: $(pp, pk) \leftarrow \mathsf{GenCom}(1^\kappa)$, with public parameters pp describing a cyclic group G of order q and public key $pk = (h_1, h_2)$ for two generators h_1, h_2 of G. (where $\mathrm{dlog}_{h_1}(h_2)$ is not known to the commiting party).
Com: For $x \in \mathbb{Z}_q$ choose $r \leftarrow_R \mathbb{Z}_q$ and compute $com = \mathsf{Com}_{pk}(x, r) = h_1^x \cdot h_2^r \in G$.
Unv: For $pk = (h_1, h_2)$, commitment com, opening (x, r): $\mathsf{Unv}_{pk}(com, x, r) = 1 \Leftrightarrow h_1^x \cdot h_2^r = com$.

Fig. 2. The Pedersen commitment.

Extended Schnorr Proof of Knowledge

Let G be a cyclic group of order q and h a generator of G. The prover wants to show to the verifier that she knows the discrete logarithms of either $R = h^r$ or $S = h^s$ without revealing its value and position. W.l.o.g. the prover knows r.

Com: The prover chooses random $v_2, w_2, z_1 \leftarrow_R \mathbb{Z}_q^*$ and sends $a_1 = h^{z_1}, a_2 = h^{w_2} \cdot S^{-v_2}$ to the verifier.
Chg: The verifier sends a random $v \leftarrow_R \mathbb{Z}_q^*$ to the prover.
Opn: The prover sends $v_1 = v - v_2, w_1 = z_1 + v_1 r, v_2$ and w_2 to the verifier.
Chk: The verifier verifies that $v = v_1 + v_2$ and that $h^{w_1} = a_1 R^{v_1}, h^{w_2} = a_2 S^{v_2}$.

Fig. 3. Schnorr proof of knowledge of one out of two secrets.

Range test

Gen: $(pp, pk) \leftarrow \text{Gen}(1^\kappa)$, with public parameters pp describing the cyclic group G of order q, a test range $[0, 2^l - 1] \subset G$ and public key $pk = (g, h)$ for generators g, h of G (the prover does not know $dlog_g(h)$).

Com: For $x = \sum_{k=0}^{l-1} x^{(k)} \cdot 2^k$ with $x^{(k)} \in \{0,1\}$ for all $k = 0, \ldots, l-1$, the prover chooses random $r^{(k)}, v_2^{(k)}, w_2^{(k)}, z_1^{(k)} \leftarrow_R \mathbb{Z}_q^*, k = 0 \ldots, l-1$, computes $r \equiv \sum_{k=0}^{l-1} r^{(k)} \cdot 2^k \bmod q$ and sends

$$com = g^x \cdot h^r \bmod q, \quad com^{(k)} = g^{x^{(k)}} \cdot h^{r^{(k)}} \bmod q$$

$$a_1^{(k)} = h^{z_1^{(k)}}, a_2^{(k)} = h^{w_2^{(k)}} \cdot (h^{r^{(k)}} g^{2x^{(k)}-1})^{-v_2^{(k)}}$$

for $k = 0, \ldots, l-1$ to the verifier.

Chg: The verifier verifies that $com \equiv \prod_{k=0}^{l-1}(com^{(k)})^{2^k} \bmod q$. The verifier sends random $v^{(k)} \leftarrow_R \mathbb{Z}_q^*$ for all $k = 0 \ldots, l-1$ to the prover.

Opn: The prover sends $v_1^{(k)} = v^{(k)} - v_2^{(k)}, w_1^{(k)} = z_1^{(k)} + v_1^{(k)} r^{(k)}, v_2^{(k)}$ and $w_2^{(k)}$ for all $k = 0 \ldots, l-1$ to the verifier.

Chk: For all $k = 0 \ldots, l-1$, the verifier verifies that $v^{(k)} = v_1^{(k)} + v_2^{(k)}$ and that either

$$(h^{w_1^{(k)}}, h^{w_2^{(k)}}) = (a_1^{(k)}(com^{(k)})^{v_1^{(k)}}, a_2^{(k)}(com^{(k)} \cdot g^{-1})^{v_2^{(k)}}) \text{ or}$$

$$(h^{w_1^{(k)}}, h^{w_2^{(k)}}) = (a_1^{(k)}(com^{(k)} \cdot g^{-1})^{v_1^{(k)}}, a_2^{(k)}(com^{(k)})^{v_2^{(k)}}).$$

Fig. 4. Range test for a commited value.

from Fig. 2 and the Range test from Fig. 4. Moreover, we provide a simple key-generation protocol for the Investcoin protocol that allows the dynamic join and leave of investors and is fault-tolerant towards investors.

3.1 Construction of Investcoin

The DIESet algorithm in Fig. 5 executes the Setup algorithms of the underlying schemes. Additionally, DIESet generates a verification parameter β_j for each project P_j (and an additional β_0 - this will be used for the security against malicious investors) which is only known to the system administration. In Sect. 3.2 we provide a simple protocol for generating the secrets. The encryption algorithm DIEEnc executes the encryption algorithm of Σ and encrypts the amounts invested by N_i into P_j. In order to prove that $C_i = \sum_{j=1}^{\lambda} x_{i,j}$, the N_i execute the commitment algorithm DIECom commiting to the amounts $x_{i,j}$ invested using the randomness $r_{i,j}$ by executing the commitment algorithm of Γ and encrypting the $r_{i,j}$ with Σ. The Range test algorithm DIETes ensures that the investments are larger or equal 0. The payment verification algorithm DIEUnvPay first verifies that the combination of the committed amounts *in the correct order* is valid for the same combination of amounts encrypted *in the correct order* by executing the verification algorithm of Γ. If the investor has not cheated, this verification will output 1 by the homomorphy of Γ and the fact that $\prod_{j=0}^{\lambda} H(t_j)^{\beta_j} = \prod_{j=1}^{\lambda} H(\tilde{t}_j)^{\beta_j} = 1$. The DIEUnvPay algorithm verifies that the combination of commitments is valid for the aggregate C_i of the investments of N_i. The decryption algorithm DIEDec then decrypts the aggregated amounts for every project by executing the decryption algorithm of Σ. After the projects are

Investcoin

Let κ be a security parameter and $n, \lambda \in \mathbb{N}$ with $n = \text{poly}(\kappa)$ and $\lambda = \text{poly}(\kappa)$. Let $\Sigma = (\text{Setup}, \text{PSAEnc}, \text{PSADec})$ be the PSA scheme from Figure 1 and let $\Gamma = (\text{GenCom}, \text{Com}, \text{Unv})$ be the Commitment scheme from Figure 2. We define Investcoin $\Omega = (\text{DIESet}, \text{DIEEnc}, \text{DIECom}, \text{DIETes}, \text{DIEUnvPay}, \text{DIEDec}, \text{DIEUnvRet})$ as follows.

DIESet:
- The system generates public parameters $\text{pp} = \{q, p, H\}$ with primes $q > m \cdot n$, $p = 2q + 1$ (where $m = 2^l - 1$ is the maximum possible amount to invest into a single project by an investor), parameters $T = \{t_0, t_1, \bar{t}_1, \ldots, t_\lambda, \bar{t}_\lambda\}$ and a hash function $H : T \rightarrow \mathcal{QR}_{p^2}$.
- The system and the investors together generate secret keys s_0 and $sk_i = (s_i, \bar{s}_i) \leftarrow_R \mathbb{Z}_{pq} \times \mathbb{Z}_{pq}$ for all $i = 1, \ldots, n$ with $s_0 = -\sum_{i=1}^{n} s_i \pmod{pq}$.
- The system generates secret parameters $\beta_0, \ldots, \beta_\lambda \leftarrow_R [-q', q']$, $q' < q/(m\lambda)$, such that $\prod_{j=0}^{\lambda} H(t_j)^{\beta_j} = \prod_{j=1}^{\lambda} H(\bar{t}_j)^{\beta_j} = 1$ (see Section 3.2 for the details). It sets $sk_0 = (s_0, \beta_0, \ldots, \beta_\lambda)$.
- System generates a public key $pk = (h_1, h_2) \in \mathbb{Z}_{p^2}^* \times \mathbb{Z}_{p^2}^*$ with $\text{ord}(h_1) = \text{ord}(h_2) = pq$.

DIEEnc: For all $j = 1, \ldots, \lambda$, each investor N_i chooses $x_{i,j} \in [0, m]$ and $x_{i,0} = 0$ and for all $j = 0, \ldots, \lambda$ N_i sends the following ciphers to the system:

$$c_{i,j} \leftarrow \text{DIEEnc}_{sk_i}(t_j, x_{i,j}) = \text{PSAEnc}_{s_i}(t_j, x_{i,j}).$$

DIECom: For all $j = 1, \ldots, \lambda$, each N_i chooses $r_{i,j} \leftarrow_R [0, m]$ and sends the following to the system:

$$(com_{i,j}, \bar{c}_{i,j}) \leftarrow \text{DIECom}_{pk, sk_i}(x_{i,j}, r_{i,j}) = (\text{Com}_{pk}(x_{i,j}, r_{i,j}), \text{PSAEnc}_{\bar{s}_i}(\bar{t}_j, r_{i,j})).$$

DIETes: For all $j = 1, \ldots, \lambda$ the algorithm from Figure 4 on $(x_{i,j}, r_{i,j})$ is executed between the system (verifier) and each investor N_i (prover) using pk. Note that there always exists a random representation $r_{i,j}^{(0)}, \ldots, r_{i,j}^{(l-1)} \leftarrow_R [0, m]$, such that $r_{i,j} \equiv \sum_{k=0}^{l-1} r_{i,j}^{(k)} \cdot 2^k \bmod p^2$.

DIEUnvPay: The system verifies for each investor N_i with commitment values $(com_{i,1}, \bar{c}_{i,1}), \ldots, (com_{i,\lambda}, \bar{c}_{i,\lambda})$ and ciphers $c_{i,1}, \ldots, c_{i,\lambda}$ that

$$\text{Unv}_{pk}\left(\prod_{j=1}^{\lambda} com_{i,j}^{\beta_j}, A_i, B_i\right) = 1,$$

where $A_i = \left(\left(\prod_{j=0}^{\lambda} c_{i,j}^{\beta_j} \bmod p^2\right) - 1\right)/p$ and $B_i = \left(\left(\prod_{j=1}^{\lambda} \bar{c}_{i,j}^{\beta_j} \bmod p^2\right) - 1\right)/p$. Then each investor N_i sends $C_i = \sum_{j=0}^{\lambda} x_{i,j}$ and $D_i = \sum_{j=1}^{\lambda} r_{i,j}$ to the system which computes

$$b_{P,i} = \text{DIEUnvPay}_{pk}\left(\prod_{j=1}^{\lambda} com_{i,j}, C_i, D_i\right) = \text{Unv}_{pk}\left(\prod_{j=1}^{\lambda} com_{i,j}, C_i, D_i\right), b_{P,i} \in \{0, 1\}.$$

and charges the bank account of N_i with the amount C_i, if $b_{P,i} = 1$.

DIEDec: For all $j = 0, \ldots, \lambda$ and ciphertexts $c_{1,j}, \ldots, c_{n,j}$, the system computes

$$X_j = \text{DIEDec}_{sk_0}(t_j, c_{1,j}, \ldots, c_{n,j}) = \text{PSADec}_{s_0}(t_j, c_{1,j}, \ldots, c_{n,j})$$

and verifies that $X_0 = 0$.

DIEUnvRet: The management of each project P_j generates a public return factor $\alpha_j \in [-q', q']$. The system charges the bank account of the management of each project P_j with the amount $\alpha_j X_j$. Each investor sends $E_i = \sum_{j=1}^{\lambda} \alpha_j x_{i,j}$ and $F_i = \sum_{j=1}^{\lambda} \alpha_j r_{i,j}$ to the system. If the verification in the DIEUnvPay algorithm has output 1, the system computes

$$b_{R,i} = \text{DIEUnvRet}_{pk}\left(\prod_{j=1}^{\lambda} com_{i,j}^{\alpha_j}, E_i, F_i\right) = \text{Unv}_{pk}\left(\prod_{j=1}^{\lambda} com_{i,j}^{\alpha_j}, E_i, F_i\right), b_{R,i} \in \{0, 1\}$$

and transfers the amount E_i to the bank account of investor N_i, if $b_{R,i} = 1$.

Fig. 5. The Investcoin protocol.

realised, each investor N_i should receive back a multiple $\alpha_j x_{i,j}$ of her amount invested in each project P_j (e.g. a ROI). The factor α_j is publicly released by the management of project P_j and denotes a rate of return, interest or similar. This value is equal for every investor, since only the investor's stake should determine how much her profit from that project is. If the first check in the DIEUnvPay algorithm has output 1, the return verification algorithm DIEUnvRet verifies that the combination of commitments and return factors is valid for the claimed aggregate E_i of the returns to receive by N_i.

We emphasize the low communication effort after the DIESet algorithm: every investor sends the messages for DIEEnc, DIECom, DIETes, DIEUnvPay in one shot to the system, later only the messages for DIEUnvRet have to be sent. Thus, there are only two communication rounds between the investors and the system. In Sect. 4.3, we provide empirical complexity measures.

Theorem 2 (Correctness of Investcoin). *Let Ω be the protocol in Fig. 5. Then the following properties hold.*

1. *For all $j = 1, \ldots, \lambda$ and $x_{1,j}, \ldots, x_{n,j} \in [0, m]$:*

$$\mathsf{DIEDec}_{sk_0}(t_j, \mathsf{DIEEnc}_{sk_1}(t_j, x_{1,j}), \ldots, \mathsf{DIEEnc}_{sk_n}(t_j, x_{n,j})) = \sum_{i=1}^{n} x_{i,j}.$$

2. *For all $i = 1, \ldots, n$ and $x_{i,1}, \ldots, x_{i,\lambda} \in [0, m]$:*

$$\mathsf{DIEUnvPay}_{pk}\left(\prod_{j=1}^{\lambda} com_{i,j}, \sum_{j=1}^{\lambda} x_{i,j}, \sum_{j=1}^{\lambda} r_{i,j}\right) = 1$$

$$\Leftrightarrow \exists\, (\tilde{c}_{i,1}, \ldots, \tilde{c}_{i,\lambda}) : (com_{i,j}, \tilde{c}_{i,j}) \leftarrow \mathsf{DIECom}_{pk,sk_i}(x_{i,j}, r_{i,j}) \,\forall\, j = 1, \ldots, \lambda.$$

3. *For all $i = 1, \ldots, n$, public integers $\alpha_1, \ldots, \alpha_\lambda$ and $x_{i,1}, \ldots, x_{i,\lambda} \in [0, m]$:*

$$\mathsf{DIEUnvRet}_{pk}\left(\prod_{j=1}^{\lambda} com_{i,j}^{\alpha_j}, \sum_{j=1}^{\lambda} \alpha_j x_{i,j}, \sum_{j=1}^{\lambda} \alpha_j r_{i,j}\right) = 1$$

$$\Leftrightarrow \exists\, (\tilde{c}_{i,1}, \ldots, \tilde{c}_{i,\lambda}) : (com_{i,j}, \tilde{c}_{i,j}) \leftarrow \mathsf{DIECom}_{pk,sk_i}(x_{i,j}, r_{i,j}) \,\forall\, j = 1, \ldots, \lambda.$$

Proof. The first correctness property is given by the correctness of the PSA scheme from Fig. 1. The second and third correctness properties are given by the correctness and the homomorphy of the Commitment scheme from Fig. 2. $\square$

By the first property, the decryption of all ciphers results in the sum of the amounts they encrypt. So the projects receive the correct investments. By the second property, the total investment amount of each investor is accepted by the system if the investor has committed to it. Thus, the investor's account will be charged with the correct amount. By the third property, the total return to each investor is accepted by the system if the investor has committed to the corresponding investment amount before. Thus, the investor will receive the correct return on investment (ROI).

3.2 Generation of Public Parameters and Secret Keys

In this section, we show how the system sets the random oracle $H : T \to \mathcal{QR}_{p^2}$ and we provide a decentralised key generation protocol for Investcoin. It supports dynamic join, dynamic leave and fault-tolarance of investors using one round of communication between the investors. The public parameters and the secret key generation protocol can be used for the security of Investcoin against maliciously behaving investors (see Sect. 4.1).

Setting the Random Oracle. Recall that we need to generate public parameters, a random oracle $H : T \to \mathcal{QR}_{p^2}$ and secret parameters $\beta_0, \dots, \beta_\lambda \leftarrow_R [-q', q']$, $q' < q/(m\lambda)$, such that for $t_0, \dots, t_\lambda, \tilde{t}_1, \dots, \tilde{t}_\lambda \in T$ the following equation holds.

$$\prod_{j=0}^{\lambda} H(t_j)^{\beta_j} = \prod_{j=1}^{\lambda} H(\tilde{t}_j)^{\beta_j} = 1. \tag{1}$$

First, for $j = 0, \dots, \lambda - 2$, on input t_j let $H(t_j)$ be random in $\mathcal{QR}_{p^2}$ and for $j = 1, \dots, \lambda - 2$, on input $\tilde{t}_j$ let $H(\tilde{t}_j)$ be random in $\mathcal{QR}_{p^2}$. The system chooses $\beta_0, \dots, \beta_{\lambda-2} \leftarrow_R [-q', q'], \beta_{\lambda-1} \leftarrow_R [-q', q'-1]$ as part of its secret key (note that choosing these values according to a different distribution gives no advantage to the system). Then it computes

$$(H(t_{\lambda-1}), H(\tilde{t}_{\lambda-1})) = \left(\prod_{j=0}^{\lambda-2} H(t_j)^{\beta_j}, \prod_{j=1}^{\lambda-2} H(\tilde{t}_j)^{\beta_j} \right),$$

$$(H(t_\lambda), H(\tilde{t}_\lambda), \beta_\lambda) = (H(t_{\lambda-1}), H(\tilde{t}_{\lambda-1}), -1 - \beta_{\lambda-i}),$$

instructs each investor N_i to set $x_{i,\lambda-1} = x_{i,\lambda} = 0$ and sets $\alpha_{\lambda-1} = \alpha_\lambda = 1$. In this way Eq. (1) is satisfied. The projects $P_{\lambda-1}, P_\lambda$ deteriorate to 'dummy-projects' (e.g. if any investor decides to set $x_{i,\lambda} > 0$, then the system simply collects $X_\lambda > 0$).

Key Generation for Investcoin. The building block for a key generation protocol is a $n - 1$ out of n secret sharing scheme between the investors and the system. It is executed before the first investment round as follows.

For all $i = 1, \dots, n$, investor N_i generates uniformly random values $s_{i,1}, \dots, s_{i,n}$ from the key space and sends $s_{i,i'}$ to $N_{i'}$ for all $i' = 1, \dots, n$ via secure channel. Accordingly, each investor $N_{i'}$ obtains the shares $s_{1,i'}, \dots, s_{n,i'}$. Then each investor N_i sets the own secret key $s_i = \sum_{i'=1}^{n} s_{i,i'}$ and each investor $N_{i'}$ sends $\sum_{i=1}^{n} s_{i,i'}$ to the system. The system then computes

$$s_0 = -\sum_{i'=1}^{n} \left(\sum_{i=1}^{n} s_{i,i'} \right) = -\sum_{i=1}^{n} \left(\sum_{i'=1}^{n} s_{i,i'} \right) = -\sum_{i=1}^{n} s_i.$$

By the secret sharing property this is a secure key generation protocol in the sense that only N_i knows s_i for all $i = 1, \ldots, n$ and only the system knows s_0.

For key generation, each investor has to send one message to every other investor and one message to the system which makes n^2 messages for the total network. As a drawback, note that the key of each single investor is controlled by the other investors together with the system: for example, if $N_1, \ldots, N_{n-1}$ (maliciously) send the shares $s_{1,n}, \ldots, s_{n-1,n}$ and $s_{n,1}, \ldots, s_{n,n-1}$ to the system, it can compute the entire key s_n of N_n.

Assume that before the start of an arbitrary investment round, new investors want to join the network or some investors want to leave the network or some investors fail to send the required ciphers. In order to be able to carry out the protocol execution, the network can make a key update that requires $O(n)$ messages (rather than $O(n^2)$ messages for a new key setup) using the established secret sharing scheme. Due to space limitations, we omit the (simple) details.

4   Investcoin: The Analysis

4.1   Security of Investcoin

The administration is honest-but-curious and may compromise investors to build a coalition for learning the values of uncompromised investors. Additionally, investors outside the coalition may try to execute the following (reasonable) malicious behaviour:

1. Use different values for $x_{i,j}$ in DIEEnc and DIECom to receive a larger profit than allowed.
2. Invest negative amounts $x_{i,j} < 0$ in order to 'steal' funds from the projects.
3. Use different parameters than generated in the Setup-phase (i.e. send inconsistent or simply random messages) to distort the whole computation.

We prevent this malicious behaviour by respectively satisfying Properties 2, 3 and 4 of the following security definition.

Theorem and Definition 3 (Security of Investcoin). *Let κ be a security parameter and $n, \lambda \in \mathbb{N}$ with $n = poly(\kappa)$ and $\lambda = poly(\kappa)$. In the random oracle model, by the DDH assumption in the group $\mathcal{QR}_{p^2}$ of quadratic residues modulo p^2 for a safe prime p, the protocol $\Omega = $ (DIESet, DIEEnc, DIECom, DIETes, DIEUnvPay, DIEDec, DIEUnvRet) as defined above is secure, i.e. it holds:*

1. *Let f_Ω be the deterministic functionality computed by Ω. Then Ω performs a secure computation (according to Definition 3).*
2. *Ω provides computational linkage, i.e. for all $i = 1, \ldots, n$, $sk_i, pk, T = \{t_1, \ldots, t_\lambda\}$, $(x_{i,j})_{j=1,\ldots,\lambda}, (\alpha_j)_{j=1,\ldots,\lambda}$ and all ppt adversaries $\mathcal{T}$ it holds:*

$$\Pr\left[c_{i,j} \leftarrow \mathsf{DIEEnc}_{sk_i}(t_j, x_{i,j}) \,\forall\, j = 1, \ldots, \lambda \wedge \right.$$

$$\mathsf{DIEUnvPay}_{pk}\left(\prod_{j=1}^{\lambda} com_{i,j}, C_i, D_i\right) = 1 \wedge$$

$$\mathsf{DIEUnvRet}_{pk}\left(\prod_{j=1}^{\lambda} com_{i,j}^{\alpha_j}, E_i, F_i\right) = 1 \wedge$$

$$\left.\left(C_i \neq \sum_{j=1}^{\lambda} x_{i,j} \vee E_i \neq \sum_{j=1}^{\lambda} \alpha_j x_{i,j}\right)\right]$$

$$\leqslant \mathsf{neg}(\kappa)$$

The probability is taken over the choices of $(c_{i,j})_{j=1,\ldots,\lambda}$, $(com_{i,j})_{j=1,\ldots,\lambda}$, C_i, D_i, E_i, F_i.

3. For all $i = 1, \ldots, n$ and $j = 1, \ldots, \lambda$: $\mathsf{DIETes}_{pk}(x_{i,j}) = 1$ iff $x_{i,j} \geqslant 0$.
4. For all $i = 1, \ldots, n$ there is a ppt distinguisher $\mathcal{D}_i$, s.t. for $c_{i,j}$ $\leftarrow$ $\mathsf{DIEEnc}_{sk_i}(t_j, x_{i,j})$, $(com_{i,j}, \tilde{c}_{i,j})$ $\leftarrow$ $\mathsf{DIECom}_{pk,sk_i}(x_{i,j}, r_{i,j})$ for a $r_{i,j} \leftarrow_R [0, m]$, where $j = 1, \ldots, \lambda$, for $c_{i,j^*}^* \leftarrow \mathsf{DIEEnc}_{sk_i^*}(t_{j^*}^*, x_{i,j^*})$ and $(com_{i,j^*}^*, \tilde{c}_{i,j^*}^*) \leftarrow \mathsf{DIECom}_{pk,sk_i^*}(x_{i,j^*}, r_{i,j^*})$, where $j^* \in \{1, \ldots, \lambda\}$ and $(sk_i^*, t_{j^*}^*) \neq (sk_i, t_{j^*})$, it holds that

$$\left| \Pr\left[\mathcal{D}_i\left(1^\kappa, c_{i,1}, com_{i,1}, \tilde{c}_{i,1}, \ldots, c_{i,j^*}, com_{i,j^*}, \tilde{c}_{i,j^*}, \ldots, c_{i,\lambda}, com_{i,\lambda}, \tilde{c}_{i,\lambda}\right) = 1\right] \right.$$
$$\left. - \Pr\left[\mathcal{D}_i\left(1^\kappa, c_{i,1}, com_{i,1}, \tilde{c}_{i,1}, \ldots, c_{i,j^*}^*, com_{i,j^*}^*, \tilde{c}_{i,j^*}^*, \ldots, c_{i,\lambda}, com_{i,\lambda}, \tilde{c}_{i,\lambda}\right) = 1\right] \right|$$
$$\geqslant 1 - \mathsf{neg}(\kappa).$$

The security theorem and definition is twofold: on the one hand it covers the security of honest investors against an honest-but-curious coalition consisting of the untrusted system administration and compromised investors (Property 1) and on the other hand it covers the security of the system against maliciously behaving investors (Properties 2, 3, 4). Note that we have to distinguish between these two requirements, since we assume different behaviours for the two groups of participants, i.e. we cannot simply give a real-world-ideal-world security proof as in the SMPC literature in the malicious model. Instead, we follow the notions of the SMPC literature [13] for the security of honest investors (only) in the honest-but-curious model and additionally provide security notions against the behaviour of malicious investors as described in the beginning of this section. For the security against an honest-but-curious coalition, the first property ensures that from the outcome of the decryption no other information than X_j can be detected for all $j = 1, \ldots, \lambda$ and that the single amounts comitted to by the investors for payment and return are hidden. For the security of the system against maliciously behaving investors, imagine the situation where an investor N_i claims to having payed amount $x_{i,j}$ to project P_j (in the DIECom algorithm) but in fact has only payed $\tilde{x}_{i,j} < x_{i,j}$ (in the DIEEnc algorithm). If the return

factor α_j is larger than 1, then N_i would unjustly profit more from her investment than she actually should and the system would have a financial damage.[1]
Therefore the second property says that for all $i = 1, \ldots, n$ with overwhelming
probability, whenever $x_{i,1}, \ldots, x_{i,\lambda}$ were send by N_i using the DIEEnc algorithm
and DIEUnvPay, DIEUnvRet accept C_i, E_i respectively, then C_i and E_i must be
the legitimate amounts that N_i respectively has invested in total and has to get
back as return in total. The third property ensures that no investor is able to
perform a negative investment. The fourth property ensures that all investors
use the correct parameters as generated by the DIESet algorithm.[2]

For the proof of Theorem 3, we first concentrate on the security against the
honest-but-curious coalition (Property 1) and then show security against malicious investors (Properties 2, 3, 4).

Proof Sketch. Due to space limitations, we omit the formal proof of Theorem
and Definition 3 and provide it in the full version. Here, we give a proof sketch.

Investcoin is the combination of the protocols described in Figs. 1, 2 and 3.
For Property 1 of Theorem 3, we can first show the security of these protocols
seperately and then use Theorem 1 in order to show composition. From the
AO security of the protocol in Fig. 1 (see [24] for the proof), formally it follows immediately that it performs a secure computation. The protocol in Fig. 2
performs a secure computation by its perfect hiding property. The protocol in
Fig. 3 performs a secure computation by its special honest verifier zero-knowledge
property. Moreover, the Investcoin protocol with calls to a trusted party computing the functionalities of the three preceeding protocols performs a secure
computation. Thus, by Theorem 1 the real Investcoin protocol performs a secure
computation.

Investcoin satisfies Property 2 of Theorem 3 by the binding-property of Γ,
the choice of $H : T \to \mathcal{QR}_{p^2}, (\beta_0, \ldots, \beta_\lambda)$ such that $\beta_0, \ldots, \beta_\lambda \in [-q', q']$, $q' <$
$q/(m\lambda)$, are pairwise different with $\prod_{j=0}^{\lambda} H(t_j)^{\beta_j} = 1$ and $\prod_{j=1}^{\lambda} H(\tilde{t}_j)^{\beta_j} = 1$
and the non-commutativity of the mapping

$$f_{\beta_0,\ldots,\beta_\lambda} : [0,m]^\lambda \to [-q,q], f_{\beta_0,\ldots,\beta_\lambda}(x_0,\ldots,x_\lambda) = \sum_{j=0}^{\lambda} \beta_j x_j.$$

The proof of Property 3 follows from the analysis in [10].

For Property 4, imagine an investor who sends inconsistent messages in the
DIEEnc algorithm (i.e. it uses different secret keys in the DIEEnc algorithm than
generated in the DIESet algorithm) in order to maliciously distort the total
amounts $X_1, \ldots, X_\lambda$ invested in the single projects. The described key generation protocol can be used for detecting such malicious investors within the

---

[1] Usually the investor cannot know if $\alpha_j > 1$ at the time of cheating, since it becomes
public in the future. However, in the scenario where a priori information about α_j
is known to some investors or where investors simply act maliciously, we need to
protect the system from beeing cheated.
[2] More precisely, it ensures that a cheating investor will be identified by the system.

Investcoin protocol as follows. First, in the DIESet algorithm, the network generates the additional public parameter t_0 and the secret $\beta_0 \in [-q', q']$ such that Eq. (1) is satisfied. Then each investor $N_{i'}$ publishes $T_{i,i'} = \mathsf{PSAEnc}_{s_{i,i'}}(t_0, 0)$ on a black board for all the key shares $s_{1,i'}, \ldots, s_{n,i'}$ received by $N_{i'}$ during the key generation. Then each N_i verifies that the other investors used the correct key shares $s_{i,1}, \ldots, s_{i,n}$ of her secret key s_i in the publications. Moreover, using the received values $\sum_{i=1}^{n} s_{i,i'}$ for all $i' = 1, \ldots, n$, the system verifies that

$$\mathsf{PSADec}_{\sum_{i=1}^{n} s_{i,i'}}(t_0, T_{1,i'}, \ldots, T_{n,i'}) = 0.$$

In this way, the system can be sure that only correct key shares were used to compute the published values $T_{i,i'}$. Now for all $i = 1, \ldots, n$, the system involves the encryption of $x_{i,0} = 0$ (i.e. the value encrypted using t_0) in order to compute the verification value A_i of N_i. Because of Eq. (1) and the linkage property, N_i has to use the same secret key s_i (and the correct $H(t_j), H(\tilde{t}_j)$) for all $j = 0, 1, \ldots, \lambda$. On the other hand, as discussed above, the encryption of $x_{i,0} = 0$ is verifiably generated with the correct secret key shares from the DIESet algorithm. This means, s_i is the correct secret key of N_i and the system knows $\mathsf{PSAEnc}_{s_i}(t_0, 0)$. Therefore the use of a different key in the DIEEnc algorithm than generated in the DIESet algorithm is not possible without being detected by the system. $\square$

4.2 Preservation of Market Liquidity

We show that it is still possible to privately exchange any part of an investment between any pair of investors within Investcoin, i.e. market liquidity is unaffected. Assume that an investment round is over but the returns are not yet executed, i.e. the system already received $c_{i,j}, (com_{i,j}, \tilde{c}_{i,j}), C_i, D_i$ for all $i = 1, \ldots, n$ and $j = 1, \ldots, \lambda$ but not E_i, F_i. Assume further that for some $i, i' \in \{1, \ldots, n\}$, investors N_i and $N_{i'}$ confidentially agree on a transfer of amount $x_{(i,i'),j}$ (i.e. a part of N_i's investment in project P_j) from investor N_i to investor $N_{i'}$. This fact needs to be confirmed by the protocol in order to guarantee the correct returns from project P_j to investors N_i and $N_{i'}$. Therefore the commitments to the invested amounts $x_{i,j}$ and $x_{i',j}$ respectively need to be updated. For the update, N_i and $N_{i'}$ agree on a value $r_{(i,i'),j} \leftarrow_R [0, m]$ via secure channel. This value should be known only to N_i and $N_{i'}$. Then N_i and $N_{i'}$ respectively compute

$$(com'_{i,j}, \tilde{c}'_{i,j}) \leftarrow \mathsf{DIECom}_{pk,sk_i}(x_{(i,i'),j}, r_{(i,i'),j}),$$
$$(com'_{i',j}, \tilde{c}'_{i',j}) \leftarrow \mathsf{DIECom}_{pk,sk_{i'}}(x_{(i,i'),j}, r_{(i,i'),j})$$

and send their commitments to the system which verifies that $com'_{i,j} = com'_{i',j}$. Then the system updates $(com_{i,j}, \tilde{c}_{i,j})$ by $(com_{i,j} \cdot (com'_{i,j})^{-1}, \tilde{c}_{i,j} \cdot (\tilde{c}'_{i,j})^{-1})$ (which is possible since DIECom is injective) and $(com_{i',j}, \tilde{c}_{i',j})$ by $(com_{i',j} \cdot com'_{i',j}, \tilde{c}_{i',j} \cdot \tilde{c}'_{i',j})$. As desired, the updated values commit to $x_{i,j} - x_{(i,i'),j}$ and to $x_{i',j} + x_{(i,i'),j}$ respectively. Moreover, N_i updates the return values (E_i, F_i) by $(E_i - \alpha_j \cdot x_{(i,i'),j}, F_i - \alpha_j \cdot r_{(i,i'),j})$ and $N_{i'}$ updates $(E_{i'}, F_{i'})$ by $(E_{i'} + \alpha_j \cdot x_{(i,i'),j}, F_{i'} + \alpha_j \cdot r_{(i,i'),j})$.

The correctness of the update is guaranteed by Property 2 and the confidentiality of the amount $x_{(i,i'),j}$ (i.e. only N_i and $N_{i'}$ know $x_{(i,i'),j}$) is guaranteed by Property 1 of Theorem 3.

Note that in general, this procedure allows a short sale for N_i when $x_{(i,i'),j} > x_{i,j}$ or for $N_{i'}$ when $x_{(i,i'),j} < 0$ and $|x_{(i,i'),j}| > x_{i',j}$ (over the integers). If this behaviour is not desired, it may also be necessary to perform a Range test for the updated commitments $com_{i,j} \cdot (com'_{i,j})^{-1}$ (between the system and N_i) and $com_{i',j} \cdot com'_{i',j}$ (between the system and $N_{i'}$) to ensure that they still commit to amounts $\geqslant 0$.

4.3 Empirical Analysis

We perform an experiment to measure the required computational running time and space of each investor and the system in one investment round of Investcoin. The Setup algorithm performs a precomputatiton for *all* investment rounds and hence is not considered here. The experiment is run on a 3 GHz CPU with $n = 1.000$ investors, $\lambda = 100$ projects to invest in (the time and space for an investor - i.e. in DIEEnc, DIECom, DIETes - are roughly linear in λ, the time for the system is roughly linear in n, λ, the space for DIEUnvPay, DIEUnvRet is linear in n and the space for DIEDec is linear in λ) and with amounts up to $m = 1.000.000$ (e.g. up to 1 million Euro can be invested in each project with only integer valued amounts). The DDH problem is considered modulo a squared safe 2048-bit prime p. The results are presented in Table 1 and show that Investcoin is indeed practicable. On the investor's side, time and space are dominated by DIETes where a large amount of Pedersen commitments has to be computed (although Boudot's scheme is not the most efficient Range test, it is the only existing one that satisfies all our security and correctness requirements *and* is based on the discrete logarithm, making the security of Investcoin dependent on only one hardness assumption, i.e. on the DDH assumption). For one investment round on a crowdfunding platform (where investment decisions are not made ad-hoc or too frequently) the measured values are reasonable. On the system's side, time and space are dominated by DIEUnvPay where several commitments for every investor must be aggregated and verified. Note that in most cases, a system running a crowdfunding platform will use more powerful CPUs.

In summary, Investcoin is a practicable and mathematically secure distributed investment encryption scheme that protects the trade secrecy of investors.

Table 1. Time and Space for $n = 1.000, \lambda = 100$.

Algorithm (investors)	Time	Space	Algorithm (system)	Time	Space
DIEEnc	2.5 s	28 KB	DIEUnvPay	32 s	1.2 MB
DIECom	2.7 s	56 KB	DIEDec	8 s	28 KB
DIETes	12 s	1.5 MB	DIEUnvRet	8 s	560 KB
Total per investor	17.2 s	1.6 MB	Total for system	48 s	1.8 MB

References

1. Abbe, E.A., Khandani, A.E., Lo, A.W.: Privacy-preserving methods for sharing financial risk exposures. Am. Econ. Rev. **102**(3), 65–70 (2012)
2. Benhamouda, F., Joye, M., Libert, B.: A new framework for privacy-preserving aggregation of time-series data. ACM Trans. Inf. Syst. Secur. **18**(3), 10 (2016)
3. Blum, A., Morgenstern, J., Sharma, A., Smith, A.: Privacy-preserving public information for sequential games. In: Proceedings of ITCS 2015, pp. 173–180 (2015)
4. Blum, M.: Coin flipping by telephone. In: Proceedings of Crypto 1981, pp. 11–15 (1981)
5. Boudot, F.: Efficient proofs that a committed number lies in an interval. In: Preneel, B. (ed.) EUROCRYPT 2000. LNCS, vol. 1807, pp. 431–444. Springer, Heidelberg (2000). doi:10.1007/3-540-45539-6_31
6. Brassard, G., Chaum, D., Crépeau, C.: Minimum disclosure proofs of knowledge. J. Comput. Syst. Sci. **37**(2), 156–189 (1988)
7. Camenisch, J., Chaabouni, R., Shelat, A.: Efficient protocols for set membership and range proofs. In: Pieprzyk, J. (ed.) ASIACRYPT 2008. LNCS, vol. 5350, pp. 234–252. Springer, Heidelberg (2008). doi:10.1007/978-3-540-89255-7_15
8. Canetti, R.: Security and composition of multiparty cryptographic protocols. J. Cryptol. J. Int. Assoc. Cryptologic Res. **13**, 143–202 (2000)
9. The Financial Crisis Inquiry Report: Final Report of the National Commission on the Causes of the Financial and Economic Crisis in the United States (2011)
10. Cramer, R., Damgård, I., Schoenmakers, B.: Proofs of partial knowledge and simplified design of witness hiding protocols. In: Desmedt, Y.G. (ed.) CRYPTO 1994. LNCS, vol. 839, pp. 174–187. Springer, Heidelberg (1994). doi:10.1007/3-540-48658-5_19
11. Fiat, A., Shamir, A.: How to prove yourself: practical solutions to identification and signature problems. In: Odlyzko, A.M. (ed.) CRYPTO 1986. LNCS, vol. 263, pp. 186–194. Springer, Heidelberg (1987). doi:10.1007/3-540-47721-7_12
12. Flood, M., Katz, J., Ong, S., Smith, A.: Cryptography and the economics of supervisory information: balancing transparency and confidentiality. Federal Reserve Bank of Cleveland, Working Paper no. 13-11 (2013)
13. Goldreich, O.: Foundations of Cryptography: Basic Applications, vol. 2. Cambridge University Press, New York (2004)
14. Jentzsch, N.: The Economics and Regulation of Financial Privacy - A Comparative Analysis of the United States and Europe (2001, submitted)
15. Joye, M., Libert, B.: A scalable scheme for privacy-preserving aggregation of time-series data. In Proceedings of FC 2013, pp. 111–125 (2013)
16. Miers, I., Garman, C., Green, M., Rubin, A.D.: Zerocoin: anonymous distributed e-cash from bitcoin. In: Proceedings of SP 2013, pp. 397–411 (2013)
17. Nakamoto, S.: Bitcoin: a peer-to-peer electronic cash system
18. Nofer, M.: The value of social media for predicting stock returns - preconditions, instruments and performance analysis. Ph.D. thesis, Technische Universität Darmstadt (2014)
19. Pedersen, T.P.: Non-interactive and information-theoretic secure verifiable secret sharing. In: Feigenbaum, J. (ed.) CRYPTO 1991. LNCS, vol. 576, pp. 129–140. Springer, Heidelberg (1992). doi:10.1007/3-540-46766-1_9
20. Peng, K., Boyd, C., Dawson, E., Okamoto, E.: A novel range test, pp. 247–258 (2006)

21. Peng, K., Dawson, E.: A range test secure in the active adversary model. In: Proceedings of ACSW 2007, pp. 159–162 (2007)
22. Schnorr, C.P.: Efficient Identification and Signatures for Smart Cards. In: Quisquater, J.-J., Vandewalle, J. (eds.) EUROCRYPT 1989. LNCS, vol. 434, pp. 688–689. Springer, Heidelberg (1990). doi:10.1007/3-540-46885-4_68
23. Elaine Shi, T.-H., Chan, H., Rieffel, E.G., Chow, R., Song, D.: Privacy-preserving aggregation of time-series data. In: Proceedings of NDSS 2011 (2011)
24. Valovich, F., Aldà, F.: Private stream aggregation revisited. CoRR abs/1507.08071 (2015)

Physical Layer Security over Wiretap Channels with Random Parameters

Ziv Goldfeld[1]([✉]), Paul Cuff[2], and Haim H. Permuter[1]

[1] Ben Gurion University of the Negev, 8499000 Beer Sheva, Israel
gziv@post.bgu.ac.il, haimp@bgu.ac.il
[2] Princeton University, Princeton, NJ 08544, USA
cuff@princeton.edu

Abstract. We study semantically secure communication over state dependent (SD) wiretap channels (WTCs) with non-causal channel state information (CSI) at the encoder. This model subsumes all other instances of CSI availability as special cases, and calls for an efficient utilization of the state sequence both for reliability and security purposes. A lower bound on the secrecy-capacity, that improves upon the previously best known result by Chen and Han Vinck, is derived based on a novel superposition coding scheme. The improvement over the Chen and Han Vinck result is strict for some SD-WTCs. Specializing the lower bound to the case where CSI is also available to the decoder reveals that it is at least as good as the achievable formula by Chia and El-Gamal, which is already known to outperform the adaptation of the Chen and Han Vinck code to the encoder and decoder CSI scenario. The results are derived under the strict semantic security metric that requires negligible information leakage for all message distributions. The proof of achievability relies on a stronger version of the soft-covering lemma for superposition codes. The lower bound is shown to be tight for a class of reversely less-noisy SD-WTCs, thus characterizing the fundamental limit of reliable a secure communication. An explicit coding scheme that includes a key extraction phase via the random state sequence is also proposed.

1 Introduction

Modern communication systems usually present an architectural separation between error correction and data encryption. The former is typically realized at the physical layer by transforming the noisy communication channel into a reliable "bit pipe". The data encryption is implemented on top of that by applying cryptographic principles. The cryptographic approach relies on restricting the computational power of the eavesdropper. The looming prospect of quantum computers (QCs) (some companies have recently reported a working prototype of a QC with over than 1000 qbits [1,2]), however, would boost computational

© Springer International Publishing AG 2017
S. Dolev and S. Lodha (Eds.): CSCML 2017, LNCS 10332, pp. 155–170, 2017.
DOI: 10.1007/978-3-319-60080-2_11

abilities, rendering some critical cryptosystems insecure and weakening others.[1] Post-QC cryptography offers partial solutions that rely on larger keys, but even now considerable efforts are made to save this expensive resource.

Physical layer security (PLS) [6], rooted in information-theoretic (IT) principles, is an alternative approach to provably secure communication that dates back to Wyner's celebrated 1975 paper on the wiretap channel (WTC) [7]. By harnessing randomness from the noisy communication channel and combining it with proper physical layer coding, PLS guarantees protection against computationally-unlimited eavesdroppers with no requirement that the legitimate parties share a secret key (SK) in advance. The eavesdroppers computational abilities are of no consequence here since the signal he/she observes from the channel carries only negligible information about the secret data. In this work we use PLS for secretly transmitting a message over state-dependent (SD) wiretap channels (WTCs) with non-causal encoder channel state information (CSI). As PLS exploits the randomness of the channel for securing the data, the considered scenario models cases where the encoder has prior knowledge of some of that randomness (i.e., the channel's state). This allows more sophisticated coding schemes that include IT secret key agreement based on the random state sequence.

1.1   SD-WTCs with Non-causal Encoder CSI

Reliably transmitting a message over a noisy SD channel with non-causal encoder CSI is a fundamental information theoretic scenario. This problem was formulated by Gelfand and Pinsker (GP) in their celebrated paper [8], where they also derived its capacity. Not only did the result from [8] have various implication for many information-theoretic problems (such as the broadcast channel), it is also the most general instance of a SD point-to-point channel in which any or all of the terminals have non-causal access to the sequence of states. Motivated by the above as well as the indisputable importance of security in modern communication systems, we study the SD wiretap channel (WTC) with non-causal encoder CSI, which incorporates the notion of security in the presence of a wiretapper into the GP channel coding problem.

First to consider a discrete and memoryless (DM) WTC with random states were Chen and Han Vinck [9], who studied the encoder CSI scenario. They established a lower bound on the secrecy-capacity based on a combination of wiretap coding with GP coding. This work was later generalized in [10] to a WTC that is driven by a pair of states, one available to the encoder and the

---

[1] More specifically, asymmetric ciphers that rely on the hardness of integer factorization or discrete logarithms can be completely broken using QCs via Shor's algorithm (or a variant thereof) [3,4]. Symmetric encryption, on the other hand, would be weakened by QC attacks but could regain its strength by increasing the size of the key [5]. This essentially follows since a QC can search through a space of size 2^n in time $2^{\frac{n}{2}}$, so by doubling the size of the key a symmetric cryptosystem would offer the same protection versus a QC attack, as the original system did versus a classic attack.

other one to the decoder. However, as previously mentioned, since CSI at the encoder is the most general setup, the result of [10] is a special of [9]. A more sophisticated coding scheme was constructed by Chia and El-Gamal for the SD-WTC with causal encoder CSI and full decoder CSI [11]. Their idea was to explicitly extract a cryptographic key from the random state, and encrypt part of the confidential message via a one-time-pad with that key. The remaining portion of the confidential message is protected by a wiretap code (whenever wiretap coding is possible). Although their code is restricted to utilize the state in a causal manner, the authors of [11] proved that it can strictly outperform the adaptations of the non-causal schemes from [9,10] to the encoder and decoder CSI setup.

1.2 This Work

We propose a novel superposition-based coding scheme for the GP WTC. The scheme results in a new lower bound on the secrecy-capacity, which recovers the previously best known achievability formulas from [9,10] as special cases. The improvement is strict for certain classes of SD-WTCs. One such interesting class in the reversely less-noisy (RLN) SD-WTC, where the channel transition probability decomposes into a WTC that given the input is independent of the state, and another channel that generates two noisy versions of the state, each observed either by the legitimate receiver or by the eavesdropper. The input dependent WTC is RLN in the sense that it produces an output to the eavesdropper that is better than this observed by the legitimate receiver. Our lower bound is tight for the RLN SD-WTC, thus characterizing its fundamental limit of reliable and secure communication. An explicit coding scheme (i.e., that does not depend on our general inner bound) that includes a key agreement protocol via the random state sequence is also proposed.

When specializing to the case where the decoder also knows the state sequence, our achievability is at least as good as the scheme from [11]. Interestingly, while the scheme from [11] relies on generating the aforementioned cryptographic key, our code construction does not involve any explicit key generation/agreement phase. Instead, we use an over-populated superposition codebook and encode the entire confidential message at the outer layer. The transmission is correlated with the state sequence by means of the likelihood encoder [12], while security is ensured by making the eavesdropper decode the inner layer codeword that contains no confidential information. Having done so, the eavesdropper is lacking the resources to extract any information about the secret message.

Our results are derived under the strict metric of semantic-security (SS). The SS criterion is a cryptographic benchmark that was adapted to the information-theoretic framework (of computationally unbounded adversaries) in [13]. In that work, SS was shown to be equivalent to a negligible mutual information (MI) between the message and the eavesdropper's observations for all message distributions. We establish SS for our superposition code via a strong soft-covering lemma (SCL) for superposition codebooks [14, Lemma 1] that produces double-exponential decay of the probability of soft-covering not happening. Since all the aforementioned secrecy results were derived under the weak-secrecy metric (i.e., a vanishing *normalized* MI with respect to a *uniformly distributed* message), our achievability outperforms the schemes from [9,10] for the SD-WTC with non-causal encoder CSI not only in terms of the achievable secrecy rate, but also in the upgraded sense of security is provides. When CSI is also available at the decoder, our result implies that an upgrade to SS is possible, without inflicting any loss of rate compared to [11].

2   Preliminaries

We use the following notations. As customary $\mathbb{N}$ is the set of natural numbers (which does not include 0), while $\mathbb{R}$ are the reals. We further define $\mathbb{R}_+ = \{x \in \mathbb{R} | x \geq 0\}$. Given two real numbers a, b, we denote by $[a : b]$ the set of integers $\{n \in \mathbb{N} | \lceil a \rceil \leq n \leq \lfloor b \rfloor\}$. Calligraphic letters denote sets, e.g., $\mathcal{X}$, while $|\mathcal{X}|$ stands for its cardinality. $\mathcal{X}^n$ denotes the n-fold Cartesian product of $\mathcal{X}$. An element of $\mathcal{X}^n$ is denoted by $x^n = (x_1, x_2, \ldots, x_n)$; whenever the dimension n is clear from the context, vectors (or sequences) are denoted by boldface letters, e.g., $\mathbf{x}$.

Let $(\mathcal{X}, \mathcal{F}, \mathbb{P})$ be a probability space, where $\mathcal{X}$ is the sample space, $\mathcal{F}$ is the σ-algebra and $\mathbb{P}$ is the probability measure. Random variables over $(\mathcal{X}, \mathcal{F}, \mathbb{P})$ are denoted by uppercase letters, e.g., X, with conventions for random vectors similar to those for deterministic sequences. The probability of an event $\mathcal{A} \in \mathcal{F}$ is denoted by $\mathbb{P}(\mathcal{A})$, while $\mathbb{P}(\mathcal{A}|\mathcal{B})$ denotes conditional probability of $\mathcal{A}$ given $\mathcal{B}$. We use $\mathbb{1}_\mathcal{A}$ to denote the indicator function of $\mathcal{A} \in \mathcal{F}$. The set of all probability mass functions (PMFs) on a finite set $\mathcal{X}$ is denoted by $\mathcal{P}(\mathcal{X})$. PMFs are denoted by the letters such as p or q, with a subscript that identifies the random variable and its possible conditioning. For example, for a two discrete correlated random variables X and Y over the same probability space, we use p_X, $p_{X,Y}$ and $p_{X|Y}$ to denote, respectively, the marginal PMF of X, the joint PMF of (X, Y) and the conditional PMF of X given Y. In particular, $p_{X|Y}$ represents the stochastic matrix whose elements are given by $p_{X|Y}(x|y) = \mathbb{P}(X = x|Y = y)$. Expressions such as $p_{X,Y} = p_X p_{Y|X}$ are to be understood to hold pointwise, i.e., $p_{X,Y}(x,y) = p_X(x)p_{Y|X}(y|x)$, for all $(x, y) \in \mathcal{X} \times \mathcal{Y}$. Accordingly, when three random variables X, Y and Z satisfy $p_{X|Y,Z} = p_{X|Y}$, they form a Markov chain, which we denote by $X - Y - Z$. We omit subscripts if the arguments of a PMF are lowercase versions of the random variables.

For a sequence of random variable X^n, if the entries of X^n are drawn in an identically and independently distributed (i.i.d.) manner according to p_X, then for every $\mathbf{x} \in \mathcal{X}^n$ we have $p_{X^n}(\mathbf{x}) = \prod_{i=1}^n p_X(x_i)$ and we write $p_{X^n}(\mathbf{x}) = p_X^n(\mathbf{x})$. Similarly, if for every $(\mathbf{x}, \mathbf{y}) \in \mathcal{X}^n \times \mathcal{Y}^n$ we have $p_{Y^n|X^n}(\mathbf{y}|\mathbf{x}) = \prod_{i=1}^n p_{Y|X}(y_i|x_i)$, then we write $p_{Y^n|X^n}(\mathbf{y}|\mathbf{x}) = p_{Y|X}^n(\mathbf{y}|\mathbf{x})$. The conditional product PMF $p_{Y|X}^n$ given a specific sequence $\mathbf{x} \in \mathcal{X}^n$ is denoted by $p_{Y|X=\mathbf{x}}^n$.

The empirical PMF $\nu_\mathbf{x}$ of a sequence $\mathbf{x} \in \mathcal{X}^n$ is $\nu_\mathbf{x}(x) \triangleq \frac{N(x|\mathbf{x})}{n}$, where $N(x|\mathbf{x}) = \sum_{i=1}^n \mathbb{1}_{\{x_i=x\}}$. We use $\mathcal{T}_\epsilon^n(p_X)$ to denote the set of letter-typical

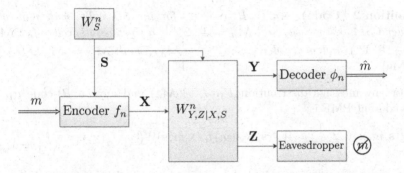

Fig. 1. The state-dependent wiretap channel with non-casual encoder channel state information.

sequences of length n with respect to the PMF p_X and the non-negative number ϵ, i.e., we have

$$T_\epsilon^n(p_X) = \left\{ \mathbf{x} \in \mathcal{X}^n \,\middle|\, |\nu_\mathbf{x}(x) - p_X(x)| \le \epsilon p_X(x), \ \forall x \in \mathcal{X} \right\}. \tag{1}$$

Definition 1 (Relative Entropy). *Let $(\mathcal{X}, \mathcal{F})$ be a measurable space, where $\mathcal{X}$ is countable, and let P and Q be two probability measures on $\mathcal{F}$, with $P \ll Q$ (i.e., P is absolutely continuous with respect to Q). The relative entropy between P and Q is*

$$D(P\|Q) = \sum_{x \in \mathrm{supp}(P)} P(x) \log\left(\frac{P(x)}{Q(x)}\right). \tag{2}$$

3   SD-WTC with Non-causal Encoder CSI

We study the SD-WTC with non-causal encoder CSI, for which we establish a new and improved achievability formula that (in some cases) strictly outperforms the previously best known coding schemes for this scenario.

3.1   Problem Setup

Let $\mathcal{S}$, $\mathcal{X}$, $\mathcal{Y}$ and $\mathcal{Z}$ be finite sets. The $\left(\mathcal{S}, \mathcal{X}, \mathcal{Y}, \mathcal{Z}, W_S, W_{Y,Z|X,S}\right)$ discrete and memoryless SD-WTC with non-causal encoder CSI is illustrated in Fig. 1. A state sequence $\mathbf{s} \in \mathcal{S}^n$ is generated in an i.i.d. manner according to W_S and is revealed in a non-causal fashion to the sender, who chooses a message m from the set $\left[1 : 2^{nR}\right]$. The sender then maps the observed state sequence $\mathbf{s}$ and the chosen message m into a sequence $\mathbf{x} \in \mathcal{X}^n$ (the mapping may be random). The sequence $\mathbf{x}$ is transmitted over the SD-WTC with transition probability $W_{Y,Z|X,S}$. The output sequences $\mathbf{y} \in \mathcal{Y}^n$ and $\mathbf{z} \in \mathcal{Z}^n$ are observed by the receiver and the eavesdropper, respectively. Based on $\mathbf{y}$, the receiver produces an estimate $\hat{m}$ of m. The eavesdropper tries to glean whatever it can about the message from $\mathbf{z}$.

Definition 2 (Code). *An* (n, R)*-code* c_n *for the SD-WTC with non-causal encoder CSI has a message set* $\mathcal{M}_n \triangleq [1 : 2^{nR}]$*, a stochastic encoder* $f_n : \mathcal{M}_n \times \mathcal{S}^n \to \mathcal{P}(\mathcal{X}^n)$ *and a decoder* $\phi_n : \mathcal{Y}^n \to \hat{\mathcal{M}}_n$*, where* $\hat{\mathcal{M}}_n = \mathcal{M}_n \cup \{e\}$ *and* $e \notin \mathcal{M}_n$*.*

For any message distribution $P_M \in \mathcal{P}(\mathcal{M}_n)$ and any (n, R)-code c_n, the induced joint PMF is:

$$P^{(c_n)}(\mathbf{s}, m, \mathbf{x}, \mathbf{y}, \mathbf{z}, \hat{m}) = W_S^n(\mathbf{s}) P_M(m) f_n(\mathbf{x}|m, \mathbf{s}) W_{Y,Z|X,S}^n(\mathbf{y}, \mathbf{z}|\mathbf{x}, \mathbf{s}) \mathbb{1}_{\{\hat{m}=\phi_n(\mathbf{y})\}}. \tag{3}$$

The performance of c_n is evaluated in terms of its rate R, the maximal decoding error probability and the SS-metric.

Definition 3 (Maximal Error Probability). *The maximal error probability of an* (n, R)*-code* c_n *is*

$$e(c_n) = \max_{m \in \mathcal{M}_n} e_m(c_n), \tag{4}$$

where $e_m(c_n) = \displaystyle\sum_{(\mathbf{s},\mathbf{x}) \in \mathcal{S}^n \times \mathcal{X}^n} W_S^n(\mathbf{s}) f_n(\mathbf{x}|m, \mathbf{s}) \sum_{\substack{\mathbf{y} \in \mathcal{Y}^n: \\ \phi_n(\mathbf{y}) \neq m}} W_{Y|X,S}^n(\mathbf{y}|\mathbf{x}, \mathbf{s}).$

Definition 4 (Information Leakage and SS Metric). *The information leakage to the eavesdropper under the* (n, R)*-code* c_n *and the message PMF* $P_M \in \mathcal{P}(\mathcal{M}_n)$ *is* $\ell(P_M, c_n) = I_{c_n}(M; \mathbf{Z})$*, where* I_{c_n} *denotes that the MI is taken with respect to the marginal* $P_{M,\mathbf{Z}}^{(c_n)}$ *of Eq. (3). The SS metric with respect to* c_n *is*

$$\ell_{\mathsf{Sem}}(c_n) = \max_{P_M \in \mathcal{P}(\mathcal{M}_n)} \ell(P_M, c_n). \tag{5}$$

Definition 5 (Achievability). *A number* $R \in \mathbb{R}_+$ *is called an achievable SS-rate for the SD-WTC with non-causal encoder CSI, if for every* $\epsilon > 0$ *and sufficiently large* n*, there exists a CR* (n, R)*-code* c_n *with* $e(c_n) \leq \epsilon$ *and* $\ell_{\mathsf{Sem}}(c_n) \leq \epsilon$*.*

Definition 6 (SS-Capacity). *The SS-capacity* C_{Sem} *of the SD-WTC with non-causal encoder CSI is the supremum of the set of achievable SS-rates.*

3.2   Main Results

The main result of this work is a novel lower bound on the SS-capacity of the SD-WTC with non-causal encoder CSI. Our achievability formula strictly outperforms the best previously known coding scheme for the considered scenario. To state our main result, let $\mathcal{U}$ and $\mathcal{V}$ be finite alphabets and for any $Q_{U,V,X|S} : \mathcal{S} \to \mathcal{P}(\mathcal{U} \times \mathcal{V} \times \mathcal{X})$ define

$$R_{\mathsf{A}}(Q_{U,V,X|S}) \triangleq \min \left\{ \begin{array}{l} I(V; Y|U) - I(V; Z|U), \\ I(U, V; Y) - I(U, V; S) \end{array} \right\}, \tag{6}$$

where the MI terms are calculated with respect to the joint PMF $W_S Q_{U,V,X|S} \times W_{Y,Z|X,S}$.

Theorem 1 (SD-WTC SS-Capacity Lower Bound). *The SS-capacity of the SD-WTC with non-causal encoder CSI is lower bounded as*

$$C_{\mathsf{Sem}} \geq R_{\mathsf{A}} \triangleq \max_{\substack{Q_{U,V,X|S}: \\ I(U;Y)-I(U;S)\geq 0}} R_{\mathsf{A}}\left(Q_{U,V,X|S}\right), \tag{7}$$

and one may restrict the cardinalities of U and V to $|\mathcal{U}| \leq |\mathcal{S}||\mathcal{X}| + 5$ and $|\mathcal{V}| \leq |\mathcal{S}|^2|\mathcal{X}|^2 + 5|\mathcal{S}||\mathcal{X}| + 3$.

An extended outline of the proof of Theorem 1 is given in Sect. 4 (see [14, Sect. 6-B] for the full proof), and is based on a secured superposition coding scheme. We encode the entire secret message in the *outer layer* of the superposition codebook, meaning no information is carried by the inner layer. The rate of the inner layer is chosen such that it is decodable by the eavesdropper. This results in the eavesdropper 'wasting' his channel resources on decoding the inner layer (which serves as a decoy), leaving it with insufficient resources to unveil the secret message. The legitimate decoder, on the other hand, decodes both layers of the codebook. The transmission is correlated with the observed state sequence by means of the likelihood encoder [12] and SS is established using the strong SCL (both the superposition version from [14, Lemma 1] and the heterogeneous version from [15, Lemma 1]).

Remark 1 (Interpretation of R_{A}). To get some intuition on the structure of R_{A} notice that $I(V;Y|U) - I(V;Z|U)$ is the total rate of secrecy resources that are produced by the outer layer of the codebook. The outer layer can achieve a secure communication rate of $I(V;Y|U) - \max\{I(V;Z|U), I(V;S|U)\}$, and it can produce secret key at a rate of $\left[I(V;S|U) - I(V;Z|U)\right]^+$, where $[x]^+ = \max(0, x)$, because some of the dummy bits needed to correlate the transmission with the state are secure for the same reason that a transmission is secure.

Also, the total amount of reliable (secured and unsecured) communication that this codebook allows is $I(U,V;Y) - I(U,V;S)$, including both the inner and outer layer. Therefore, one interpretation of our encoding scheme is that secret key produced in the outer layer (if any) is applied to the non-secure communication in the inner layer. In total, this achieves a secure communication rate that is the minimum of the total secrecy resources $I(V;Y|U) - I(V;Z|U)$ (i.e. secure communication and secret key) and the total communication rate $I(U,V;Y) - I(U,V;S)$, corresponding to the statement of R_{A}. Of course, this effect happens naturally by the design of the superposition code, without the need to explicitly extract a key and apply a one-time pad.

Remark 2 (Relation to Past Results). Our achievability result recovers the previously best known scheme for the SD-WTC with non-causal encoder CSI from [9,10] as a special case. If the state sequence **S** is also known at the legitimate receiver (obtained by replacing Y with (Y, S) in the considered SD-WTC), our result is at least as good as the best known lower bound by Chia and El-Gamal from [11, Theorem 1]. The latter work considered the case where the encoder learns the channel's state in a causal manner. Nonetheless, the authors of [11] show that using their causal scheme even when the CSI is available non-causally to the encoder can strictly outperform the schemes from [9, 10] when $Y = (Y, S)$. Replacing Y with (Y, S) in R_A from Eq. (7), the auxiliary random variables U and V can be chosen to recover the rate bounds from [11, Theorem 1]. In addition, since our scheme is tailored for the non-causal CSI scenario, our joint distribution allows correlation between the auxiliary random variable and the state, while in [11, Theorem 1] they are uncorrelated.

3.3 Reversely Less Noisy SD-WTC

An interesting special case for which the result of Theorem 1 is tight is the RLN SD-WTC. Let $\mathcal{S}_1$ and $\mathcal{S}_2$ be finite sets and consider a SD-WTC $W_{\tilde{Y}, \tilde{Z}|X,S}$ with non-causal encoder CSI, where $\tilde{Y} = (Y, S_1)$, $\tilde{Z} = (Z, S_2)$ and $W_{S_1, S_2, Y, Z|X, S} = W_{S_1, S_2|S} W_{Y, Z|X}$. Namely, the transition probability $W_{S_1, S_2, Y, Z|X, S}$ decomposes into a product of two WTCs, one being independent of the state given the input, while the other one depends only on it. The legitimate receiver (respectively, the eavesdropper) observes not only the output $\mathbf{Y}$ (respectively, $\mathbf{Z}$) of the WTC $W_{Y, Z|X}^n$, but also $\mathbf{S}_1$ (respectively, $\mathbf{S}_2$) - a noisy version of the state sequence drawn according to the marginal of $W_{S_1, S_2|S}^n$. We characterize the SS-capacity of this setting when the WTC $W_{Y, Z|X}$ is RLN, i.e., when $I(U; Y) \leq I(U; Z)$, for every random variable U with $U - X - (Y, Z)$.

To state the SS-capacity result let $\mathcal{A}$ and $\mathcal{B}$ be finite sets and for any $P_X \in \mathcal{P}(\mathcal{X})$, $P_{A|S} : \mathcal{S} \to \mathcal{P}(\mathcal{A})$ and $P_{B|A} : \mathcal{A} \to \mathcal{P}(\mathcal{B})$ define

$$R_{\text{RLN}}\left(P_X, P_{A|S}, P_{B|A}\right) = \min\left\{I(A; S_1|B) - I(A; S_2|B), I(X; Y) - I(A; S|S_1)\right\}, \tag{8}$$

where the mutual information terms are calculated with respect to the joint PMF $W_S P_{A|S} P_{B|A} P_X W_{S_1, S_2|S} W_{Y, Z|X}$, i.e., where (X, Y, Z) is independent of (S, S_1, S_2, A, B) and $A - S - (S_1, S_2)$ and $B - A - (S, S_1, S_2)$ form Markov chains (as well as the Markov relations implied by the channels).

Theorem 2 (RLN SD-WTC SS-Capacity). *The SS-capacity of the RLN SD-WTC with full encoder and noisy decoder and eavesdropper CSI is*

$$C_{\text{RLN}} = \max_{P_X, P_{A|S}, P_{B|A}} R_{\text{RLN}}\left(P_X, P_{A|S}, P_{B|A}\right). \tag{9}$$

A proof of Theorem 2, where the direct part is established based on Theorem 1, is given in Sect. 5. Instead, one can derive an explicit achievability for Eq. (9) via a coding scheme based on a key agreement protocol via multiple blocks and a one-time-pad operation. To gain some intuition, an outline of the scheme for the simplified case where $S_2 = 0$ is described in the following remark. This scenario is fitting for intuitive purposes since the absence of correlated observations with S at the eavesdropper's site allows to design an explicit secured protocol over a single transmission block. We note however, the even when S_2 is not a constant, a single-block-based coding scheme is feasible via the superposition code construction in the proof of Theorem 1.

Remark 3 (Explicit Achievability for Theorem 2). It is readily verified that when $S_2 = 0$, setting $B = 0$ into Eq. (9) is optimal. The resulting secrecy rate $\tilde{R}_{\mathrm{RLN}}(P_X, P_{A|S}) \triangleq \min \left\{ I(A; S_1), I(X; Y) - I(A; S|S_1) \right\}$, for any fixed P_X and $P_{A|S}$ as before, is achieved as follows:

1. Generate 2^{nR_A} a-codewords as i.i.d. samples of P_A^n.
2. Partition the set of all a-codewords into $2^{nR_{\mathrm{Bin}}}$ equal sized bins. Accordingly, label each a-codeword as $\mathbf{a}(b, k)$, where $b \in \left[1 : 2^{nR_{\mathrm{Bin}}}\right]$ and $k \in \left[1 : 2^{n(R_A - R_{\mathrm{Bin}})}\right]$.
3. Generate a point-to-point codebook that comprises $2^{n(R+R_{\mathrm{Bin}})}$ codewords $\mathbf{x}(m, b)$, where $m \in \mathcal{M}_n$ and $b \in \left[1 : 2^{nR_{\mathrm{Bin}}}\right]$, drawn according to P_X^n.
4. Upon observing the state sequence $\mathbf{s} \in \mathcal{S}^n$, the encoder searches the entire a-codebook for an a-codeword that is jointly-typical with $\mathbf{s}$, with respect to their joint PMF $W_S P_{A|S}$. Such a codeword is found with high probability provided that

$$R_A > I(A; S). \tag{10}$$

Let $(b, k) \in \left[1 : 2^{nR_{\mathrm{Bin}}}\right] \times \left[1 : 2^{n(R_A - R_{\mathrm{Bin}})}\right]$ be the indices of the selected a-codeword. To sent the message $m \in \mathcal{M}_n$, the encoder one-time-pads m with k to get $\tilde{m} = m \oplus k \in \mathcal{M}_n$, and transmits $\mathbf{x}(\tilde{m}, b)$ over the WTC. The one-time-pad operation restricts the rates to satisfy

$$R \leq R_A - R_{\mathrm{Bin}}. \tag{11}$$

5. The legitimate receiver first decodes the x-codeword using it's channel observation $\mathbf{y}$. An error-free decoding requires the total number of x-codewords to be less than the capacity of the sub-channel $W_{Y|X}$, i.e.,

$$R + R_{\mathrm{Bin}} < I(X; Y). \tag{12}$$

Denoting the decoded indices by $(\hat{\tilde{m}}, \hat{b}) \in \mathcal{M}_n \times \left[1 : 2^{nR_{\mathrm{Bin}}}\right]$, the decoder then uses the noisy state observation $\mathbf{s}_1 \in \mathcal{S}_1^n$ to isolate the exact a-codeword from the $\hat{b}$-th bin. Namely, it searches for a unique index $\hat{k} \in \left[1 : 2^{n(R_A - R_{\mathrm{Bin}})}\right]$, such that $\left(\mathbf{a}(\hat{b}, \hat{k}), \mathbf{s}_1\right)$ are jointly-typical with respect to the PMF P_{A,S_1} the marginal of $W_S W_{S_1|S} P_{A|S}$. The probability of error in doing so is arbitrarily small with the blocklength provided that

$$R_A - R_{\mathrm{Bin}} < I(A; S_1). \tag{13}$$

Having decoded $(\hat{\tilde{m}}, \hat{b})$ and $\hat{k}$, the decoder declares $\hat{m} \triangleq \hat{\tilde{m}} \oplus \hat{k}$ as the decoded message.
6. For the eavesdropper, note that although the it has the correct $(\tilde{m}, b)$ (due to the less noisy condition), it cannot decode k since it has no observation that is correlated with the A, S and S_1 random variables. Security of the protocol is implies by the security of the one-time-pad operation.
7. Putting the aforementioned rate bounds together establishes the achievability of $\tilde{R}_{\mathrm{RLN}}(P_X, P_{A|S})$.

To the best of our knowledge, the result of Theorem 2 was not established before. It is, however, strongly related to [16], where a similar model was considered for the purpose of key generation (rather than the transmission of a confidential message). In particular, [16] established lower and upper bounds on the secret-key capacity of the RLN WTC with noisy decoder and eavesdropper CSI. The code construction proposed in [16] is reminiscent of this described in Remark 3 (with the proper adjustments for the key-agreement task).

Remark 4 (Strict Improvement over Past Results). This secrecy-capacity result from Theorem 2 cannot be achieved from the previously known achievable schemes from [9–11]. For [11], this conviction is straightforward since the considered setting falls outside the framework of a SD-WTC with full (non-causal) encoder and decoder CSI. The sub-optimality of [9,10] follows by furnishing an explicit example of a RLN SD-WTC, for which our scheme achieves strictly higher secrecy rates. Due to space limitation, the example is omitting from this work. The reader is referred to [14, Sect. 5-C] for the details.

4 Outline of Proof of Theorem 1

We give a detailed description of the codebook construction and of the encoding and decoding processes. Due to space limitation, the analysis of reliability and SS is omitted and only the required rate bounds accompanied by broad explenations are provided (see [14, Sect. 6-B] for the full details). Fix $\epsilon > 0$ and a conditional PMF $Q_{U,V,X|S}$ with $I(U;Y) \geq I(U;S)$.

Codebook $\mathcal{B}_n$: We use a superposition codebook where the outer layer also encodes the confidential message. The codebook is constructed independently of $\mathbf{S}$, but with sufficient redundancy to correlate the transmission with $\mathbf{S}$.

Let I and J be two independent random variables uniformly distributed over $\mathcal{I}_n \triangleq [1 : 2^{nR_1}]$ and $\mathcal{J}_n \triangleq [1 : 2^{nR_2}]$, respectively. Let $\mathcal{B}_U^{(n)} \triangleq \{\mathbf{u}(i)\}_{i \in \mathcal{I}_n}$ be an inner layer codebook generated as i.i.d. samples of Q_U^n. For every $i \in \mathcal{I}_n$, let $\mathcal{B}_V^{(n)}(i) \triangleq \{\mathbf{v}(i,j,m)\}_{(j,m) \in \mathcal{J}_n \times \mathcal{M}_n}$ be a collection of $2^{n(R_2+R)}$ vectors of length n drawn according to the distribution $Q_{V|U=\mathbf{u}(i)}^n$. We use $\mathcal{B}_n$ to denote our superposition codebook, i.e., the collection of the inner and all the outer layer codebooks. The encoder and decoder are described next for a fixed superposition codebook $\mathcal{B}_n$.

Encoder $f_n^{(\mathcal{B}_n)}$: The encoding phase is based on the likelihood-encoder [12], which, in turn, allows us to approximate the (rather cumbersome) induced joint distribution by a much simpler distribution which we use for the analysis.

Given $m \in \mathcal{M}_n$ and $\mathbf{s} \in \mathcal{S}^n$, the encoder randomly chooses $(i,j) \in \mathcal{I}_n \times \mathcal{J}_n$ according to

$$P_{\mathsf{LE}}^{(\mathcal{B}_n)}(i,j|m,\mathbf{s}) = \frac{Q_{S|U,V}^n(\mathbf{s}|\mathbf{u}(i),\mathbf{v}(i,j,m))}{\sum_{(i',j')} Q_{S|U,V}^n(\mathbf{s}|\mathbf{u}(i'),\mathbf{v}(i',j',m))}, \tag{14}$$

where $Q_{S|U,V}$ is the conditional marginal of $Q_{S,U,V}$ defined by $Q_{S,U,V}(s,u,v) = \sum_{x \in \mathcal{X}} W_S(s) Q_{U,V,X|S}(u,v,x|s)$, for every $(s,u,v) \in \mathcal{S} \times \mathcal{U} \times \mathcal{V}$. The channel input sequence is then generated by feeding the chosen u- and v-codewords along with the state sequence into the DMC $Q_{X|U,V,S}^n$.

Decoder $\phi_n^{(\mathcal{B}_n)}$: Upon observing $\mathbf{y} \in \mathcal{Y}^n$, the decoder searches for a unique triple $(\hat{i}, \hat{j}, \hat{m}) \in \mathcal{I}_n \times \mathcal{J}_n \times \mathcal{M}_n$ such that $\left(\mathbf{u}(\hat{i}), \mathbf{v}(\hat{i}, \hat{j}, \hat{m}), \mathbf{y} \right) \in \mathcal{T}_\epsilon^n(Q_{U,V,Y})$.
If such a unique triple is found, then set $\phi_n^{(\mathcal{B}_n)}(\mathbf{y}) = \hat{m}$; otherwise, $\phi_n^{(\mathcal{B}_n)}(\mathbf{y}) = e$.

The triple $(\mathcal{M}_n, f_n^{(\mathcal{B}_n)}, \phi_n^{(\mathcal{B}_n)})$ defined with respect to the codebook $\mathcal{B}_n$ constitutes an (n, R)-code c_n.

Main Idea Behind Analysis: The key step is to approximate (in total variation) the joint PMF induced by the above encoding and decoding scheme, say $P^{(\mathcal{B}_n)}$, by a new distribution $\Gamma^{(\mathcal{B}_n)}$, which lands itself easier for the reliability and security analyses. For any $P_M \in \mathcal{P}(\mathcal{M}_n)$, $\Gamma^{(\mathcal{B}_n)}$ is

$$\Gamma^{(\mathcal{B}_n)}(m,i,j,\mathbf{u},\mathbf{v},\mathbf{s},\mathbf{x},\mathbf{y},\mathbf{z},\hat{m}) = P_M(m) \frac{1}{|\mathcal{I}_n||\mathcal{J}_n|} \mathbb{1}_{\left\{ \mathbf{u} = \mathbf{u}(i) \right\}} \mathbb{1}_{\left\{ \mathbf{v} = \mathbf{v}(i,j,m) \right\}}$$
$$\times Q_{S|U,V}^n(\mathbf{s}|\mathbf{u},\mathbf{v}) Q_{X|U,V,S}^n(\mathbf{x}|\mathbf{u},\mathbf{v},\mathbf{s}) W_{Y,Z|X,S}^n(\mathbf{y},\mathbf{z}|\mathbf{x},\mathbf{s}) \mathbb{1}_{\left\{ \phi_n^{(\mathcal{B}_n)}(\mathbf{y}) = \hat{m} \right\}}, \quad (15)$$

Namely, with respect to $\Gamma^{(\mathcal{B}_n)}$, the indices $(i,j) \in \mathcal{I}_n \times \mathcal{J}_n$ are uniformly drawn from their respective ranges. Then, the sequence $\mathbf{s}$ is generated by feeding the corresponding u- and v-codewords into the DMC $Q_{S|U,V}^n$. Based on the superposition SCL from [14, Lemma 1], it can be shown the with respect to a random superposition codebook $\mathcal{B}_n$, $P^{(\mathcal{B}_n)}$ and $\Gamma^{(\mathcal{B}_n)}$ are close in total variation in several senses (both in expectation and with high probability), if

$$R_1 > I(U;S) \tag{16a}$$
$$R_1 + R_2 > I(U,V;S)). \tag{16b}$$

Having this, both the reliability and the security analysis are preformed with respect to $\Gamma^{(\mathcal{B}_n)}$ instead of $P^{(\mathcal{B}_n)}$. Standard joint-typicality decoding arguments for superposition codes show that reliability follows provided that

$$R + R_2 < I(V;Y|U), \tag{17a}$$
$$R + R_1 + R_2 < I(U,V;Y). \tag{17b}$$

Using the heterogeneous strong SCL from [15, Lemma 1], SS is ensured if

$$R_2 > I(V;W|U). \tag{18}$$

The rate bound in Eq. 18 ensures that the distribution of the eavesdropper's observation given the inner layer codeword and each secret message is asymptotically indistinguishable form random noise. This asymptotic independence, in turn, implies SS. Finally, applying the Fourier-Motzkin Elimination on Eqs. (16a), (16b), (17a), (17b) and (18) shows that $R_A\left(Q_{U,V,X|S}\right)$ is achievable.

5   Proof of Corollary 2

5.1   Direct

We use Theorem 1 to establish the achievability of Theorem 2. For any $Q_{U,V,X|S}$:
$S \to \mathcal{U} \times \mathcal{V} \times \mathcal{X}$, replacing Y and Z in $R_A\left(Q_{U,V,X|S}\right)$ with $(Y.S_1)$ and (Z, S_2), respectively, gives that

$$R_A^{\mathsf{RLN}}(Q_{U,V,X|S}) = \min \left\{ I(V; Y, S_1|U) - I(V; Z, S_2|U), I(U, V; Y, S_1) - I(U, V; S) \right\}, \tag{19}$$

where the joint PMF $W_S Q_{U,V,X|S} W_{S_1,S_2|S} W_{Y,Z|X}$ satisfies

$$I(U; Y, S_1) - I(U; S) \geq 0 \tag{20}$$

is achievable.

To properly define the choice of $Q_{U,V,X|S}$ that achieves Eq. (9), recall the P distribution stated after Eq. (8) that factors as $W_S P_{A|S} P_{B|A} P_X W_{S_1,S_2|S} W_{Y,Z|X}$ and let $\tilde{P}$ be a PMF over $\mathcal{S} \times \mathcal{A} \times \mathcal{B} \times \mathcal{X} \times \mathcal{Y} \times \mathcal{Z} \times \mathcal{S}_1 \times \mathcal{S}_2 \times \mathcal{B} \times \mathcal{X}$, such that

$$\tilde{P}_{S,A,B,X,S_1,S_2,Y,Z,\tilde{B},\tilde{X}} = P_{S,A,B,X,S_1,S_2,Y,Z} \mathbb{1}_{\{\tilde{B}=B\} \cap \{\tilde{X}=X\}}. \tag{21}$$

Now, fix $P_{S,A,B,X,S_1,S_2,Y,Z}$ and let $Q_{U,V,X|S}$ in Eq. (6) be such that $V = (A, B)_{\tilde{P}}$, $U = (\tilde{B}, \tilde{X})_{\tilde{P}}$ and $Q_{X|S,U,V} = \tilde{P}_X = P_X$, where the subscript $\tilde{P}$ means that the random variables on the RHS are distributed according to their marginal from Eq. (21). Consequently, $Q_{U,V,X|S} W_{S_1,S_2|S} W_{Y,Z|X}$ equals to the RHS of Eq. (21).

Using the statistical relations between the random variable in Eq. (21), one observes that the mutual information term in R_A^{RLN} from Eq. (6) coincide with those from Eq. 9. Namely, we have

$$I_Q(V; Y, S_1|U) - I_Q(V; Z, S_2|U) = I_P(A; S_1|B) - I_P(A; S_2|B) \tag{22}$$
$$I_Q(U, V; Y, S_1) - I_Q(U, V; S) = I_P(X; Y) - I_P(A; S|S_1). \tag{23}$$

Finally, we show that Eq. (20) is satisfied by any PMF $Q_{U,V,X|S}$ of the considered structure for which $R_A^{\mathsf{RLN}}(Q_{U,V,X|S}) \geq 0$. This follows because

$$I_Q(U; Y, S_1) - I_Q(U; S) = I_P(X; Y) - I_P(B; S|S_1) \overset{(a)}{\geq} I_P(X; Y) - I_P(A, B; S|S_1), \tag{24}$$

where (a) is by the non-negativity of conditional mutual information. Thus, any PMF $Q_{U,V,X|S}$ such that $I_Q(U; Y, S_1) - I_Q(U; S) < 0$, induces that the RHS of Eq. (24) is also negative (in which case $R_A^{\mathsf{RLN}}(Q_{U,V,X|S}) < 0$). Since taking $U = V = X = 0$ achieves a higher rate (namely, zero) we may consider only input distribution that satisfy Eq. (20). The achievability result from Theorem 1 establishes the direct part of Theorem 2.

5.2   Converse

Let $\{c_n\}_{n\in\mathbb{N}}$ be a sequence of (n,R) semantically-secure codes for the SD-WTC with a vanishing maximal error probability. Fix $\epsilon > 0$ and let $n \in \mathbb{N}$ be sufficiently large so that the achievability requirements from Definition 5 are satisfied. Since both requirements hold for any message distribution $P_M \in \mathcal{P}(\mathcal{M})$, in particular, they hold for a uniform $P_M^{(U)}$. All the following multi-letter mutual information and entropy terms are calculated with respect to the induced joint PMF from Eq. (3), where the channel $W_{Y,Z|X,S}$ is replaced with $W_{S_1,S_2,Y,Z|X,S}$ defined in Sect. 3.3. Fano's inequality gives

$$H(M|S_1^n, Y^n) \le 1 + n\epsilon R \triangleq n\epsilon_n, \tag{25}$$

where $\epsilon_n = \frac{1}{n} + \epsilon R$.

The security criterion from Definition 5 and the RLN property of the channel $W_{Y,Z|X}$ (that, respectively, justify the two following inequalities) further gives

$$\epsilon \ge I(M; S_2^n, Z^n) = I(M; S_2^n) + I(M; Z^n|S_2^n) \ge I(M; S_2^n, Y^n). \tag{26}$$

Having Eqs. (25) and (26), we bound R as

$$nR = H(M)$$
$$\overset{(a)}{\le} I(M; S_1^n, Y^n) - I(M; S_2^n, Y^n) + n\delta_n$$
$$= I(M; S_1^n|Y^n) - I(M; S_2^n|Y^n) + n\delta_n$$
$$\overset{(b)}{=} \sum_{i=1}^n \left[I(M; S_{1}^i, S_{2,i+1}^n|Y^n) - I(M; S_1^{i-1}, S_{2,i}^n|Y^n) \right] + n\delta_n$$
$$\overset{(c)}{=} \sum_{i=1}^n \left[I(M; S_{1,i}|B_i) - I(M; S_{2,i}|B_i) \right] + n\delta_n$$
$$\overset{(d)}{=} n \sum_{i=1}^n P_T(i) \left[I(M; S_{1,T}|B_T, T=i) - I(M; S_{2,T}|B_T, T=i) \right] + n\delta_n$$
$$= n \left[I(M; S_{1,T}|B_T, T) - I(M; S_{2,T}|B_T, T) \right] + n\delta_n$$
$$\overset{(e)}{=} n \left[I(A; S_1|B) - I(A; S_2|B) \right] + n\delta_n \tag{27}$$

where:

(a) is by Eqs. (25) and (26) while setting $\delta_n \triangleq \epsilon_n + \frac{\epsilon}{n}$;
(b) is a telescoping identity [17, Eqs. (9) and (11)];
(c) defined $B_i \triangleq (S_1^{i-1}, S_{2,i+1}^n, Y^n)$, for all $i \in [1:n]$.
(d) is by introducing a time-sharing random variable T that is uniformly distributed over the set $[1:n]$ and is independent of all the other random variables in $P^{(c_n)}$;
(e) defines $S \triangleq S_T$, $S_1 \triangleq S_{1,T}$, $S_2 \triangleq S_{2,T}$, $X \triangleq X_T$, $Y \triangleq Y_T$, $Z \triangleq Z_T$, $B \triangleq (B_T, T)$ and $A \triangleq (M, B)$.

Another way to bound R is

$$nR = H(M)$$

$$\overset{(a)}{\leq} I(M; S_1^n, Y^n) + n\epsilon_n$$

$$= I(M; S_1^n, Y^n, S^n) - I(M; S^n | S_1^n, Y^n) + n\epsilon_n$$

$$\overset{(b)}{=} I(M; Y^n | S_1^n, S^n) - I(M, Y^n; S^n | S_1^n) + I(S^n; Y^n | S_1^n) + n\epsilon_n$$

$$= I(M, S^n; Y^n | S_1^n) - I(M, Y^n; S^n | S_1^n) + n\epsilon_n$$

$$\overset{(c)}{\leq} I(M, S^n; Y^n) - I(M, Y^n; S^n | S_1^n) + n\epsilon_n$$

$$\overset{(d)}{\leq} I(X^n; Y^n) - I(M, Y^n; S^n | S_1^n) + n\epsilon_n$$

$$\overset{(e)}{\leq} \sum_{i=1}^{n} \left[I(X_i; Y_i) - I(M, Y^n; S_i | S_1^n, S^{i-1}) \right] + n\epsilon_n$$

$$\overset{(f)}{\leq} \sum_{i=1}^{n} \left[I(X_i; Y_i) - I(M, Y^n, S_1^{n \setminus i}, S^{i-1}; S_i | S_{1,i}) \right] + n\epsilon_n$$

$$\overset{(g)}{\leq} \sum_{i=1}^{n} \left[I(X_i; Y_i) - I(M, B_i; S_i | S_{1,i}) \right] + n\epsilon_n$$

$$\overset{(h)}{=} n \sum_{i=1}^{n} P_T(i) \left[I(X_T; Y_T | T = i) - I(M, B_T; S_T | S_{1,T}, T = i) \right] + n\epsilon_n$$

$$\overset{(i)}{\leq} n \left[I(X_T; Y_T) - I(M, B_T, T; S_T | S_{1,T}) \right] + n\epsilon_n$$

$$\overset{(j)}{\leq} n \left[I(X; Y) - I(A; S | S_1) \right] + n\epsilon_n \tag{28}$$

where:

(a) is by Eq. (25);

(b) uses the independence of M and (S_1^n, S^n) (1st term);

(c) is because conditioning cannot increase entropy and since $Y^n - (M, S^n) - S_1^n$ forms a Markov chain (1st term);

(d) uses the Markov relation $Y^n - X^n - (M, S^n)$;

(e) follows since conditioning cannot increase entropy and by the discrete and memoryless property of the WTC $W_{Y,Z|X}^n$;

(f) is because $P_{S^n, S_1^n, S_2^n}^{(c_n)} = W_{S,S_1,S_2}^n$, i.e., the marginal distribution of (S^n, S_1^n, S_2^n) are i.i.d.;

(g) is by the definition of B_i;

(h) follows for the same reason as step (d) in the derivation of Eq. (27);

(i) is because conditioning cannot increase entropy and the Markov relation $Y_T - X_T - T$ (1st term), and because $\mathbb{P}(S_T = s, S_{1,T} = s_1, T = t) = W_{S,S_1}(s, s_1) P_T(t)$, for all $(s, s_1, t) \in \mathcal{S} \times \mathcal{S}_1 \times [1:n]$ (2nd term);

(j) reuses the definition of the single-letter random variable from step (e) in the derivation of Eq. (27).

It can be verified that the joint distribution of the defined random variables factors in accordance to the statement of Theorem 2. Thus, we have the following bound on the achievable rate

$$R \leq \frac{\min\left\{I(A;S_1|B) - I(A;S_2|B), I(X;Y) - I(A;S|S_1)\right\}}{1-\epsilon} + \frac{1}{(1-\epsilon)n} + \frac{\epsilon}{1-\epsilon}, \tag{29}$$

where the mutual information terms are calculated with respect to the joint PMF $W_S W_{S_1,S_2|S} P_{A|S} P_{B|A} P_{X|S,S_1,S_2,A,B} W_{Y,Z|X}$. However, noting that in none of the mutual information terms from Eq. (29) do X and (S, S_1, S_2, A, B) appear together, we may replace $P_{X|S,S_1,S_2,A,B}$ with P_X without affecting the expressions. Taking $\epsilon \to 0$ and $n \to \infty$ completes the proof of the converse.

References

1. Johnson, M.W., et al.: Quantum annealing with manufactured spins. Nature **473**(7346), 194–198 (2011)
2. Jones, N.: Google and NASA snap up quantum computer D-Wave Two (2013). http://www.scientificamerican.com/article.cfm?id=google-nasa-snap-up-quantum-computer-dwave-two
3. Shor, P.W.: Polynomial-time algorithms for prime factorization and discrete logarithms on a quantum computer. SIAM Rev. **41**(2), 303–332 (1999)
4. Bernstein, D.J.: Introduction to post-quantum cryptography. In: Bernstein, D.J., Buchmann, J., Dahmen, E. (eds.) Post-quantum Cryptography, pp. 1–14. Springer, Heidelberg (2009)
5. Perlner, R.A., Cooper, D.A.: Quantum resistant public key cryptography: a survey. In: Proceedings of Symposium Identity and Trust on the Internet (IDtrust), Gaithersburg, Maryland, pp. 85–93. ACM, April 2009
6. Bloch, M., Barros, J.: Physical-Layer Security: From Information Theory to Security Engineering. Cambridge University Press, Cambridge (2011)
7. Wyner, A.D.: The wire-tap channel. Bell Syst. Technol. **54**(8), 1355–1387 (1975)
8. Gelfand, S.I., Pinsker, M.S.: Coding for channel with random parameters. Probl. Pered. Inform. (Probl. Control Inf. Theor.) **9**(1), 19–31 (1980)
9. Chen, Y., Han Vinck, A.J.: Wiretap channel with side information. IEEE Trans. Inf. Theor. **54**(1), 395–402 (2008)
10. Liu, W., Chen, B.: Wiretap channel with two-sided state information. In: Proceedings of 41st Asilomar Conference Signals, System and Computer, Pacific Grove, CA, US, pp. 893–897 (2007)
11. Chia, Y.-K., El Gamal, A.: Wiretap channel with causal state information. IEEE Trans. Inf. Theor. **58**(5), 2838–2849 (2012)
12. Song, E., Cuff, P., Poor, V.: The likelihood encoder for lossy compression. IEEE Trans. Inf. Theor. **62**(4), 1836–1849 (2016)
13. Bellare, M., Tessaro, S., Vardy, A.: A cryptographic treatment of the wiretap channel. In: Proceedings of Advance Cryptology, (CRYPTO 2012), Santa Barbara, CA, USA (2012)

14. Goldfeld, Z., Cuff, P., Permuter, H.H.: Wiretap channel with random states non-causally available at the encoder. IEEE Trans. Inf. Theor. (2016, submitted)

15. Goldfeld, Z., Cuff, P., Permuter, H.H.: Arbitrarily varying wiretap channels with type constrained states. IEEE Trans. Inf. Theor. **62**(12), 7216–7244 (2016)

16. Khisti, A., Diggavi, S.N., Wornell, G.W.: Secret-key generation using correlated sources and channels. IEEE Trans. Inf. Theor. **58**(2), 652–670 (2012)

17. Kramer, G.: Teaching IT: an identity for the Gelfand-Pinsker converse. IEEE Inf. Theor. Soc. Newslett. **61**(4), 4–6 (2011)

Assisting Malware Analysis with Symbolic Execution: A Case Study

Roberto Baldoni, Emilio Coppa, Daniele Cono D'Elia[(✉)],
and Camil Demetrescu

Software Analysis and Optimization Laboratory,
Department of Computer, Control, and Management Engineering,
Cyber Intelligence and Information Security Research Center,
Sapienza University of Rome, Rome, Italy
{baldoni,coppa,delia,demetres}@dis.uniroma1.it

Abstract. Security analysts spend days or even weeks in trying to understand the inner workings of malicious software, using a plethora of manually orchestrated tools. Devising automated tools and techniques to assist and speed up the analysis process remains a major endeavor in computer security. While manual intervention will likely remain a key ingredient in the short and mid term, the recent advances in static and dynamic analysis techniques have the potential to significantly impact the malware analysis practice. In this paper we show how an analyst can use symbolic execution techniques to unveil critical behavior of a remote access trojan (RAT). Using a tool we implemented in the ANGR framework, we analyze a sample drawn from a well-known RAT family that leverages thread injection vulnerabilities in the Microsoft Win32 API. Our case study shows how to automatically derive the list of commands supported by the RAT and the sequence of system calls that are activated for each of them, systematically exploring the stealthy communication protocol with the server and yielding clues to potential threats that may pass unnoticed by a manual inspection.

Keywords: Malware · RAT · APT · Symbolic execution · ANGR

1 Introduction

The unprecedented spread of network-connected devices and the increasing complexity of operating systems is exposing modern ICT infrastructures to malicious intrusions by different threat actors, which can steal sensitive information, gain unauthorized access, and disrupt computer systems. Attacks are often perpetrated in the context of targeted or broad-spectrum campaigns with different scopes, including hacktivism, cyber warfare, cyber crime, and espionage. One of the most common form of intrusion is based on malicious software, or malware, which can exploit vulnerabilities in applications and operating systems to infect, take over, or disrupt a host computer without the owner's knowledge and consent. Sustained by the explosion of messaging applications and social

© Springer International Publishing AG 2017
S. Dolev and S. Lodha (Eds.): CSCML 2017, LNCS 10332, pp. 171–188, 2017.
DOI: 10.1007/978-3-319-60080-2_12

networks, malware can nowadays affect virtually any device connected to the Internet including unconventional targets such as network printers, cooling systems, and Web-based vehicle immobilization systems. Malware infections can cause potentially significant harm by exfiltrating sensitive data, tampering with databases and services, and even compromising critical infrastructures.

According to [17], malware is responsible for the most frequent and costly attacks on public and private organizations. ICT infrastructures are not the only targets: Kindsight Security reports that at least 14% of private home networks were infected with malware in April–June 2012 [16]. One of the main vectors of malware spreading remain emails and infected websites, where unsuspecting users are daily hijacked by inadvertently opening seemingly benign attachments or lured into browsing deceitful links or click-baits that stealthily download and install malicious software. Malware scammers often resort to social engineering techniques to trick their victims and infect them with a variety of clever approaches including backdoors, trojans, botnets, rootkits, adware, etc.

The job of a professional malware analyst is to provide quick feedback on security incidents that involve malicious software, identifying the attack vectors and the proper actions to secure and repair the infected systems. In many cases involving critical compromised services, time is of the essence. Analysts seek clues to the parts of the system that were disrupted and attempt to reconstruct and document the chain of events that led to the attack. Often, intrusions are carried out by variants of previously encountered malware. In other cases, malware is based on zero-day vulnerabilities or novel attack strategies that may require days or even weeks to be identified and documented. Analysts usually combine and relate the reports generated by a wide range of dedicated static and dynamic analysis tools in a complex manual process and are often required to sift through thousands or even millions of lines of assembly code.

A skilled professional is able to glance over irrelevant details and follow the high-level execution flow, identifying any stealthy API calls that can compromise the system. However, some tasks may keep even the most experienced analysts busy for days or even weeks. For instance, malware such as backdoors or trojans provide a variety of hidden functionalities that are activated based on unknown communication protocols with remote servers maintained by the threat actors. Identifying the supported commands and reconstructing how the protocols work may require exploring a wide range of possible execution states and isolating the input data packets that make the execution reach them. While this can be amazingly difficult without automated software analysis techniques, the state of the art of modern binary reverse engineering tools still requires a great deal of manual investigation by malware analysts.

Contributions. In this paper, we argue that the significant advances in software analysis over the last decade can provide invaluable support to malware analysis. In particular, we describe how symbolic execution [1], a powerful analysis technique pioneered in software testing, can be applied to malware analysis by devising a prototype tool based on the ANGR symbolic executor [26]. The tool automatically explores the possible execution paths of bounded length starting from a given entry point. The analysis is static and the code is not concretely executed. As output, the tool produces a report that lists for each explored execution path the sequence of encountered API calls and their arguments, along with properties of the malware's input for which the path is traversed, e.g., the actual data values read from a socket that would trigger the path's execution.

We evaluate our tool on a sample taken from a well-known family of remote access trojans [31], showing how to automatically reconstruct its communication protocol and the supported commands starting from initial hints by the malware analysts on the portions of code to analyze. We estimate the reports our tools generates to be worth hours of detailed investigation by a professional analyst.

Paper organization. This paper is organized as follows. In Sect. 2 we address the background and the features of our case study. Section 3 discusses our symbolic analysis tool and how we used it to analyze the sample. Section 4 puts our results into the perspective of related work, while Sect. 5 concludes the paper with final thoughts and ideas for further investigations.

2   Features of the RAT

In this section, we discuss the background and the features of the malware instance we use as our case study. The RAT executable can be downloaded from VirusTotal and its MD5 signature is 7296d00d1ecfd150b7811bdb010f3e58. It is drawn from a family of backdoor malware specifically created to execute remote commands in Microsoft Windows platforms.

Variants of this RAT date as far back as 2004 and have successfully compromised thousands of computers across more than 60 different countries. This constantly morphing malware is known under different aliases such as Enfal and GoldSun, and its instances typically contain unique identifiers to keep track of which computers have been compromised by each campaign [31].

The RAT gathers information on the infected computer, and communicates with a command-and-control (C&C) server. Once activated, the malware allows a remote attacker to take over the infected machine by exchanging files with the server and executing shell commands. It then copies itself in a number of executable files of the Windows system folder and modifies the registry so that it is automatically launched at every startup.

The malware uses thread injection to activate its payload in Windows Explorer. The payload connects to http://mse.vmnat.com, sending a sprintf-formatted string with information on the system retrieved using the Netbios API. At the core of the malware is a loop that periodically polls the server for encrypted remote commands and decrypts them by applying a character-wise XOR with the 0x45 constant. The malware version we analyzed supports 17 commands, which provide full control over the infected host by allowing remote attackers to list, create, and delete directories, move, delete, send, and receive files, execute arbitrary commands, terminate processes, and other more specific tasks. Communication with the server is carried out over HTTP 1.1 on port 80, user "reader", and password "1qazxsw2".

174   R. Baldoni et al.

A high-level picture of the control flow graph of the command processing thread, automatically reconstructed by the IDA disassembler, is shown in Fig. 1. The thread code starts at address 0x402BB0 and spans over 4 KiB of machine code, not including the code of the called subroutines.

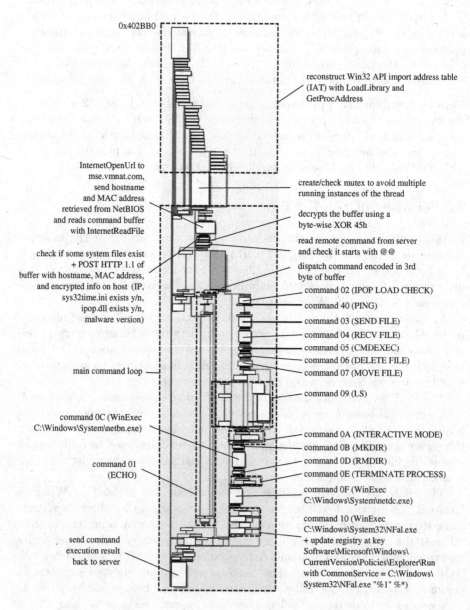

Fig. 1. Control flow graph of the RAT's command processing thread.

3 Analyzing the RAT with Symbolic Execution

In this section, we report our experience in the analysis of our RAT sample using ANGR, a popular symbolic executor. After a brief description of symbolic execution techniques, we discuss the practical challenges in using these tools for the malware realm, and present a number of domain-specific optimizations we adopted. We describe the setup we used to dissect the core of the sample, and discuss our findings and their value from a malware analyst's perspective.

3.1 Introducing Symbolic Execution

Symbolic execution is a popular program analysis technique for testing a property in program against multiple execution paths at a time. Rather than executing a program on a specific input – thus exploring a single control flow path at a time – symbolic execution can concurrently handle multiple paths that would be exercised under different inputs.

In a nutshell, a program is allowed to take on *symbolic* rather than concrete input values, while an execution engine collects across each explored path a set of constraints that are combined into a formula describing the path. When an assignment instruction is evaluated, the formula is simply updated to reflect it. When a branching instruction is encountered and its outcome depends on one or more symbolic values, the execution is forked by creating two states described by two distinct formulas, derived from the current formula by adding to it the branch condition or its negation, respectively. A constraint solver - typically one suited for satisfiability modulo theories (SMT) - is used to evaluate expressions involving symbolic values, as well as for generating concrete inputs that can be used to run concretely the program along the desired path.

We have implemented our ideas in ANGR [26], an open source framework for symbolic execution developed at UC Santa Barbara. ANGR ships as a collection of Python libraries for binary analysis and dissection. It has been employed in a number of research works [25,29], as well as by the Shellphish team from UCSB that recently participated in the DARPA Cyber Grand Challenge, a two-year competition seeking to create automatic systems for vulnerability detection, exploitation, and patching in near real-time [26]. The simplicity of its Python APIs and the support provided by the community make it an ideal playground for prototyping research ideas in a powerful framework.

3.2 Addressing Challenges from the Malware Domain

Symbolic execution techniques have largely been employed in the software testing domain, usually with the goal of automatically generating test inputs that yield a high degree of code coverage. Albeit a few works have explored security applications of these techniques (Sect. 4), the majority of currently available symbolic executors are not well equipped for analyses in the malware realm.

Firstly, most research tools target Linux as a platform, while Windows is by far the most malware-ridden platform. The few tools that support Windows

APIs such as BitBlaze [28] commonly resort to a concrete execution for API calls, asking the constraint solver to provide a valid model for a path formula when a symbolic argument is provided for a call. This *concretization* process typically causes a loss of accuracy in the analysis; also, it does not solve scalability problems that might arise when analyzing real-world malware [30]. Secondly, techniques and heuristics designed for software testing purposes might not fit malware well. While in software testing it is profitable to explore paths capturing behaviors unexpected for a standard usage session, such as system or library call failures, in the malware domain an analyst is rather interested in behaviors commonly exposed by a sample, provided that the right triggering inputs (e.g., valid commands from a C&C server) are received and/or the environmental conditions (e.g., a specific version of Windows) are met.

Extending ANGR. To make the dissection of the RAT possible, we had to devise a number of extensions to the original ANGR framework, tackling both of the above-mentioned problems. In order to support Windows APIs, we had to implement 57 models of commonly used functions, such as `GetProcAddress`, `LoadLibrary`, `HttpOpenRequest`, and `CreateFile`. A *model* is a summary for a function that simulates its effects by propagating symbolic data in the same way that the original function would have [3], requiring a significantly shorter amount of time than in a symbolic exploration. Models are commonly employed in symbolic executors when dealing with the environment (e.g., to simulate filesystem and network I/O) and also to speed up the analysis of classes of functions, such as those for string manipulation. The lack of models for Win32 APIs in ANGR was thus the first obstacle we had to overcome, along with the missing support for the `stdcall` calling convention employed by these APIs.

Writing a model can be a time-consuming and error-prone task [2]. We thus developed a tool that fetches API prototypes from Windows header files and retrieves from the MSDN documentation not only a description of the function's behavior, but also of the directional attribute (i.e., *in, out,* or *both*) of each argument. The output of the tool is thus an ANGR model stub annotated with the input-output relationships, so that a developer can more easily encode the logic of the function in terms of symbolic data manipulation.

Domain-Specific Optimizations. We then had to assess the practicality of our methodology with respect to the so-called *path explosion* problem, which haunts every symbolic execution implementation and can prevent it from scaling to real-world software. Indeed, as a symbolic executor may fork states at every input-dependent branch, the total number of paths to explore might be exponential. This impacts both space and time, and a common approach is to employ search strategies that can limit the exploration to a subset of paths that look appealing for a given goal (e.g., identifying null-pointer dereferences).

The adoption of domain-specific optimizations and search heuristics can mitigate the path explosion problem in the analysis of malicious binaries, making their symbolic analysis feasible. A recent work [30] explores this approach for binaries packed with sophisticated code packers that reveal pages only when about to execute code in them. We thus devised a number of heuristics aiming at revealing information useful to an analyst for the dissection, discarding instead from the exploration paths that are unlikely to.

For a number of models, we restricted the number of possible outcomes by discarding error paths, or put a limit on the length of the symbolic buffer returned by a method. For instance, in our case study we found fairly reasonable to assume that having Win32 methods such as `HttpSendRequest` or `InternetReadFile` to succeed should fairly reflect the run-time behavior expected for a malware. Shall an adversary put a number of faulty function invocations in the code as a decoy, the analyst can still revert the model to an exhaustive configuration (either at specific call sites or for any invocation) and restart the symbolic execution. Exploring error-handling paths might become necessary for malware, especially nation-state APTs, that conceals interesting behavior for the security research to explore, e.g., attempting to spread the infection to other computers until Internet access is found on one of them. Selectively enabling error-handling paths provides the analyst with the flexibility needed to explore such scenarios as well.

Limiting the length of symbolic buffers is a common practice in symbolic executors, as exploring all possible lengths would quickly lead to the path explosion phenomenon [1]. We devised an optimization that targets symbolic buffers originating in Win32 models and processed by tight loops. A *tight loop* is a code sequence ending in a conditional backward jump within a short reach; for our setting we empirically chose a 45-byte threshold. When the loop trip count depends on the content of a symbolic buffer originating in a Win32 model, we do as follows: if the loop has not already exited within k iterations, we force it to. The rationale behind this heuristic is that while such buffers are typically large, the amount of data they usually contain is much smaller. For instance, in the analysis of our RAT sample we found out that most buffers have length 0x1000 but are filled only for the first few bytes. Tight loops for string processing are quite frequent in the sample, especially in the form of REP instructions. We also provided an optimization for ANGR that speeds up the processing of REP when the trip count can be determined statically: rather than symbolically iterating over the loop, we compute its effects and update the state at once.

Finally, we encoded a simple iterative deepening search (IDS) strategy to discriminate which alternative should be explored first when a branching instruction is encountered (Sect. 3.1). As our goal is to reconstruct which strings exercise the different commands supported by a RAT and possibly discover any dependencies between them, exploring sequences of commands of increasing length might provide insights to an analyst in a timely manner. We also favored IDS over breadth-first search (BFS) for its lower memory consumption. In fact, ANGR currently lacks a mature checkpointing mechanism to automatically suspend the exploration for a set of states and release their resources when too many paths are being concurrently explored. While this might not be an issue when executing the framework on a cluster, an analyst might also want to perform a preliminary analysis on commodity hardware such as the laptop we used in our experiments.

3.3   Dissecting the RAT with ANGR

In this section, we provide an overview of how the dissection of the RAT sample can be carried out in our tool for ANGR.

When a malware analyst first disassembles the executable, they can determine from the inspection of the `WinMain` method that the RAT - after running a number of checks and collecting preliminary information about the attacked machine (such as the Windows version and the locale in use) - resorts to the Win32 `CreateRemoteThread` function to inject and run three threads in the virtual address space of `explorer.exe`. A quick look at their code reveals that the first thread carries out the command-and-control logic and executes the remote commands (Fig. 1), while the remaining two do not reveal any interesting behavior and can be discarded from the analysis for the moment.

Execution Context. Ideally, an analyst would like to set the symbolic entry point (SEP) to the entry point of the thread and start the symbolic execution from there. ANGR treats as symbolic any data fetched from uninitialized memory locations during the exploration. We thus define the *execution context* as the set of memory location holding a value initialized when reaching SEP from the program's entry point, and that any path starting at SEP might then need later[1].

Providing the engine with information on the context can be profitable depending on the application being analyzed. For instance, one might employ symbolic execution to find out how the context should look like in order for the execution to reach a specific instruction (e.g., to discover which input string defuses a logic bomb). For other applications, however, a fully symbolic context might quickly lead to an explosion of the paths, as too little information is available when dealing with assignments and branching instructions.

In our case study, the execution context for the command processing thread consists of a large array provided as argument to the thread. This array contains gathered information describing the attacked machine, the addresses of the `LoadLibrary` and `GetProcAddress` functions in the address space of the process to inject, and a number of strings describing the names of the Win32 APIs that will be used in the thread. In fact, when executing in the injected process the RAT will solve the address of each Win32 function it needs to invoke in a block of code, constructing on the stack a minimal import table for the block on demand.

It is unrealistic to think that in general a constraint solver can guess which API a malware writer requires at specific points in the program. The analyst in this scenario is thus expected to provide the symbolic executor with portions of the context, i.e., the API-related strings added to the argument array in the early stage of a concrete execution. This should not be surprising for an analyst, as they often have to fix the program state when manipulating the stack pointer in a debugger to advance the execution and skip some code portions. We have explored two ways to perform this task. The first is to take a memory dump of a concrete execution of the program right before it reaches the starting point SEP

---

[1] A context can formally be defined in terms of live variables, i.e., the set of locations that execution paths starting at SEP might read from before writing to.

for the symbolic execution. A script then processes it and fills the context for the execution with concrete values taken from the dump. Optionally, portions of the context can be overridden and marked as symbolic: for instance, turning into symbolic a buffer containing the IP address or the Windows version for the machine can reveal more execution paths if a malware discriminates its behavior according to the value it holds.

The problem with a dump-based approach is that in some cases it might not be simple for an analyst to have a concrete execution reach the desired SEP. A more general alternative is to employ symbolic execution itself to fill a portion of the context, by moving the SEP backward. In our RAT case study we symbolically executed the instructions filling the argument array, obtaining a context analogous to what we extracted from a memory dump.

We believe that in general a combination of the two approaches might help an analyst reconstruct the context for even more sophisticated samples, if it is required by the executor to produce meaningful results. Additionally, reverse engineering practitioners that use symbolic execution often perform simple adjustments on an initial fully symbolic context in order to explore complex portions of code. We believe this approach can effectively be applied to samples from the malware domain as well. In fact, such adjustments are common in the ANGR practice as part of a trial-and-error process, and do not require intimate knowledge of the inner workings of the framework.

Starting the Exploration. Our RAT dissection tool ships as an ANGR script that we devise in two variants: one takes on a concrete instantiation of the argument array for the thread, while the other constructs it symbolically. Nevertheless, once the entry point of the injected thread is reached, their behavior is identical: from here on we will thus use the term *script* to refer to both of them. From the thread's entry point the symbolic execution follows a single execution path until a `while` cycle is reached. As we will discuss in the next section, this cycle is responsible for command processing. The analyst would then observe in the run-time log of the script that a symbolic buffer is created and manipulated in Internet-related Win32 API calls (e.g., `InternetReadFile`), and different execution paths are then explored depending on its content.

The analyst can ask the tool to display how the branch condition looks like. The language used for expressing the condition is the one used by the Z3 SMT solver employed by ANGR. If the condition is hardly readable by the analyst, the script can also query the constraint solver and retrieve one or more concrete instances that satisfy the condition. Our sample initially checks whether the first two bytes in a symbolic buffer are equal to a constant that when subsequently XOR-ed with the encryption key yields the sequence "@@".

As the exploration proceeds, paths in which the condition is not satisfied will quickly return to the beginning of the `while` cycle. This might suggest that the constraints on the content of the symbolic buffer do not match the syntax of the command processing core, which would then wait for a new message.

The analyst can also employ similar speculations or findings to speed up the symbolic execution process. In particular, ANGR allows users to mark certain addresses as to avoid, i.e., the engine will not follow paths that bring the instruction pointer to any of them. For the sample we dissected this was not a compelling issue: the number of paths that our iterative deepening search (IDS) strategy would explore would still be small. Nonetheless, this optimization can become very valuable in a scenario where the number of paths is much larger due to a complex sequence of format controls. A direct consequence of the optimization is that paths spanning sequences with at least one invalid command in it are not reported to the analyst. We believe that such paths would not normally reveal valuable insights into the command processing logic and protocol, and thus can safely be discarded.

While attempting to dissect the command processing loop, the IDS strategy used for path exploration (Sect. 3.2) has proved to be very convenient. In fact, an IDS allows us to explore possible sequences of commands of increasing length k, which also corresponds to the number of times the first instruction in the while cycle is hit again. The script will produce a report every time a sequence of k commands is fully explored, and once all possible sequences have been analyzed the symbolic executor proceeds by exploring a sequence of $k + 1$ commands, producing in turn incremental reports for the updated length. As the number of iterations of a command processing loop is typically unbounded, the analyst has to explicitly halt the exploration once they have gained sufficient insights for the objective of the analysis. For our sample, all the accepted commands were already revealed for $k = 3$.

3.4   The RAT Dissected

The reports generated from our tool capture a number of relevant and useful facts to an analyst regarding each execution path. Each report is a polished version of the execution trace revealing the sequence of Win32 API invocations performed inside each x86 subroutine in the sample. A report also captures control flow transfers across subroutines, showing their call sites and return addresses along with the API calls they perform in between.

```
...
[0x4030a0] InternetOpenA(<BV32 0x0>, <BV32 0x0>, <BV32 0x0>, <BV32 0x0>, <BV32 0x0>)
    => <BV32 hInternet_39_32>
[0x4030bc] InternetOpenUrlA(<BV32 hInternet_39_32>, <BV32 0xabcd161c>,
    <BV32 0x0>, <BV32 0x0>, <BV32 0x84000100>, <BV32 0x0>,
    'http://mse.vmnat.com/httpdocs/mm/$machine_host_name:$mac_address/Cmwhite')
    => <BV32 hInternet_url_40_32>
[0x4030d5] InternetReadFile(<BV32 hInternet_url_40_32>, <BV32 0x7ffd4c00>,
    <BV32 0x1000>, <BV32 0x7ffd4a60>) => <BV32 0x1>
    SO: <BV32768 InternetReadFile_buffer_41_32768> @ 0x7ffd4c00
    SO: <BV32 InternetReadFile_buffer_written_42_32> @ 0x7ffd4a60
[0x4030dc] InternetCloseHandle(<BV32 hInternet_url_40_32>) => <BV32 0x1>
[0x4030df] InternetCloseHandle(<BV32 hInternet_39_32>) => <BV32 0x1>
...
```

Fig. 2. Fragment of detailed report automatically generated for one execution path in the RAT's command processing thread.

Figure 2 shows an excerpt from a report. For each API the call site, the list of arguments and the return value are shown. Constant string arguments are printed explicitly, while for other data types we resort to the BVxx notation, which in the ANGR domain describes a *bitvector* (i.e., a sequence of consecutive bits in memory) of xx bits. BV32 objects occur frequently on a 32-bit architecture as they can hold pointers or primitive data types. When a bitvector holds a concrete value, the value is explicitly listed in the report. For symbolic values a string indicating a data type (e.g., hInternet_url) or the name of the API that created the buffer (e.g., InternetReadFile_buffer) is displayed, followed by a suffix containing a numeric identifier for the buffer and the buffer size in bits. Observe that while the contents of a bitvector can be symbolic, it will normally be allocated at a concrete address: when such an address is passed as argument to an API, the report will contain a row starting with SO: that describes the symbolic buffer the address points to.

We remark that the sequence of Win32 API calls performed by a malware typically provides valuable information in the malware analysis practice, as the analyst can find out which are the effects of a possible execution path in a black-box fashion. We will further discuss this aspect in Sect. 4.

Further insights can be revealed once the constraint solver produces for the report instances of symbolic buffers matching the path formula (Sect. 3.1), i.e., inputs can steer the execution across the analyzed path. For instance, this allowed us to find out which strings were required to exercise the 17 commands accepted by our sample (Fig. 1).

Dependencies between commands can instead surface from the analysis of sequences of increasing length k. We found that each valid sequence of commands should always start with two invocations of the 01 command, used to implement a handshaking protocol with the command and control server.

From a concrete instantiation of the symbolic buffers we then found out the RAT checks for the presence of a magic sequence in the server's response during the handshaking phase. In particular, the response has to contain "8040$(" starting at byte 9 in order for the malware to update its state correctly and eventually unlock the other commands. Constraint solving techniques thus proved to be valuable in the context of message format reconstruction.

Sequences of three commands reveal sufficient information for an analyst to discover all the commands accepted by the RAT. Due to the particular structure of the handshaking protocol, our tool explored (and thus reported) as many paths as supported commands. Figure 3 provides a compact representation of the logic of the command processing thread that we could infer from the reports. The sample starts by creating a mutex to ensure that only one instance of the RAT is running in the system. The internal state of the malware, represented by the two variables c_1 and c_2, is then initialized.

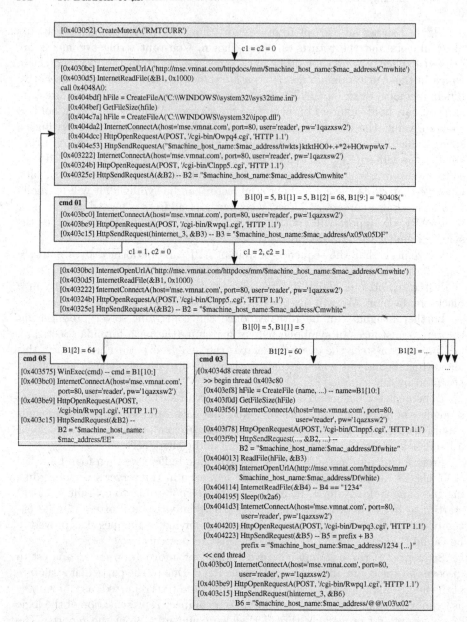

Fig. 3. Compact report for the RAT's command processing thread.

The subroutine starting at address 0x4048A0 is invoked to collect further information on the machine (specifically, the presence of two files), and a connection is established with the server in order to transmit the identity of the infected machine. This step is performed twice, resulting in different increments to c1 and c2. Edges between code blocks have been annotated with the conditions on the symbolic bytes in the response from the server that should be met in order to make the transitions possible.

Once $c1 = 2$ and $c2 = 1$, the whole set of commands for the sample is unlocked. Figure 3 reports a high-level description of two commands: namely, command 05 executes the *cmd* application on the machine, while command 03 spawns a thread to transmit a file to the server. Both command handlers end with a sequence that notifies the server of the reception of the command. We extracted from the reports the sequence of Win32 API calls performed in each command, thus identifying all the actions available to the attackers.

4   Related Work

Malware Detection. Anti-malware vendors receive every day thousands of samples, many of which are unknown. A large body of works have explored automatic techniques to determine whether a sample is malicious and, if so, whether it is a variation of a previously analyzed threat or it requires a closer inspection from a malware analyst. Solutions based on static techniques analyze the code without actually executing it, with the advantage of covering it in its entirety. For instance, [8] relies on model checking to defy simple obfuscations employed by malicious code writers to subvert detection tools such as anti-virus products. [9] extends this technique by supporting a wider range of morphing techniques common in polymorphic and metamorphic malware.

The major weakness of static solutions is that they can be defeated by resorting to self-modifying code, as in packer programs, or to techniques designed to foil static disassembly. Dynamic solutions are thus a necessary complement to static detection techniques [18]. Dynamic techniques typically execute a sample in a contained environment and verify the action it performs, providing analysts with a report. For instance, GFI Sandbox [33] and Norman Sandbox [27] are popular tools among security professionals. Dynamic analyses can monitor a number of aspects including function calls and their arguments, information flow, and instruction traces. We refer the interested reader to [12] for a recent survey of this literature. The main drawbacks of dynamic solutions are that only a single program execution is observed, and that a malware might hide its behavior once it detects it is running in a contained environment.

Automatic Code Analysis. A few works have attempted to automatically explore multiple execution paths for a malware sample. In [18] Moser *et al.* present a system that can identify malicious actions carried out only when certain conditions along the execution paths are met. Their results show that on a set of 308 real-world malicious samples many of them show different behavior depending on the inputs from the environment. Brumley *et al.* [3,4] have designed similar systems aiming at identifying trigger-based behavior in malware. In particular, [3] discovers all commands in a few simple DDoS zombies and botnet programs.

These approaches employ mixed concrete and symbolic execution to explore multiple paths, starting the execution from the very beginning of the program. In this paper, we leverage symbolic execution to dissect a portion of a sample that is of interest for an analyst, provided they have sufficient knowledge to set up a minimal execution context for the exploration to start. Automatic systems suffer from known limitations that hinder the analysis of complex malware, such as the inherent cost of constraint solving and difficulties in handling self-modifying code and obfuscated control flow [20]. They are thus not generally used for real-scale malware analysis [30]. We believe manual symbolic execution as devised in this work can help get around these issues, as an analyst can provide the engine with insights to refine and guide the exploration (on a possibly limited scope compared to a whole-code automatic analysis) as part of a trial-and-error process.

In [30] Ugarte-Pedrero *et al.* show that by leveraging a set of domain-specific optimizations and heuristics multi-path exploration can be used to defeat complex packers. They also present an interesting case study on Armadillo, which is very popular among malware writers. Ad-hoc techniques and heuristics, including even simple ones as those we describe in this paper, can indeed be very effective in the malware domain.

Of a different flavor is the framework presented in [20]. X-Force is a binary analysis engine that can force a binary to execute requiring no input or proper environment. By trading precision for practicality, branch outcomes are forced in order to explore multiple paths inconsistently, while an exception recovery mechanism allocates memory and updates pointers accordingly. In one of the case studies the authors discuss the discovery of hidden malicious behaviors involving library calls.

Symbolic Execution. Symbolic execution techniques have been pioneered in the mid 1970s to test whether certain properties can be violated by a piece of software [15]. Symbolic techniques have been largely employed in software testing, with the goal of finding inputs that exercise certain execution paths or program points (e.g., [6,13]). A number of security applications have been discussed as well, e.g., in [25,26,28,29]. The reader can refer to previous literature (e.g., [1,23]) for a better understanding of the challenges that affect the efficiency of symbolic execution and when it might become impractical.

To the best of our knowledge, symbolic execution tools are not commonly employed yet by malware analysts. However, the 2013 DARPA announcement regarding the Cyber Grand Challenge competition has raised a lot of interest among security professionals. For instance, the 2016 Hex-Rays plugin contest for IDA Pro was won by Ponce [14], which provides support for taint analysis [23] and symbolic execution: Ponce allows the user to control conditions involving symbolic registers or memory locations, in order to steer the execution as desired. Symbolic execution is employed also in several open-source projects such as Triton [22] for binary analysis, reverse engineering, and software verification.

Obfuscation. Obfuscation techniques can be used also with the specific goal of thwarting symbolic execution. In particular, [24] discusses how to use cryptographic hash functions to make it hard to identify which values satisfy a branch condition in a malware, while [32] relies on unsolved mathematical conjectures to deceive an SMT solver. [34] addresses the limitations of symbolic execution in the face of three generic code obfuscations and describes possible mitigations.

Botnet Analysis. In this paper, we show how to derive the sequence of commands for a specific RAT. Many works have tackled the more general problem of automatic protocol reconstruction and message format reverse engineering (e.g., [5, 7, 10, 11]). We refer the interested reader to [19] for a survey of protocol reverse engineering techniques, and to [21] for a taxonomy of botnet research.

5   Conclusions

In this paper we have shown a successful application of symbolic execution techniques to malware analysis. A prototype tool we designed based on ANGR was able to automatically derive the list of commands supported by a well-known RAT and its communication protocol with the C&C server. To design our tool we had to overcome a number of complex issues. A primary requirement for symbolically executing a Windows binary is the availability of API models. Unfortunately, the current release of ANGR does not provide any Win32 API model, forcing us to develop them when needed for executing our RAT sample. An interesting research direction is how we can extend our tool for generating API stubs in order to minimize the implementation effort required to transform these stubs into working API models.

The most common issue when performing symbolic execution of a complex binary is the well-known *path explosion* problem. Indeed, a large number of paths could be generated during a program's execution, making the analysis hardly scalable. To mitigate this problem, we have implemented several domain-specific optimizations as well as a variety of *common-sense* heuristics. Although these tweaks may harm the efficacy of an automatic analysis by discarding potentially interesting paths, they can be easily disabled or tuned by a malware analyst whenever few useful reports are generated by our tool.

The case study presented in this paper has shown how communication protocols used by RATs could be potentially nontrivial. In the examined RAT sample, our tool was able to highlight that a specific handshaking phase is required to activate the majority of commands. While an analyst may spend hours trying to understand this protocol, our tool could reveal it without any manual intervention. However, our prototype still lacks support for mining the reports. Ideally, our tool should continuously evaluate the generated reports and provide the analyst with a clear summary of the findings, possibly highlighting and clustering reports based on common features. Visualizing the flow of concrete and symbolic data across API calls would provide valuable information to analysts as well.

While our prototype has been tested only on a single RAT sample, we believe our approach is rather general and replicable on other well-known RAT families.

We plan to address this topic in future work. One obstacle to a large-scale validation is that each sample may need a different setup, i.e., a different symbolic entry point and execution context. It remains an interesting open question how to minimize the amount of manual intervention required for malware analysts.

Another challenging issue that is likely to emerge when approaching other RAT samples is the use of strongly encrypted commands. Indeed, if a RAT resorts to a robust crypto function to decrypt the command, the constraint solver may be unable to provide a concrete model, i.e., break the encryption schema. In this scenario, our tool may fail to fully reconstruct the communication protocol of the RAT, but may still provide useful hints for the analyst. Although this may seem a critical limitation of our tool, we remark that, when performing a manual dissection, the analyst will face the same issue. Common crypto attacks, such as dictionary-based and brute-force attacks, could be integrated in our tool to attempt to defeat the encryption when the solver fails.

Acknowledgments. We are grateful to the anonymous CSCML 2017 referees for their many useful comments. This work is partially supported by a grant of the Italian Presidency of Ministry Council and by CINI Cybersecurity National Laboratory within the project "FilieraSicura: Securing the Supply Chain of Domestic Critical Infrastructures from Cyber Attacks" (www.filierasicura.it) funded by CISCO Systems Inc. and Leonardo SpA.

References

1. Baldoni, R., Coppa, E., D'Elia, D.C., Demetrescu, C., Finocchi, I.: A survey of symbolic execution techniques. CoRR, abs/1610.00502 (2016)
2. Ball, T., Bounimova, E., Cook, B., Levin, V., Lichtenberg, J., McGarvey, C., Ondrusek, B., Rajamani, S.K., Ustuner, A.: Thorough static analysis of device drivers. In: Proceedings of the 1st ACM SIGOPS/EuroSys European Conference on Computer Systems, EuroSys 2006, pp. 73–85. ACM, New York (2006)
3. Brumley, D., Hartwig, C., Kang, M.G., Liang, Z., Newsome, J., Poosankam, P., Song, D., Yin, H.: Bitscope: automatically dissecting malicious binaries. Technical report, CMU-CS-07-133 (2007)
4. Brumley, D., Hartwig, C., Liang, Z., Newsome, J., Song, D., Yin, H.: Automatically identifying trigger-based behavior in malware. In: Lee, W., Wang, C., Dagon, D. (eds.) Botnet Detection, pp. 65–88. Springer, Boston (2008)
5. Caballero, J., Yin, H., Liang, Z., Song, D.: Polyglot: automatic extraction of protocol message format using dynamic binary analysis. In: Proceedings of the 14th ACM Conference on Computer and Communications Security, CCS 2007, pp. 317–329. ACM, New York (2007)
6. Cadar, C., Dunbar, D., Engler, D.: KLEE: unassisted and automatic generation of high-coverage tests for complex systems programs. In: Proceedings of the 8th USENIX Conference on Operating Systems Design and Implementation, OSDI 2008, pp. 209–224. USENIX Association, Berkeley (2008)
7. Cho, C.Y., Shin, E.C.R., Song, D.: Inference and analysis of formal models of botnet command and control protocols. In: Proceedings of the 17th ACM Conference on Computer and Communications Security, CCS 2010, pp. 426–439. ACM, New York (2010)

8. Christodorescu, M., Jha, S.: Static analysis of executables to detect malicious patterns. In: Proceedings of the 12th Conference on USENIX Security Symposium, SSYM 2003, vol. 12. USENIX Association, Berkeley (2003)
9. Christodorescu, M., Jha, S., Seshia, S.A., Song, D., Bryant, R.E.: Semantics-aware malware detection. In: Proceedings of the 2005 IEEE Symposium on Security and Privacy, SP 2005, pp. 32–46. IEEE Computer Society, Washington, DC (2005)
10. Cui, W., Kannan, J., Wang, H.J.: Discoverer: automatic protocol reverse engineering from network traces. In: Proceedings of 16th USENIX Security Symposium on USENIX Security Symposium, SS 2007, pp. 14:1–14:14. USENIX Association, Berkeley (2007)
11. Cui, W., Peinado, M., Chen, K., Wang, J.H. and Irun-Briz, L.: Automatic reverse engineering of input formats. In: Proceedings of the 15th ACM Conference on Computer and Communications Security, CCS 2008, pp. 391–402. ACM, New York (2008)
12. Egele, M., Scholte, T., Kirda, E., Kruegel, C.: A survey on automated dynamic malware-analysis techniques and tools. ACM Comput. Surv. 44(2), 6:1–6:42 (2008)
13. Godefroid, P., Levin, M.Y., Molnar, D.A.: Automated whitebox fuzz testing. In: Proceedings of the Network and Distributed System Security Symposium, NDSS 2008 (2008)
14. Illera, A.G., Oca, F.: Introducing ponce: one-click symbolic execution. http://research.trust.salesforce.com/Introducing-Ponce-One-click-symbolic-execution/. Accessed Mar 2017
15. King, J.C.: Symbolic execution and program testing. Commun. ACM 19(7), 385–394 (1976)
16. Kindsight Security Labs: Malware report - Q2 2012 (2012). http://resources.alcatel-lucent.com/?cid=177650. Accessed Mar 2017
17. RSA Security LLC: Current state of cybercrime (2016). https://www.rsa.com/content/dam/rsa/PDF/2016/05/2016-current-state-of-cybercrime.pdf. Accessed Mar 2017
18. Moser, A., Kruegel, C., Kirda, E.: Exploring multiple execution paths for malware analysis. In: Proceedings of the 2007 IEEE Symposium on Security and Privacy, SP 2007, pp. 231–245 (2007)
19. Narayan, J., Shukla, S.K., Clancy, T.C.: A survey of automatic protocol reverse engineering tools. ACM Comput. Surv. 48(3), 40:1–40:26 (2015)
20. Peng, F., Deng, Z., Zhang, X., Xu, D., Lin, Z., Su, Z.: X-force: force-executing binary programs for security applications. In: Proceedings of the 23rd USENIX Conference on Security Symposium, SEC 2014, pp. 829–844. USENIX Association, Berkeley (2014)
21. Rodríguez-Gómez, R.A., Maciá-Fernández, G., García-Teodoro, P.: Survey and taxonomy of botnet research through life-cycle. ACM Comput. Surv. 45(4), 45:1–45:33 (2013)
22. Saudel, F., Salwan, J.: Triton: a dynamic symbolic execution framework. In: Symposium sur la sécurité des technologies de l'information et des communications, SSTIC, Rennes, France, pp. 31–54. SSTI, 3–5 June 2015
23. Schwartz, E.J., Avgerinos, T., Brumley, D.: All you ever wanted to know about dynamic taint analysis and forward symbolic execution (but might have been afraid to ask). In: Proceedings of the 2010 IEEE Symposium on Security and Privacy, SP 2010, pp. 317–331. IEEE Computer Society, Washington, DC (2010)
24. Sharif, M.I., Lanzi, A., Giffin, J.T., Lee, W.: Impeding malware analysis using conditional code obfuscation. In: Proceedings of the Network and Distributed System Security Symposium, NDSS 2008 (2008)

25. Shoshitaishvili, Y., Wang, R., Hauser, C., Kruegel, C., Vigna, G.: Firmalice - automatic detection of authentication bypass vulnerabilities in binary firmware. In: 22nd Annual Network and Distributed System Security Symposium, NDSS 2015 (2015)
26. Shoshitaishvili, Y., Wang, R., Salls, C., Stephens, N., Polino, M., Dutcher, A., Grosen, J., Feng, S., Hauser, C., Krügel, C., Vigna, G.: SOK: (state of) the art of war: offensive techniques in binary analysis. IEEE Symposium on Security and Privacy, SP 2016, pp. 138–157 (2016)
27. Norman Solutions: Norman sandbox analyzer. http://download01.norman.no/product_sheets/eng/SandBox_analyzer.pdf. Accessed Mar 2017
28. Song, D., et al.: BitBlaze: a new approach to computer security via binary analysis. In: Sekar, R., Pujari, A.K. (eds.) ICISS 2008. LNCS, vol. 5352, pp. 1–25. Springer, Heidelberg (2008). doi:10.1007/978-3-540-89862-7_1
29. Stephens, N., Grosen, J., Salls, C., Dutcher, A., Wang, R., Corbetta, J., Shoshitaishvili, Y., Kruegel, C., Vigna, G.: Driller: augmenting fuzzing through selective symbolic execution. In: 23nd Annual Network and Distributed System Security Symposium, NDSS 2016 (2016)
30. Ugarte-Pedrero, X., Balzarotti, D., Santos, I., Bringas, P.G.: RAMBO: run-time packer analysis with multiple branch observation. In: Caballero, J., Zurutuza, U., Rodríguez, R.J. (eds.) DIMVA 2016. LNCS, vol. 9721, pp. 186–206. Springer, Cham (2016). doi:10.1007/978-3-319-40667-1_10
31. Villeneuve, N., Sancho, D.: The "Lurid" downloader. Trend Micro Incorporated (2011). http://la.trendmicro.com/media/misc/lurid-downloader-enfal-report-en.pdf. Accessed Mar 2017
32. Wang, Z., Ming, J., Jia, C., Gao, D.: Linear obfuscation to combat symbolic execution. In: Atluri, V., Diaz, C. (eds.) ESORICS 2011. LNCS, vol. 6879, pp. 210–226. Springer, Heidelberg (2011). doi:10.1007/978-3-642-23822-2_12
33. Willems, C., Holz, T., Freiling, F.: Toward automated dynamic malware analysis using CWSandbox. IEEE Secur. Priv. 5(2), 32–39 (2007)
34. Yadegari, B., Debray, S.: Symbolic execution of obfuscated code. In: Proceedings of the 22nd ACM SIGSAC Conference on Computer and Communications Security, CCS 2015, pp. 732–744. ACM (2015)

Brief Announcement: A Consent Management Solution for Enterprises

Abigail Goldsteen, Shelly Garion$^{(\boxtimes)}$, Sima Nadler, Natalia Razinkov, Yosef Moatti, and Paula Ta-Shma

IBM Research Haifa, Haifa University Campus, Mount Carmel, 3498825 Haifa, Israel
{abigailt,shelly,sima,natali,moatti,paula}@il.ibm.com

Abstract. Technologies such as cloud, mobile and the Internet of Things (IoT) are resulting in the collection of more and more personal data. While this sensitive data can be a gold mine for enterprises, it can also constitute a major risk for them. Legislation and privacy norms are becoming stricter when it comes to collecting and processing personal data, requiring the informed consent of individuals to process their data for specific purposes. However, IT solutions that can address these privacy issues are still lacking. We briefly outline our solution and its main component called "Consent Manager", for the management, automatic enforcement and auditing of user consent. We then describe how the Consent Manager was adopted as part of the European FP7 project COSMOS.

1 Introduction

Enterprises today are collecting personal data, and even if they have the best of intentions, they are challenged to abide by privacy laws, norms and policies, such as the US Health Insurance Portability and Accountability Act (HIPAA)[1] or the European Union General Data Protection Regulation (GDPR)[2], which are designed to protect all personal data collected about EU residents. Current IT systems are lacking the infrastructure support to automate the process of personal data collection. There is no standardized way to share privacy options and their implications with end users. Consent, once obtained, is not correlated with the data collected. When data is accessed, there is no way to know for what purpose it is being accessed nor whether the data subject agreed to using it for the given purpose. Privacy audits are, for the most part, manual checklists, and are very subjective and error prone.

Our main contribution is a comprehensive and relatively easy to implement solution for the automatic enforcement and auditing of user consent preferences, including for legacy applications and systems that cannot easily be changed. This gives both users and enterprises more control and confidence in what is actually being done with personal data within the system; it also gives the enterprise

---

[1] www.hhs.gov/hipaa/.
[2] www.eugdpr.org/.

© Springer International Publishing AG 2017
S. Dolev and S. Lodha (Eds.): CSCML 2017, LNCS 10332, pp. 189–192, 2017.
DOI: 10.1007/978-3-319-60080-2_13

the ability to better comply (and prove compliance) with data protection regulations. The proposed solution provides tools for modeling consent, a repository for storing it, and a data access management component to enforce consent and log the enforcement decisions.

The main component of our solution is the *Consent Manager*. It is responsible for the collection, storage, and maintenance of user consent. For each application the organization wishes to provide to its users, the relevant consent parameters must first be defined. The Consent Manager uses the concept of *consent templates* to define the parameters of consent for a specific purpose for which the data is used. The consent template contains general information about the purpose, such as its name, description, legal terms, etc. It also contains a list of data items collected and/or processed for that purpose that are subject to consent. Once such a template is in place, it can be used to request specific users' consent to supply their personal data for that purpose. Figure 1 shows an example of a possible consent template and in Fig. 2 we show one example of how to collect consent from users that doesn't confuse them with too many options but provides information about what data is collected in each the option.

The next pivotal piece of the solution is the *Data Access Manager*, which is responsible for controlling access to the data and enforcing consent. This is the component that enforces the decisions made by the Consent Manager and actually controls the release of data by filtering/masking/blocking the response to the requesting application; this is done based on the purpose of the request and the allowed usage of the requested data. There are several options for the location of the Data Access Manager in the enterprise architecture: either at the application level, which is easy to implement but less secure; or at the data storage level which is more secure, but must be implemented for each different type of data store; or as part of a separate security suite which requires non-trivial integration.

Purpose information

Properties

Validity (days)	354
Privacy officer ID	3111
Location City	New York
Location State	NY
Location Country	USA

Purpose data

ID	Data	Data type	Description	Sensitivity type	Mandatory	Duration	Anonymization method	Dest. Organization
	Data	Data type		Sensitivity type	Mandatory	Duration	Anonymization method	
	First Name	text		PII	Yes	365	none	
	Last Name	text		PII	No	365	none	
	Phone number	text		PII	Yes	365	none	
	Email	text		PII	No	365	none	
	Birth Year	date		PII	No	365	none	
	Gender	text		PII	Yes	365	none	
	User ID	text		PII	No	365	none	
	Activity Type ID	text		PII	No	365	none	

Fig. 1. Possible consent template UI

Fig. 2. Possible UI for collecting consent from end users

2   Consent Management in the COSMOS Project

COSMOS (Cultivate resilient smart Objects for Sustainable city applicatiOnS)[3] is a European FP7 project whose goal was to enhance the sustainability of smart city applications by enabling IoT "things" to evolve and act in a more autonomous way, becoming smarter and more reliable by learning from others' experience. In the COSMOS smart heating application, data is collected from various IoT devices (such as window/door activity sensors, temperature and humidity sensors) installed in public housing in Camden, London. This helps residents manage their heating schedule efficiently and control their consumption, taking into consideration various factors such as their targeted heating budget, comfort level and health, to reduce energy waste and identify operational issues. A flat can learn from its own experience or from the experience of the other flats.

However, sharing smart energy data leads to certain privacy risks. As an illustration, [1] describes one household's electricity demand profile over a 24-hour period, where one can clearly observe the usage patterns of common electricity devices such as a refrigerator, washing machine, oven or kettle. From such a pattern one can learn the daily habits, occupancy level and even the employment status and religion of the household residents. Naturally, it is extremely important to enhance security and privacy for such IoT applications. In particular, the residents' consent should be obtained before analyzing their data.

Therefore, we integrated consent management into the COSMOS smart heating application. We implemented enforcement for a SQL-based data access layer based on OpenStack Swift[4] object store and Apache Spark[5] analytics engine. However, to simplify the implementation, the Data Access Manager was implemented as part of the data access application and not at the storage level. Two purposes were defined in the Consent Manager for COSMOS. The first, is that the resident's personal energy data is used to provide him personal recommendations; and the second is that the data is shared with others to provide crowd-based recommendations. Each resident can choose from one of these options or opt-out from both.

The current design of the Consent Manager is a result of several refinements we did while improving the APIs and the scale and performance of the solution. We observed several shortcomings that we have already improved upon or plan to improve in future versions of the Consent Manager. The first issue was that the Consent Manager did not handle batch queries. As a result we added to the Consent Manager the option to request approval for access to multiple data items and multiple data subjects in a single request. Another change we plan to make resulted from a customer requirement about how to deal with changes in a purpose definition. For example, if the terms and conditions change or if a new type of data needs to be processed. This question was also raised in the

---

[3] http://iot-cosmos.eu/.

[4] http://docs.openstack.org/developer/swift/.

[5] http://spark.apache.org/.

COSMOS project. As a result we decided to add a new feature for versioning of purposes, with the option to determine which changes require gathering new consent from the end-users.

Acknowledgments. The research leading to these results was supported by the EU FP7 project COSMOS under grant ICT-609043.

Reference

1. Quinn, E.L.: Privacy and the New Energy Infrastructure, Social Science Research Network (SSRN), February 2009

Brief Announcement: Privacy Preserving Mining of Distributed Data Using a Trusted and Partitioned Third Party

Nir Maoz and Ehud Gudes[✉]

Department of Mathematics and Computer Science, The Open University,
1 University Road, 43537 Ra'anana, Israel
ntaizi@yahoo.com

1 Introduction

We like to discuss the usability of new architecture of partitioned third party, offered in [1] for conducting a new protocols for data mining algorithms over shared data base between multiple data holders. Current solution for data mining over partitioned data base are: Data anonimization [4], homomorphic encryption [5], trusted third party [2] or secure multiparty computation algorithms [3]. Current solutions suffer from different problems such as expensive algorithms in terms of computation overhead and required communication rounds, revealing private information to third party. The new architecture offered by Sherman et al. allow the data holders to use simple masking techniques that are not expensive in computation nor assume trust in the third party, yet allow to perform simple and complex data mining algorithms between multiple data owners while private data is not revealed. That come with the assumption of no collude between the two parts of the PTTP. In the PTTP architecture offered by Sherman et al. [1] the trusted third party is divided into two parts CE the Computer Engine which does the data mining and mathematical calculation on behalf of the participants and to the Randomizer R which generates random numbers and permutations needed for the protocol, and share them securely with the participants. All communication between data base holders and the CE or R is assumed to be private e.g. using symmetric encryption with private key shared between the two sides broadcasting each other. In this paper, we show one basic data mining algorithms for calculating union/intersection, to show the power of this architecture. We developed few more basic and complex algorithms, for calculating aggregation functions, Min/Max and association rules. Although, some of these operations like union/intersection were discussed in [1], we developed different and simpler protocols than those suggested there.

2 Intersection/Union

In this section we describe a PTTP protocol for computing intersection and union of private sets. The set is separated horizontally between the different

S. Dolev and S. Lodha (Eds.): CSCML 2017, LNCS 10332, pp. 193–195, 2017.
DOI: 10.1007/978-3-319-60080-2_14

databases holders. In the setting that we consider here, each of the private databases includes a column in which the entries are taken from some groundset Ω. For example, if the column includes the age of the data subject, in whole years, then we may take $\Omega = \{0, 1, \ldots, 120\}$. Our protocol enables the different database holders to compute the union or intersection of their columns; The protocol uses a PTTP. The main idea of the protocol, is to create for each DB_i two Boolean vectors. Both are of the length of the groundset. All DB_i will sort the groundset in the same order (e.g. alphabetic order). Each DB_i will create one vector that have 1 for each index in the vector that represent the position of the item it has in its DB and 0 for the rest of the indexes. Then each DB creates another vector which has a 0 for each index in the vector that represent the position of the item it has in its DB and 1 for the rest of the indexes (Exactly the inverse vector). Than it will concatenate both vectors, and apply a permutation, shared by all participants, on the result vector. After applying the permutation, no one can know, without knowing the permutation, which item a specified DB holds or not even how many items there are in each DB. The reason is, that in the way we build the concatenated vector, there is exactly the same number of 1's and 0's (Ω times) which prevent the knowledge of how many items each DB holds. And since the vector is permuted, there is no way to know which 1's belong to the first half of the vectors (index of item the participant hold) and which 1's belong to the second half (index of item the participants doesn't hold). The parties in our protocol are as follows:

- Q is the querier who issues the query to be answered.
- D_i, $1 \leq i \leq M$, are the databases.
- CE (Computation Engine) and R (Randomizer) are the two parts of the PTTP.

In the protocol, $V = V_\Omega$ is a vector that includes all values in the groundset Ω. Protocol 1 shows the algorithm in detail.

3 Conclusions

We show in this paper how one can use the new PTTP architecture proposed by Sherman et al. to create simple algorithms like union/intersection over distributed database, without the need for strong cryptography techniques or the use of hash functions. Using PTTP also results in less communication rounds and in some cases also reduces the size of the messages. In most cases the new algorithms can also protect against malicious coalitions. That said, more research is still needed in order to shift more privacy preserving responsibilities from the two parts of the the PTTP back to the database holders, so that even coalition which involve both parts of the PTTP won't allow to reveal significant information. Finally, more research is also needed to show the utility of the PTTP architecture for other data mining tasks such as: clustering or decision trees construction.

Protocol 1. A PTTP protocol for Computing Set Operations.

1: Q sends the query to DB_i, $1 \leq i \leq M$, and the query type (either intersection or union) to R.
2: Q sends R which private column participate in the query (e.g. the age column).
3: R generates a random permutation σ on the set of integers $\{1, \ldots, 2|\Omega|\}$ and sends it to DB_i, $1 \leq i \leq M$.
4: **for all** $1 \leq i \leq M$ **do**
5:    DB_i sets a Boolean vector $V_i = \{v_{i,1}, \ldots, v_{i,|\Omega|}\}$ where

$$v_{i,j} = \begin{cases} 1 \ if\, DB_i \ hold \ private \ value \ j \\ 0 \ \text{otherwise} \end{cases},$$

6:    DB_i sets a Boolean vector $\neg V_i = \{\neg v_{i,1}, \ldots, \neg v_{i,|\Omega|}\}$ where

$$\neg v_{i,j} = \begin{cases} 0 \ if\, DB_i \ hold \ private \ value \ j \\ 1 \ \text{otherwise} \end{cases},$$

7:    DB_i sets a new vector that is the concatenation of the two previous mentioned vectors, $CV = V_i \parallel \neg V_i$
8:    DB_i calculate $PV_i = \sigma(CV) = \sigma(V_i \parallel \neg V_i) = \sigma(\{v_{i,1}, \ldots, v_{i,|\Omega|}, \neg v_{i,1}, \ldots, \neg v_{i,|\Omega|}\})$
9:    DB_i sends its vector to CE.
10: CE computes the intersection or union of all vectors received from the M databases and send the result vector RV to R.
11: R calculate the final vector $FV = \sigma^{-1}(RV)$
12: R throws away the second half of the vector and output it.

Acknowledgment. The authors would like to thank Tassa Tamir, for providing very helpful comments on the algorithms presented here.

References

1. Chow, S.S.M., Lee, J.-H., Subramanian, L.: Two-party computation model for privacy-preserving queries over distributed databases. In: NDSS 2009
2. Ghosh, J., Reiter, J.P., Karr, A.F.: Secure computation with horizontally partitioned data using adaptive regression splines. Comput. Stat. Data Anal. **51**(12), 5813–5820 (2007)
3. Tassa, T.: Secure mining of association rules in horizontally distributed databases. IEEE Trans. Knowl. Data Eng. **26**(4), 970–983 (2014)
4. Tassa, T., Gudes, E.: Secure distributed computation of anonymized views of shared databases. ACM Trans. Database Syst. (TODS) **37**(2), 11 (2012)
5. Zhong, S.: Privacy-preserving algorithms for distributed mining of frequent itemsets. Inf. Sci. **177**(2), 490–503 (2007)

Brief Announcement: A Technique for Software Robustness Analysis in Systems Exposed to Transient Faults and Attacks

Sergey Frenkel[(✉)] and Victor Zakharov

Federal Research Center "Computer Science and Control"
Russian Academy of Sciences, Moscow, Russia
fsergei51@gmail.com, VZakharov@ipiran.ru

1 Introduction

At present, the problem of accounting of possible failures effect that can occur in the program memory area both due to some physical effects and some malicious attacks, and can distort values of variables, operations, the codes, etc., is solved by applying the widely used Fault Injection (FI) simulation technique. Main drawback of the FI is necessity to have different expensive software that can not be used to solve other design problems, in particular verification and testing. In this paper we present an approach to estimate the robustness to faults caused externally during a program execution, which in contrast to FI can be implemented by well-known design tools for testing and verification, which are mandatory in the CS designing. It can decrease essentially the total cost of the development. We consider a fault manifestation model which is based on a product of two Markov chains corresponding to FSMs (Finite State Machine) modeling of a program. One of them operates in normal conditions, while another FSM operates with a momentary failure occurs (e.g., within the execution time of one operation). This model has previously been proposed for probabilistic verification of hardware systems robustness [1, 2].

2 Model of Program Under Transient Faults and Attacks

The application program model considered in this paper is the FSM of Mealy type, corresponding to the algorithm implemented by this program. The failure of a program is a computational errors caused by random changes in any of the bits of a certain word from the binary image of the program, which can occur due to various causes. We assume that such a change leads to erroneous modification in a state transition of the FSM representation of the program, but does not lead to the appearance of new states.

Let $\{M_t, t \geq 0\}$ be a Markov chain (MC) describing the target behavior of target fault-free FSM with n states under random input, that is, functioning without any effect on transient faults, and $\{F_t, t \geq 0\}$ is the MC based on the same FSM but exposed to some transient fault. Let $Z_t = \{(M_t, F_t, t \geq 0\}$ corresponding to the behavior of the

S. Dolev and S. Lodha (Eds.): CSCML 2017, LNCS 10332, pp. 196–199, 2017.
DOI: 10.1007/978-3-319-60080-2_15

MCs pairs that are two-dimensional MC with space $S^2 = S \times S$ of pairs $(a_i, a_j) \in S$. The size of the MC state space will be $n(n-1) + 2$. The matrix of transition probabilities of these MCs are calculated from the given FSM transitions table and the probabilities of Boolean input binary variables of the FSM as well. Along with the states, MC Z_t has two absorbing states A_0 and A_1, where A_0 is the event "the output variables are not distorted before the moment when the FSM's trajectory will be restored", and A_1 means "malfunctioning has already manifested itself in the output signal". The pairs of (a_i, a_j) states enable representation of any transient faults as "the FSM instead of the state a_i, in which it should be on this clock after the transition at the previous time cycle, as a result of the malfunction was in the state a_j".

We characterize the program *robustness* as the probability of an event where the trajectories (states, transitions and outputs) of M_t and F_t will be coincided after t time slots (e.g., the clocks) after the termination of an external factor causing a transient fault, before than outputs of both FSMs (underlying these MCs) become mismatched.

This probability that the FSM returns to correct functioning after this number t of time slots can be computed by Chapman-Kolmogorov equation

$$\vec{p}(t) = \vec{p}(t-1)P^* = \vec{p}(0)(P^*)^t, \tag{1}$$

where the initial distribution $\vec{p}(0)$ is determined by the initial states of the fault-free and faulty automata, and P^* is the transition matrix of this two-dimensional Markov chain. If the valid automaton at the initial moment 0 is in the state i_0, but the faulty state (say, due to an attack effect) is in the state $j_0 \neq i_0$, then $p_{i_0 j_0}(0) = 1$, and the remaining coordinates of the vector $\vec{p}(0)$ are zero.

The components of the vector $\vec{p}(t)$ are the probabilities $p_0(t)$, $p_1(t)$ of getting into the absorbing state A_0 and A_1 mentioned above, and the probability of transitions to the rest (transient) states of the MC. The sum equals to $1 - p_0(t) - p_1$ A detailed description and mathematical analysis of the model is carried out in [1]. In dependence of the absorbing states definition, we may consider two modifications of the model above, which can be chosen depending on design goal.

The model 1. The MC gets into the absorbing state A_1 in all cases of the mismatch between the outputs of the fault-free and faulty FSM.

The model 2. The occurrence of the Markov chain in the absorbing state A_1 only in the case of a mismatch between the outputs of the fault-free and failing FSM when the failing FSM hits a state that coincides with the state of the faulty FSM.

3   Estimation of Robustness to Malicious Attacks

We can adapt the above models to estimate the likelihood of incorrect program behavior due to malicious attacks, where the system calls trace corresponding to the program execution is modeled the FSM. We consider that every event in system finally results in state transition in the mentioned above FSM. Due to this injected fault, an erroneous result may be observed in the output. Thereby the attack may lead to the states, in which fault-free and faulty FSMs will have mismatched outputs, like in the models discussed earlier. We can describe this behavior by the product of two normal and faulty ("malicious") FSMs, and corresponding two-dimensional Markov chain as well as mentioned in the models earlier.

If we consider binary outputs Y and states A, then the error can be expressed as an XOR operation ($\tilde{Y} = Y \oplus e$), states $A = A + e$, e is a given Boolean vector, describing the bits corruption. Thereby, if we consider a malware with obfuscated codes, Model 2 could be more relevant, as the states coinciding can be a result of the obfuscation. The system calls may be considered as input sequence of the FSM and the probabilities of the different transitions may be based on the frequency of certain calls during a normal (fault-free) execution.

Let us consider a utility program which should remove the spaces from words flow, such, that (i) leading spaces are suppressed, (ii) if the word ends with spaces, then they are also deleted, (iii) within a word there can not be more than one space, extra spaces are deleted (in more details, the program is described in [3]).

Let the programmer have a functional test set with a suitable coverage. Then the model can be represented (at algorithmic level) by the FSM (Fig. 1), where $(x_1, .. x_6)$ are the inputs, $(y_1, ... y_{15})$ are outputs, and a_i, a_j are the previous and next states correspondingly. The inputs are the FSM transition conditions, and the outputs correspond to the program's operations.

a_1	a_{13}	1	y_{16}
a_2	a_{11}	1	y_{13}
a_3	a_9	1	y_{12}
a_4	a_5	x_1	y_3
a_4	a_3	$\sim x_1$	y_2
a_5	a_2	x_2	y_1
a_5	a_4	$\sim x_2$	y_4
a_6	a_7	1	y_{11}
a_7	a_8	x_3	y_{10}
a_8	a_4	$\sim x_3$	y_5
a_8	a_1	x_1	y_8
a_8	a_1	$\sim x_1$	y_9
a_9	a_6	1	y_6
a_{10}	a_7	x_4	y_7
a_{10}	a_9	$\sim x_4$	y_{12}
a_{11}	a_{10}	x_5	y_{12}
a_{11}	a_8	$\sim x_5$	y_{10}
a_{12}	a_2	x_6	y_{14}
a_{12}	a_{12}	$\sim x_6$	y_{15}
a_{13}	a_{12}	1	y_{15}

Fig. 1. Mealy automaton as a model of the program.

For example, the state a_4 is a state where a space is found in a word, followed by a non-blank space. If this occurs at the beginning of a word, then this space is not written to the word being formed, otherwise it is written.

To construct the Markov chain Z_t, the probability of the Boolean unit for the inputs $(x_1, ..., x_6)$ must be assigned, e.g., if the input corresponds to the condition the input "word has non-zero length"), then Prob $(x_6 = 1)$ is evaluated as N_{emp}/N_T, where N_{emp} is the number of empty words in the functional test, N_T is the total number of words in the functional test set. For the considered example, the total number of states $\{(i, j)\}$ of the MC Z_t, where i, j are the indices of the states a_i, a_j of the fault-free FSM and the FSM subjected to failure (the "failed" FSM), is 158. The states are ordered as follows: $\{A_0$, (1, 2), (1, 3), (1, 4), (1, 5), (1, 6), (1, 7), (1, 8), (1, 9),(1, 10), (1, 11), (1, 12), (1, 13), (2, 1), ... $A_1\}$.

Let the programmer evaluate the probability that the number of program cycles before the program's self-healing after the termination of a failure (attack) does not exceed a certain value. The Model 2 is more adequate than Model 1 in this case, as it considers self-healing only in the case when all states of both FSM are coincided.

Consider failure (10, 8) "the FSM instead of the state a_{10} was in the state a_8", and y_{10} executed instead of y_{12}.

It may be due to the fact that the formation of the resulting string of words (without any spaces except for one space between the words) is completed before the exhaustion of the characters in the input string. For example, if we have estimated some functional test set where the probabilities of Boolean 1 for the $(x_1, x_2, .. x_6)$ are: 0.2, 0.7, 0.2, 0.7, 0.01, 0.2, then Chapmen-Kolmogorov equation for the MC, corresponding to FSM of Fig. 1 gives that the probability of the correct functioning will be rather high (more than 0.9) at the 18-th clock after the failure.

Acknowledgements. This work was supported by RFBR grants no. 15-07-05316 and 16-07-01028.

References

1. Frenkel, S., Pechinkin, A.: Estimation of self-healing time for digital systems under transient faults. Inf. Appl. (Informatika i ee Primeneniya) **4**(3), 2–8 (2010)
2. Frenkel, S.: Some measures of self-repairing ability for fault-tolerant circuits design. In: Second Workshop MEDIAN 2013, Avignon, France, pp. 57–60, May 30–31, 2013
3. Frenkel, S., et al.: Technical report (2017). http://www.ipiran.ru/publications/Tech_report.pdf

Symmetric-Key Broadcast Encryption:
The Multi-sender Case

Cody Freitag[1], Jonathan Katz[2(✉)], and Nathan Klein[3]

[1] University of Texas, Austin, TX, USA
cody.freitag@utexas.edu
[2] University of Maryland, College Park, MD, USA
jkatz@cs.umd.edu
[3] Oberlin College, Oberlin, OH, USA
nklein@oberlin.edu

Abstract. The problem of (stateless, symmetric-key) *broadcast encryption*, in which a central authority distributes keys to a set of receivers and can then send encrypted content that can be decrypted only by a designated subset of those receivers, has received a significant amount of attention. Here, we consider a generalization of this problem in which *all* members of the group must have the ability to act as both sender and receiver. The parameters of interest are the number of keys stored per user and the bandwidth required per transmission, as a function of the total number of users n and the number of excluded/revoked users r.

As our main result, we show a multi-sender scheme allowing revocation of an arbitrary number of users in which users store $O(n)$ keys and the bandwidth is $O(r)$. We prove a matching lower bound on the storage, showing that for schemes that support revocation of an arbitrary number of users $\Omega(n)$ keys are necessary for *unique predecessor* schemes, a class of schemes capturing most known constructions in the single-sender case. Previous work has shown that $\Omega(r)$ bandwidth is needed when the number of keys per user is polynomial, even in the single-sender case; thus, our scheme is optimal in both storage and bandwidth.

We also show a scheme with storage polylog(n) and bandwidth $O(r)$ that can be used to revoke any set of polylog(n) users.

1 Introduction

In the classical setting of *broadcast encryption* [16], there is a group of n users to which a sender periodically transmits encrypted data. At times, the sender requires that only some designated subset S of the users should be able to decrypt the transmission and recover the original plaintext; the remaining users R—who should be unable to learn anything about the underlying plaintext, even if they all collude—are said to be *revoked* from that transmission. We are interested here in *symmetric-key* schemes that use no public-key operations, and which are also *stateless*, i.e., in which the keying material stored by each user remains fixed even as different subsets of users are revoked. This problem is motivated

© Springer International Publishing AG 2017
S. Dolev and S. Lodha (Eds.): CSCML 2017, LNCS 10332, pp. 200–214, 2017.
DOI: 10.1007/978-3-319-60080-2_16

by applications to secure content distribution, but has applications to secure multicast communication in distributed systems more generally.

To the best of our knowledge, all previous considerations of broadcast encryption explicitly consider the case in which there is one, designated sender, and each of the n users acts only as a (potential) receiver. (A case that *has* been considered previously is the "point-to-point" setting in which each user should be able to communicate securely with every other user. See further discussion in Sect. 1.1.) But in the setting of multicast communication it makes sense to assume that each of the n users might need to communicate with any subset of the others; that is, each of the n users might sometimes act as a sender and sometimes as a receiver. We refer to this as the *multi-sender* setting. Multi-sender broadcast encryption is applicable in military settings or ad-hoc networks, or whenever there is some group of users all of whom wish to jointly communicate, yet from time-to-time some users' devices are compromised and so those users must be revoked. Or, users in the group may each have different access privileges, and so the set of revoked users for any particular transmission (being made by any one of the n users) may vary depending on the context. We initiate the study of multi-sender broadcast encryption in this paper.

As in the single-sender case, the main parameters of interest are the storage per user and the bandwidth overhead per transmission, as a function of the total number of parties n and the number of revoked users r. There is a trivial solution in which each user shares a key with every other user, and uses the appropriate keys to encrypt to any desired subset. This solution requires each user to store $n - 1$ keys and has bandwidth $n - r - 1$. The natural questions are whether it is possible to achieve storage and/or bandwidth sublinear in n. (We remark that traditionally $r \ll n$ is considered the interesting case, as it is assumed that the number of revoked users will be small in normal operation of the scheme.)

As our main results, we show:

- There is a multi-sender scheme supporting revocation of arbitrarily many users, in which each user stores $O(n)$ keys and the bandwidth is $O(r)$. Moreover, we prove a lower bound (when revocation of arbitrarily many users must be supported) showing that $\Omega(n)$ storage is necessary, regardless of the bandwidth, for *unique predecessor* schemes [2], a class capturing all state-of-the-art constructions in the single-sender setting [19,20,29].
  Austrin and Kreitz [2] have previously shown that the bandwidth must be $\Omega(r)$, even in the single-sender case, when polynomially many keys are used; thus, our scheme is asymptotically optimal in both storage and bandwidth.
- There is a multi-sender scheme that can support revocation of any set of $r \leq \text{polylog}(n)$ users, having storage $\text{polylog}(n)$ and bandwidth $O(r)$.

We refer to Sect. 1.2 for a more complete discussion of our results.

1.1   Prior Work

As noted earlier, to the best of our knowledge all prior work treating symmetric-key broadcast encryption focuses only on the case of a single sender.

(Nevertheless, as we discuss below, some prior work is applicable to the multi-sender setting.) We briefly survey this work here, without intending to be exhaustive.

We remark that in some formulations of broadcast encryption, security is defined to hold with respect to all coalitions $R \subseteq [n] \backslash S$ containing at most r' users, for some bound r', rather than with respect to $R \stackrel{\text{def}}{=} [n] \backslash S$ as here. The former offers reduced security, but (potentially) allows for security/efficiency tradeoffs depending on the assumed number of colluding users. For simplicity, and following recent work in this area (e.g., [18–20,29]), we assume $R = [n] \backslash S$ in the discussion below and throughout the paper.

Single-sender broadcast encryption. The work of Blundo et al. [7], which extends the work of Blom [6], can be used to construct a scheme in which a group of n users is given keying material that allows any subset S' of size t to compute a shared key that is information-theoretically hidden from the $r = n - t$ other users. Their work implies a multi-sender broadcast encryption scheme with bandwidth 1: user i can transmit to a set of $n - r - 1$ other users S by encrypting with the key shared by users in $S' = S \cup \{i\}$. Unfortunately, the storage per user in their scheme is $\binom{n-1}{n-r-1}$, which they prove is optimal for their setting. Blundo et al. [8] also consider a more careful application of these ideas to the problem of broadcast encryption, trading off higher bandwidth for lower storage. For most interesting settings of the parameters, however, this work is subsumed by the schemes below.

Fiat and Naor [16] introduce the term "broadcast encryption," and show a scheme that simultaneously has storage $O(r_{\max} \log r_{\max} \log n)$ and bandwidth $O(r_{\max}^2 \log^2 r_{\max} \log n)$, where $r_{\max}$ denotes a pre-determined upper bound on the size of r. Further improvements were given by [1,17,18,24,25]. In all these schemes both the storage and the bandwidth depend on $r_{\max}$, so either $r_{\max}$ must be small or the parameters of the scheme are high even if only few users are actually revoked.[1]

Most recent work has focused on schemes that directly have the flexibility to communicate with arbitrary subsets of users while revoking all others. The following general approach can be used for constructing such schemes: Fix a set of keys K held by the sender. Each user i is given some subset $K_i \subset K$ of these keys. For the sender to securely send a message to a group S, it suffices if there is a set $K_S \subset K$ of keys such that (1) each user in S knows at least one key in K_S, and (2) no user in the revoked set $R = [n] \backslash S$ knows any of the keys in K_S. This implies a solution with bandwidth $|K_S|$ in which the sender encrypts the content independently using each of the keys in K_S, and each intended receiver decrypts the appropriate ciphertext using a key they know. Following [26], we refer to this as the *OR approach*. Naor, Naor, and Lotspiech [29] propose the

---

[1] Another possibility is to run $\log r_{\max}$ independent copies of the scheme using powers of two for the maximum size of the revoked set. This allows the bandwidth to depend on the actual number of revoked users r, though increases storage by a factor of $\log r_{\max}$.

complete subtree (CS) scheme that uses this approach, and has storage $\log n$ and bandwidth $r \log n/r$.

All the schemes described so far have information-theoretic security. One can use *key derivation* to reduce the per-user storage, at the expense of achieving only computational security. When using key derivation, roughly speaking, users need not explicitly store all the keys they have access to; instead, they may derive one key from another, or derive multiple keys from a single predecessor, using a hash function (possibly modeled as a random oracle), a pseudorandom generator, or a pseudorandom function. Naor, Naor, and Lotspiech [29] present the *subset difference (SD) scheme* that uses the OR approach and key derivation, and achieves storage $O(\log^2 n)$ and bandwidth $O(r)$. This was improved in subsequent work [4,5,20], culminating in the *SSD scheme* of Goodrich et al. [19] that achieves storage $O(\log n)$ and bandwidth $O(r)$, though at the expense of requiring computation linear in n. (Hwang et al. [21] show how to improve the computation to $O(\log n)$ at the expense of a small increase in bandwidth.) Jho et al. [22] show a scheme with storage $O(c^p)$ and bandwidth $O(\frac{r}{p} + \frac{n-r}{c})$, where c, p are parameters; the scheme fares best (and beats [20,29] in terms of both storage and bandwidth) when r is a large constant fraction of n. This scheme was further improved in [21], but even in that case either the bandwidth is $\Omega(\sqrt{n})$ when $r = O(1)$ or else the storage is $\Omega(\sqrt{n})$. Other relevant work in the single-sender case includes [12,31].

Lower bounds for broadcast encryption schemes following the OR approach have been studied in both the information-theoretic setting [18,26,29] and when key derivation is used [2].

Secure point-to-point communication. Motivated by achieving secure point-to-point communication, Dyer et al. [14] (see also [28]) consider the setting in which each user i holds a subset $K_i \subset K$ of keys, and each pair of users i, j has a set of keys $K_{i,j} = K_i \cap K_j$ in common that are not *all* known to any set of at most $r_{\max}$ other users. Although Dyer et al. do not explicitly treat the case of multi-sender broadcast encryption, we observe that their scheme can be used to solve that problem if the number of revoked users is bounded: Consider a user i who wishes to securely transmit a message to some set S of users. Let $R = [n] \backslash (S \cup \{i\})$, where $|R| \leq r_{\max}$. Then i and each user in S must share at least one key not known to any user in R; user i can encrypt its content using all such keys. This results in a multi-sender broadcast encryption scheme with[2] storage and bandwidth $O(r_{\max}^2 \log n)$.

Other related work. The case of *stateful* broadcast encryption has also received extensive attention (for the single-sender case), in terms of both constructions [3,10,11,30,32] and lower bounds [11,27]. Here, some set of authorized users S is continually maintained by the sender; the authorized users always

---

[2] These parameters are not stated explicitly by Dyer et al., who report only the total number of keys. However, Corollary 1 and the proof of Theorem 4 in their paper show that the per-user storage is $O(r_{\max}^2 \log n)$; the bandwidth is bounded by the number of keys held by any user acting as a sender.

Table 1. Constructions of multi-sender broadcast encryption schemes. Scheme 1 is described in Appendix A.

Security	Storage	Bandwidth	Scheme
Info. theoretic	$O(r_{\max}^2 \log n)$	$O(r_{\max}^2 \log n)$	Follows from prior work [14]
Info. theoretic	$n \log n$	$r \log \frac{n}{r}$	Result 1 applied to CS scheme [29]
Info. theoretic	$O(n^{1+1/k})$	$O(kr)$	Result 1 applied to Scheme 1
Info. theoretic	$O(r_{\max}^4 \, n^{1/2} \log n)$	$2r_{\max}$	Result 3 applied to [18]
Computational	$O(n)$	$O(r)$	Result 2 applied to Scheme 1
Computational	$O(r_{\max}^2 \log^2 n)$	$O(r)$	Result 3 applied to SSD scheme [19]

share a single key under which the sender encrypts its communication. From time to time, the sender *revokes* a user i, thus changing the set of authorized users. When this happens, the sender transmits rekeying information that allows all users in $S \backslash \{i\}$ to both compute a new, shared key as well as to update their individual keying material.

Broadcast encryption has also been studied in the *public-key* setting [9,13]. Of course, there the single-sender and multi-sender cases are equivalent. Public-key schemes inherently require stronger assumptions than symmetric-key schemes, and generally incur higher computational costs.

1.2   Our Results

We show general transformations from single-sender broadcast encryption (BE) schemes to multi-sender ones. Fix a single-sender BE scheme Π with storage s, bandwidth b, and where the sender stores s^* keys.[3] We show:

Result 1: There is a multi-sender BE scheme with storage $(n-1) \cdot s + s^*$ and bandwidth b that supports the same number of revoked users as Π does; if Π is information-theoretic then so is the derived scheme.

Result 2: If Π is information-theoretic, there is a multi-sender BE scheme with storage s^* and bandwidth b that supports the same number of revoked users as Π does. The scheme uses key derivation, and so is no longer information-theoretic.

Result 3: For any bound $r_{\max}$ on the number of revoked users, there is a multi-sender BE scheme with storage $O((s \cdot r_{\max}^2 + s^* \cdot r_{\max}) \cdot \log n)$ and bandwidth b; moreover, if Π is information-theoretic then so is the derived scheme.

Applying the above to known single-sender schemes (cf. Appendix A) gives the results in Table 1. Particularly interesting in practice, where computational security suffices, are:

---

[3] When computational security suffices, any single-sender BE scheme can be modified to have $s^* = 1$ by having the sender use a PRF to derive all the keys in the system. In the information-theoretic setting that is not the case.

- A scheme supporting revocation of arbitrarily many users, where each user stores $O(n)$ keys and the bandwidth is $O(r)$. (Although the storage may seem high, we prove that it is optimal for schemes of a certain class allowing arbitrary revoked sets.)
- A scheme with a pre-determined bound $r_{\max}$ on the number of revoked users that has storage $O(r_{\max}^2 \log^2 n)$ and bandwidth $O(r)$.

As noted earlier, we also prove a lower bound on the key storage for multi-sender BE schemes that support revocation of an arbitrary number of users, and that are constructed in a certain way. Specifically, we focus on so-called *unique predecessor* schemes [2], which are schemes that follow the OR approach and in which keys are derived from secret values by applying a hash function (possibly modeled as a random oracle), a pseudorandom generator, or a pseudorandom function to those values individually. To the best of our knowledge, this class includes all known computationally secure, single-sender schemes that improve on information-theoretic schemes, and lower bounds for unique predecessor schemes (in the single-sender case) were previously studied by Austrin and Kreitz [2]. Our bound shows that, in the multi-sender setting, any such scheme requires at least one user to store at least $\frac{n-1}{2}$ keys. Interestingly, we also show that this bound is tight, as there is a multi-sender BE scheme in which all users hold this many keys. (The bandwidth in this scheme is $n - r$, which is why we do not include it in Table 1.)

Austrin and Kreitz [2] also show that any (unique predecessor) single-sender BE scheme with polynomially many keys per user has bandwidth $\Omega(r)$ for small r, showing that our computationally secure scheme supporting unbounded revocation is asymptotically optimal in terms of both storage and bandwidth.

2   Definition of the Problem

We consider multi-sender broadcast encryption schemes, in which there is a set of users $[n] = \{1, \ldots, n\}$, each of whom is given some keying material by a trusted authority. Subsequently, each user $i \in [n]$ should be able to send a message to any desired subset of users $S \subseteq [n] \setminus \{i\}$ such that the revoked users $R = [n] \setminus (S \cup \{i\})$ cannot recover the message even if they all collude. We let $r = |R|$ denote the number of revoked users. Some schemes support revocation of an arbitrary number of users, whereas others impose an *a priori* bound $r_{\max}$ on the number of users who can be revoked. We consider two classes of schemes—information-theoretic and computational—both following the OR approach described in the previous section.

Information-theoretic schemes. In the information-theoretic schemes we consider, there is a set K of keys chosen uniformly and independently from some key space. Each user i is assigned a set $K_i \subset K$, and the *per-user storage* of the scheme is defined as $\max_i |K_i|$. When user i wishes to send a message to a subset S, it finds the smallest $K_{i,S} \subseteq K_i$ such that (1) each user $j \in S$ holds at least one of the keys in $K_{i,S}$ (i.e., $K_j \cap K_{i,S} \neq \emptyset$ for $j \in S$) and (2) no user

$j \in R$ holds any of the keys in $K_{i,S}$ (i.e., $K_j \cap K_{i,S} = \emptyset$ for $j \in R$). (For schemes supporting a bounded number $r_{\max}$ of revoked users, such a $K_{i,S}$ is only required to exist if $n - |S| - 1 \leq r_{\max}$.) User i can then encrypt its message using[4] each key in $K_{i,S}$. The *bandwidth* of the scheme for a given number of revoked users r is defined to be $\max_{i,S:n-|S|-1\leq r} |K_{i,S}|$.

Computationally secure schemes. We consider *unique predecessor* schemes, as defined by Austrin and Kreitz [2], that can offer reduced storage but only achieve computational security. In such schemes, we have sets K, K_i, and $K_{i,S}$ satisfying the same conditions as above, and the bandwidth is defined in the same way. Now, however, users need not store their keys explicitly. Instead, they may derive their keys from secret values they store, with the canonical example of this being the use of a single secret value v to derive keys $k_1 = F_v(1), \ldots, k_\ell = F_v(\ell)$ for F a pseudorandom function. Following [2], we model this by a set $V \supseteq K$ of "secret values" along with a directed graph G (a *key-derivation graph*) whose nodes are in one-to-one correspondence with the elements of V and such that each node has in-degree 0 or 1 (hence the name "unique predecessor"). To instantiate such a scheme, a uniform value is chosen for each node with in-degree 0; then, for every $v' \in V$ that is the ℓth child of some node $v \in V$, we set the value of v' equal to $F_v(\ell)$.

Nodes labeled with elements of K are called "keys"; we say $k \in K$ can be derived from $v \in V$ if v is an ancestor of k in the graph G (this includes the case $v = k$). Each user i is now given a subset $V_i \subset V$ of secret values, and we define K_i to be the set of keys that can be derived from V_i. The per-user storage is now $\max_i |V_i|$.

We remark that the information-theoretic setting is a special case of the above, where $V = K$ and all nodes have in-degree 0.

3   Constructions

We first consider two transformations of single-sender BE schemes to multi-sender BE schemes that are applicable for any number of revoked users. Then, we look at the special case where there is an *a priori* bound $r_{\max}$ on the number of users to be revoked.

3.1   A Trivial Construction

Let Π be a single-sender BE scheme for n users. The construction described here applies regardless of how Π works, but for simplicity we assume Π is a unique predecessor scheme as defined in Sect. 2 (adapted appropriately for the single-sender case). Thus, we let $\bar{V}$ denote the set of secret values in Π, let $\bar{V}_i \subset \bar{V}$ denote the values given to user i, and let $\bar{V}_0 \subset \bar{V}$ denote the values with in-degree 0 (these are the only values the sender needs to store). We construct a

---

[4] For long messages, user i can encrypt the message using a fresh key k and encrypt k using each key in $K_{i,S}$.

multi-sender scheme for n users by simply running Π in parallel n times, with each user acting as the sender in an instance of Π. Our set of secret values will be $V = [n] \times \bar{V}$, and user i will be given

$$V_i = \big\{(i,v) \mid v \in \bar{V}_0\big\} \cup \big\{(j,v) \mid j \neq i, \ v \in \bar{V}_i\big\};$$

that is, user i will be given the values that the sender would store in the ith instance of Π, and the values that user i would store (as a receiver) in all other instances of Π. For a user i to send a message to a designated subset S, that user will simply act as the sender would in the ith instance of Π when sending to S.

It is easy to see that this multi-sender scheme is secure if Π is. Consider any sender i and designated subset of receivers S. Since only the ith instance of Π will be used, we can focus our attention on values of the form $\{(i,v)\}_{v \in \bar{V}}$. But then security of Π implies that even if all the users in R collude, they will not be able to decrypt the message sent by user i. We thus have:

Theorem 1. *Let Π be a single-sender BE scheme with s^* sender storage, per-user storage s, and bandwidth b. Then the multi-sender BE scheme described above supports the same number of revoked users as Π does, and has per-user storage $(n-1) \cdot s + s^*$ and the same bandwidth as Π. Moreover, if Π is information-theoretic then so is the derived scheme.*

3.2   An Improved Construction

We now give an improved construction that uses key derivation applied to an information-theoretic, single-sender scheme Π. Let $\bar{K}$ denote the set of keys used by Π, and let $\bar{K}_i \subset \bar{K}$ denote the keys stored by user i in that scheme. Conceptually, in our multi-sender scheme we will again have n instances of Π, with each user acting as a sender in one of the schemes. Now, however, the keys in the various schemes will be *correlated*. Specifically, the keys used in the ith instance of Π will be $K^{(i)} = \{F_{\bar{k}}(i) \mid \bar{k} \in \bar{K}\}$. In our multi-sender scheme, each user i is given all the keys that the sender would store in the ith instance of Π (namely, $K^{(i)}$), as well as the values $\bar{K}_i \subset \bar{K}$ that can be used to derive the keys that user i would store (as a receiver) in all other instances of Π. Note that user i need not store $F_{\bar{k}}(i)$ for $\bar{k} \in \bar{K}_i$; hence the storage of user i is exactly $|\bar{K}|$.

More formally, we now have a set of keys $K = \{k_{i,j} \mid i \in [n], \ j \in \bar{K}\}$ and additional values $V_0 = \{k_{0,j} \mid j \in \bar{K}\}$; define $V = K \cup V_0$. The keys satisfy $k_{i,j} = F_{k_{0,j}}(i)$; in terms of the underlying key-derivation graph, all nodes corresponding to V_0 have in-degree 0, and node $k_{0,j}$ is a parent of all nodes of the form $k_{i,j}$. User i is given

$$K_i = \big\{k_{0,j} \mid j \in \bar{K}_i\big\} \cup \big\{k_{i,j} \mid j \in \bar{K}\big\}.$$

(We can also use the optimization mentioned above to reduce the storage slightly.) If we let $K^{(i)} = \{k_{i,j} \mid j \in \bar{K}\}$ and $K_\ell^{(i)} = \{k_{i,j} \mid j \in \bar{K}_\ell\} \subset K^{(i)}$, then the key observations are: (1) for each i, the sets $K^{(i)}, K_1^{(i)}, \ldots, K_n^{(i)}$ correspond to $\bar{K}, \bar{K}_1, \ldots, \bar{K}_n$, and we thus have n instances of Π; moreover, (2) user i can derive both $K^{(i)}$ as well as $K_i^{(j)}$ for all j. Put differently, user i can act as a sender in the ith instance of Π, and as a receiver in any other instance of Π. Thus, for a user i to send a message to some designated subset S, that user simply acts as the sender using keys $K^{(i)}$; each receiver $j \in S$ derives the keys $K_j^{(i)}$ and uses those to decrypt.

Security follows in a straightforward manner based on security of Π and the assumption that F is a pseudorandom function. We thus have:

Theorem 2. *Let Π be an information-theoretic, single-sender BE scheme with s^* total keys, per-user storage s, and bandwidth b. Then the multi-sender BE scheme described above is computationally secure, supports the same number of revoked users as Π, and has per-user storage s^* and the same bandwidth as Π.*

3.3   A Construction Supporting Bounded Revocation

In this section we explore an approach for constructing multi-sender BE schemes supporting a bounded number of revoked users. Our construction uses the notion of r-cover-free families [15,23]:

Definition 1. *Fix a universe K. A family of sets $\mathcal{F} = \{K_1, \ldots, K_n\}$, where $K_i \subset K$, is called r-**cover free** if $K_j \not\subseteq K_{i_1} \cup \cdots \cup K_{i_r}$ for any distinct $j, i_1, \ldots, i_r \in [n]$.*

Kumar et al. [24] show, for any r, n, an explicit construct of an r-cover-free family of size n with $|K| \leq 16r^2 \log n$ and $|K_i| \leq 4r \log n$ for all i. We remark that $r_{\max}$-cover-free families immediately imply single-sender broadcast encryption schemes supporting up to $r_{\max}$ revoked users. In general, however, the bandwidth of the resulting construction may be high.

We now show how to use an $r_{\max}$-cover-free family in conjunction with any single-sender broadcast encryption scheme Π supporting up to $r_{\max}$ revoked users to construct a *multi-sender* scheme supporting up to $r_{\max}$ revoked users.

Fix some $r_{\max}$, and let $\{T_1, \ldots, T_n\}$ be an $r_{\max}$-cover-free family over a set T of size t. The construction described below applies regardless of how Π works, but for simplicity we assume Π is a unique predecessor scheme as defined in Sect. 2 (adapted appropriately for the single-sender case). Thus, we let $\bar{V}$ denote the set of secret values in Π, let $\bar{V}_i \subset \bar{V}$ denote the values given to user i, and let $\bar{V}_0 \subset \bar{V}$ denote the values with in-degree 0 (these are the values the sender stores).

Our construction of a multi-sender scheme works by generating t independent instances of Π, and giving each user i (1) the values that the sender would store in the jth instance of Π, for all $j \in T_i$, and (2) the values that user i would store in all instances of Π. That is, our set of values is now $V = T \times \bar{V}$, and user i is given

$$V_i = \big\{(j, v) \mid j \in T_i,\ v \in \bar{V}_0\big\} \cup \big\{(i, v) \mid i \in T,\ v \in \bar{V}_i\big\}.$$

Say user i wants to send a message to some designated subset S, where $R = [n]\backslash(S \cup \{i\})$ has size at most $r_{\max}$. User i first finds an $i^* \in T_i$ such that $i^* \notin \bigcup_{j \in R} T_j$; such an i^* exists by the properties of the cover-free family. It then acts as the sender in instance i^* of Π, revoking the users in R. Security follows since Π is secure for at most $r_{\max}$ revoked users. Using [24], we thus have:

Theorem 3. *Let Π be a single-sender BE scheme supporting up to $r_{\max}$ revoked users, and having s^* sender storage, per-user storage s, and bandwidth $b(r)$ when revoking $r \leq r_{\max}$ users. Then the multi-sender BE scheme described above supports up to $r_{\max}$ revoked users, and has per-user storage $O(s^*\, r_{\max}\, \log n + s\, r_{\max}^2\, \log n)$ and the same bandwidth as Π. If Π is information-theoretic then so is the derived scheme.*

4   Lower Bounds on Per-User Storage

In this section we consider bounds on the per-user storage s for multi-sender broadcast encryption schemes. We first observe a storage/communication trade-off for information-theoretic schemes. Say there is a scheme with per-user storage s and bandwidth b when revoking r users. Consider some sender i storing keys K_i with $|K_i| = s$. There are $\binom{n-1}{r}$ different authorized subsets $S \subset [n]/\{i\}$ that exclude r users, and for each one the set of keys $K_{i,S} \subseteq K_i$ used by user i to encrypt must be different and non-empty. Moreover, $|K_{i,S}| \leq b$ for all S. Thus, we must have

$$\sum_{j=1}^{b} \binom{s}{j} \geq \binom{n-1}{r}.$$

Simplifying, this gives $s \geq \binom{n-1}{r}^{1/b}$. If r is small, the above gives (asymptotically) $b \geq r \frac{\log(n-1)}{\log s}$. The most relevant consequence is that if r is constant, and the per-user storage is polylogarithmic in n, then $b = \omega(r)$. It is interesting to note (cf. Table 1) that when r is constant there is a computationally secure scheme with polylogarithmic storage and $b = O(r)$.

Can the storage be improved in computationally secure schemes? Unfortunately, the following theorem shows that any unique predecessor scheme supporting arbitrarily many revoked users must have per-user storage $\Omega(n)$.

Theorem 4. *Any unique predecessor scheme for n-user, multi-sender broadcast encryption supporting arbitrarily many revoked users must have per-user storage at least $\lceil \frac{n-1}{2} \rceil$.*

Proof. The ability to revoke $r = n - 2$ users implies that each pair of distinct users i, j must be able to derive a shared key $k_{\{i,j\}}$ that cannot be derived by any other user. Call this the *pairwise key* for i and j. We claim that for each such i, j, either user i or user j (or possibly both) explicitly stores a value $v_{\{i,j\}}$ such that the *only* pairwise key that can be derived from $v_{\{i,j\}}$ is $k_{\{i,j\}}$. This implies that there are at least $\binom{n}{2}$ values $v_{\{i,j\}}$ that are stored overall, and hence some user must store at least $\binom{n}{2}/n = \frac{n-1}{2}$ values.

To prove the claim, let v_i (resp., v_j) denote the stored value used by user i (resp., user j) to derive $k_{\{i,j\}}$. Assume toward a contradiction that user i derives some other pairwise key (say, $k_{\{i,j'\}}$ with $j' \neq j$) from v_i, and that user j derives some other pairwise key (say, $k_{\{i',j\}}$ with $i' \neq i$) from v_j. (Security implies that user i cannot derive the pairwise key $k_{\{i',j'\}}$ if $i \notin \{i', j'\}$, and similarly for user j.) The unique predecessor property implies that v_i and v_j must lie on the same path in the underlying graph, and hence one must be an ancestor of the other. Without loss of generality, say v_i is an ancestor of v_j. But then user i can derive v_j and hence $k_{\{i',j\}}$, violating security.

This lower bound is essentially tight, as we now show an n-user scheme in which each user stores exactly $\lceil \frac{n-1}{2} \rceil + 1$ values. For notational convenience, define $H(x) = F_x(0)$ where F is a pseudorandom function, and let $H^{(i)}(\cdot)$ denote the i-fold iteration of H. Number the users from 0 to $n-1$. Each user i stores:

- A value v_i. Define $k_{\{i,j\}} = H^{(j)}(v_i)$ for $j = i+1, \ldots, i + \lfloor \frac{n-1}{2} \rfloor$ (taken modulo n).
- Keys $k_{i,j}$ for $j = i + \lfloor \frac{n+1}{2} \rfloor, \ldots, i + n - 1$ (taken modulo n).

Each user stores $1 + \left(n - \lfloor \frac{n+1}{2} \rfloor\right) = \lceil \frac{n-1}{2} \rceil + 1$ values. Note that each key $k_{\{i,j\}}$ can be derived only by users i and j. Any user i can thus securely send a message to any designated subset S by using the set of keys $\{k_{\{i,j\}} \mid j \in S\}$.

As described, key derivation requires $O(n)$ invocations of H. Using a tree-based construction, however, this can be improved to $O(\log n)$.

5   Conclusion

We have introduced the problem of *multi-sender* broadcast encryption, a natural generalization of symmetric-key broadcast encryption, and explored upper- and lower bounds on such schemes.

The most interesting open question is whether or not there exists a computationally secure scheme with storage $o(n)$ using an altogether different paradigm. It would also be interesting to find an information-theoretic scheme with storage $O(n)$ and bandwidth better than the trivial $n - r - 1$, or to show that doing asymptotically better is not possible.

Acknowledgments. This research was supported in part by the NSF REU-CAAR program, award #1262805; we thank Bill Gasarch for organizing that program. We thank Daniel Apon, Seung Geol Choi, Jordan Schneider, and Arkady Yerukhimovich for discussing various aspects of this problem with us.

A   Information-Theoretic Single-Sender Schemes

In this section we describe various single-sender schemes that, to the best of our knowledge, have not appeared previously in the literature. The parameters of the schemes presented here do not beat the parameters of the best known single-sender schemes, but they have the advantage of having information-theoretic security.

We begin with a simple scheme that revokes exactly one user (i.e., $r_{\max} = 1$). Fix some b, and identify the n users with b-tuples whose coordinates range from 1 to $n^{1/b}$. The sender holds a set of keys $K = \{k_{i,w}\}_{i \in [b], w \in [n^{1/b}]}$ of size $b \cdot n^{1/b}$. The user associated with tuple $(w_1, \ldots, w_b)$ is given the set of keys $\{k_{i,w}\}_{i \in [b], w \neq w_i}$; in other words, key $k_{i,w}$ is held by all users whose ith coordinate is not w. To revoke the single user $(w_1, \ldots, w_b)$, the sender encrypts the message using the b keys $k_{1,w_1}, \ldots, k_{b,w_b}$ not held by that user. It follows that:

Theorem 5. *For any b, there is an information-theoretic, single-sender BE scheme with $r_{\max} = 1$ having per-user storage $b \cdot n^{1/b} - b$, bandwidth b, and $b \cdot n^{1/b}$ total keys.*

Gentry et al. [18] show that in any information-theoretic, single-sender scheme with $r_{\max} = 1$, storage s, and bandwidth b, it holds that $n \leq s^b$. The above scheme shows this bound is tight within a constant factor.

We now show how to build an information-theoretic scheme Π^* revoking any number of users based on any scheme Π revoking a single user. The high-level idea is to apply the SD approach [29] but to schemes rather than keys. In the SD approach, users are arranged at the leaves of a binary tree, and for each pair of nodes i, j in the tree with i a parent of j, we let $S_{i,j}$ denote the users who are descendants of i but not descendants of j. Naor et al. show that any set of users S can be partitioned into $O(r)$ such sets, where $r = n - |S|$ is the number of revoked users. In the SD scheme, for all i, j as above there is a single key $k_{i,j}$ that is known exactly to those users in $S_{i,j}$; hence, the bandwidth of the scheme is $O(r)$. Here, we generalize the approach so that there is a *set* of keys allowing only those users in $S_{i,j}$ to decrypt.

We again arrange the users at the leaves of a binary tree. In this tree, let T_i denote the sub-tree rooted at some node i. For each such sub-tree T_i of height h, we associate the root node i of that sub-tree with h instances of Π (recall, Π is a single-sender scheme supporting revocation of a single user) corresponding to the h levels of T_i not including the root node itself. The "virtual users" of instance $\ell \in \{0, \ldots, h-1\}$ of Π correspond to the nodes at height ℓ in T_i, and we imagine giving each node the keys it would receive as a virtual user in all instances of Π in which it is involved. The real users, at the leaves, store the keys that would be given to its ancestors.

To send a message to a subset S of the users, the sender partitions S into a collection of subsets $S_{i,j}$ as in the SD scheme. To encrypt a message such that only the users in $S_{i,j}$ can read it, the sender uses the instance of Π in which node i is the sender and the nodes on the same level as j are the receivers, and revokes user j.

Rather than analyzing the above in the general case, we compute the bandwidth and storage when applied to the single-sender scheme Π from Theorem 5. Naor et al. showed that any set of S users can be partitioned into at most $2r - 1$

subsets $S_{i,j}$, where $r = n - |S|$ is the number of revoked users. Since the scheme Π from Theorem 5 has fixed bandwidth b independent of the number of users, we conclude that the bandwidth of our scheme here is at most $b \cdot (2r - 1)$. The storage per user is given by $\sum_{h=1}^{\log n} \sum_{\ell=0}^{h-1} (n/2^{h-\ell})^{1/b} = O(n^{1/b})$. Similarly, one can show that the total number of keys is $O(n)$. Summarizing:

Theorem 6 (Scheme 1). *For any b, there is an information-theoretic, single-sender BE scheme supporting arbitrarily many revoked users having per-user storage $O(n^{1/b})$, bandwidth $O(b \cdot r)$, and $O(n)$ total keys.*

Specifically, there is an information-theoretic, single-sender BE scheme supporting arbitrarily many revoked users having per-user storage $O(\sqrt{n})$, bandwidth $O(r)$, and $O(n)$ total keys.

References

1. Aiello, W., Lodha, S., Ostrovsky, R.: Fast digital identity revocation. In: Krawczyk, H. (ed.) CRYPTO 1998. LNCS, vol. 1462, pp. 137–152. Springer, Heidelberg (1998). doi:10.1007/BFb0055725
2. Austrin, P., Kreitz, G.: Lower bounds for subset cover based broadcast encryption. In: Vaudenay, S. (ed.) AFRICACRYPT 2008. LNCS, vol. 5023, pp. 343–356. Springer, Heidelberg (2008). doi:10.1007/978-3-540-68164-9_23
3. Balenson, D., McGrew, D., Sherman, A.: One-way function trees and amortized initialization. Internet Draft, Key management for large dynamic groups (1999)
4. Bhattacharjee, S., Sarkar, P.: Reducing communication overhead of the subset difference scheme. IEEE Trans. Comput, to appear. https://eprint.iacr.org/2014/577
5. Bhattacherjee, S., Sarkar, P.: Concrete analysis and trade-offs for the (complete tree) layered subset difference broadcast encryption scheme. IEEE Trans. Comput. **63**(7), 1709–1722 (2014)
6. Blom, R.: An optimal class of symmetric key generation systems. In: Beth, T., Cot, N., Ingemarsson, I. (eds.) EUROCRYPT 1984. LNCS, vol. 209, pp. 335–338. Springer, Heidelberg (1985). doi:10.1007/3-540-39757-4_22
7. Blundo, C., De Santis, A., Herzberg, A., Kutten, S., Vaccaro, U., Yung, M.: Perfectly secure key distribution for dynamic conferences. Inf. Comput. **146**(1), 1–23 (1998)
8. Blundo, C., Mattos, L.A.F., Stinson, D.R.: Trade-offs between communication and storage in unconditionally secure schemes for broadcast encryption and interactive key distribution. In: Koblitz, N. (ed.) CRYPTO 1996. LNCS, vol. 1109, pp. 387–400. Springer, Heidelberg (1996). doi:10.1007/3-540-68697-5_29
9. Boneh, D., Gentry, C., Waters, B.: Collusion resistant broadcast encryption with short ciphertexts and private keys. In: Shoup, V. (ed.) CRYPTO 2005. LNCS, vol. 3621, pp. 258–275. Springer, Heidelberg (2005). doi:10.1007/11535218_16
10. Canetti, R., Garay, J.A., Itkis, G., Micciancio, D., Naor, M., Pinkas, B.: Multicast security: a taxonomy and some efficient constructions. In: IEEE INFOCOM, pp. 708–716 (1999)
11. Canetti, R., Malkin, T., Nissim, K.: Efficient communication-storage tradeoffs for multicast encryption. In: Stern, J. (ed.) EUROCRYPT 1999. LNCS, vol. 1592, pp. 459–474. Springer, Heidelberg (1999). doi:10.1007/3-540-48910-X_32

12. Cheon, J.H., Jho, N.-S., Kim, M.-H., Yoo, E.S.: Skipping, cascade, and combined chain schemes for broadcast encryption. IEEE Trans. Inf. Theor. **54**(11), 5155–5171 (2008)
13. Dodis, Y., Fazio, N.: Public-key broadcast encryption for stateless receivers. In: Security and Privacy in Digital Rights Management (ACM CCS Workshop), pp. 61–80. ACM (2002)
14. Dyer, M., Fenner, T., Frieze, A., Thomason, A.: On key storage in secure networks. J. Cryptol. **8**(4), 189–200 (1995)
15. Erdös, P., Frankl, P., Füredi, Z.: Families of finite sets in which no set is covered by the union of r others. Israeli J. Math. **51**(1–2), 79–89 (1985)
16. Fiat, A., Naor, M.: Broadcast encryption. In: Stinson, D.R. (ed.) CRYPTO 1993. LNCS, vol. 773, pp. 372–387. Springer, Heidelberg (1994)
17. Gafni, E., Staddon, J., Yin, Y.L.: Efficient methods for integrating traceability and broadcast encryption. In: Wiener, M. (ed.) CRYPTO 1999. LNCS, vol. 1666, pp. 372–387. Springer, Heidelberg (1999). doi:10.1007/3-540-48405-1_24
18. Gentry, C., Ramzan, Z., Woodruff, D.P.: Explicit exclusive set systems with applications to broadcast encryption. In: 47th Annual Symposium on Foundations of Computer Science (FOCS), pp. 27–38. IEEE (2006)
19. Goodrich, M.T., Sun, J.Z., Tamassia, R.: Efficient tree-based revocation in groups of low-state devices. In: Franklin, M. (ed.) CRYPTO 2004. LNCS, vol. 3152, pp. 511–527. Springer, Heidelberg (2004). doi:10.1007/978-3-540-28628-8_31
20. Halevy, D., Shamir, A.: The LSD broadcast encryption scheme. In: Yung, M. (ed.) CRYPTO 2002. LNCS, vol. 2442, pp. 47–60. Springer, Heidelberg (2002). doi:10.1007/3-540-45708-9_4
21. Hwang, J.Y., Lee, D.H., Lim, J.: Generic transformation for scalable broadcast encryption schemes. In: Shoup, V. (ed.) CRYPTO 2005. LNCS, vol. 3621, pp. 276–292. Springer, Heidelberg (2005). doi:10.1007/11535218_17
22. Jho, N.-S., Hwang, J.Y., Cheon, J.H., Kim, M.-H., Lee, D.H., Yoo, E.S.: One-way chain based broadcast encryption schemes. In: Cramer, R. (ed.) EUROCRYPT 2005. LNCS, vol. 3494, pp. 559–574. Springer, Heidelberg (2005). doi:10.1007/11426639_33
23. Kautz, W.H., Singleton, R.C.: Nonrandom binary superimposed codes. IEEE Trans. Inf. Theor. **10**(4), 363–377 (1964)
24. Kumar, R., Rajagopalan, S., Sahai, A.: Coding constructions for blacklisting problems without computational assumptions. In: Wiener, M. (ed.) CRYPTO 1999. LNCS, vol. 1666, pp. 609–623. Springer, Heidelberg (1999). doi:10.1007/3-540-48405-1_38
25. Kumar, R., Russell, A.: A note on the set systems used for broadcast encryption. In: 14th Annual Symposium on Discrete Algorithms (SODA), pp. 470–471. ACM-SIAM (2003)
26. Luby, M., Staddon, J.: Combinatorial bounds for broadcast encryption. In: Nyberg, K. (ed.) EUROCRYPT 1998. LNCS, vol. 1403, pp. 512–526. Springer, Heidelberg (1998). doi:10.1007/BFb0054150
27. Micciancio, D., Panjwani, S.: Optimal communication complexity of generic multicast key distribution. In: Cachin, C., Camenisch, J.L. (eds.) EUROCRYPT 2004. LNCS, vol. 3027, pp. 153–170. Springer, Heidelberg (2004). doi:10.1007/978-3-540-24676-3_10
28. Mitchell, C.J., Piper, F.C.: Key storage in secure networks. Discrete Appl. Math. **21**(3), 215–228 (1988)

29. Naor, D., Naor, M., Lotspiech, J.: Revocation and tracing schemes for stateless receivers. In: Kilian, J. (ed.) CRYPTO 2001. LNCS, vol. 2139, pp. 41–62. Springer, Heidelberg (2001). doi:10.1007/3-540-44647-8_3
30. Wallner, D.M., Harder, E.J., Agee, R.C.: Key management for multicast: issues and architectures. Internet Draft, RFC 2627 (1999)
31. Wang, S.-Y., Yang, W.-C., Lin, Y.-J.: Balanced double subset difference broadcast encryption scheme. Secur. Commun. Netw. **8**(8), 1447–1460 (2015)
32. Wong, C.K., Gouda, M., Lam, S.S.: Secure group communications using key graphs. In: Proceedings of ACM SIGCOMM, pp. 68–79 (1998)

A Supervised Auto-Tuning Approach for a Banking Fraud Detection System

Michele Carminati$^{(\boxtimes)}$, Luca Valentini, and Stefano Zanero

Dipartimento di Elettronica, Informazione e Bioingegneria,
Politecnico di Milano, Milan, Italy
{michele.carminati,stefano.zanero}@polimi.it,
luca.valentini@mail.polimi.it

Abstract. In this paper, we propose an extension to Banksealer, one of the most recent and effective banking fraud detection systems. In particular, until now Banksealer was unable to exploit analyst feedback to self-tune and improve its performance. It also depended on a complex set of parameters that had to be tuned by hand before operations.

To overcome both these limitations, we propose a supervised evolutionary wrapper approach, that considers analyst's feedbacks on fraudulent transactions to automatically tune feature weighting and improve Banksealer's detection performance. We do so by means of a multi-objective genetic algorithm.

We deployed our solution in a real-world setting of a large national banking group and conducted an in-depth experimental evaluation. We show that the proposed system was able to detect sophisticated frauds, improving Banksealer's performance of up to 35% in some cases.

Keywords: Internet banking · Fraud detection · Genetic algorithm · Supervised learning

1 Introduction

Nowadays, Internet banking has become one of the major target of fraudulent cyber-attacks such as phishing, malware, and trojan infections, and has brought to a worldwide loss of billions of dollars every year [4, 36]. According to Kaspersky, in 2016 financial malware infected about 2,8 millions personal devices, a 40% increase since 2015 [1].

To contrast fraudulent cyber-attacks, banks developed fraud analysis and detection systems that aim at identifying unauthorized activities as quickly as possible. These systems monitor and scrutinize transactions, scoring suspicious ones for analyst verification. In spite of the importance of the subject, very little research is openly carried out, because of privacy restrictions and difficulties in obtaining real-world data.

Thanks to the collaboration with a major banking group, we were able to develop Banksealer [7,8], a novel, unsupervised fraud analysis system that

© Springer International Publishing AG 2017
S. Dolev and S. Lodha (Eds.): CSCML 2017, LNCS 10332, pp. 215–233, 2017.
DOI: 10.1007/978-3-319-60080-2_17

automatically ranks frauds and anomalies in banking transactions. While the experiments presented in [8] showed that Banksealer is an effective approach in identifying frauds and seems to provide a meaningful support to the banking analysts in fraud investigation, one of its main limitations is the inability to collect and exploit banking analyst's feedback. It also depended on a complex set of parameters that had to be tuned by hand before operations.

To overcome these limitations, in this paper we propose a general supervised evolutionary wrapper approach that considers analyst feedback on fraudulent transactions to find an optimal tuning of Banksealer's parameters. Our approach implements the Non-dominated Sorting Genetic Algorithm II (NSGA-II) to find a configuration of parameters that optimizes the ranking of potential frauds at runtime. Most of the parameters are feature weights, which makes the task particularly challenging, since we are confronted with large and unbalanced datasets in which there are multiple variants of frauds that are, overall, extremely rare (i.e. less than 1% with respect to legitimate transactions), and dynamically evolve over the time.

We deployed our solution in a real-world setting of a large national banking group and conducted an in-depth experimental evaluation. Thanks to collaboration with this bank and leveraging the domain expert's knowledge, we reproduced frauds (in a controlled environment) performed against online banking users, and recorded the resulting fraudulent transactions. We show that the proposed system was able to detect sophisticated frauds improving Banksealer's performance up to a factor of 35%.

In summary, in this paper we make the following novel contributions:

- We propose a supervised learning module based on Multi-Objective-Genetic-Algorithm (MOGA) able to automatize the feature weighting task and to improve detection performances of Banksealer.
- We free banking analysts from the manual job of the feature weighting task and exploit their knowledge analyzing their feedback.
- We improve Banksealer's ability to evolve over the time and to adapt itself to changes in both threats and user behavior.

2   Overview of Banksealer and Goals

In this section we will recall the main concepts underlying the existing Banksealer system, insofar as they are needed to explain the motivation of the present work. We refer the interested reader to the original paper [8] for additional details.

Banksealer characterizes the customers of the bank by means of a local, a global, and a temporal *profile*, which are built during a training phase taking as input a list of *transactions*. Each type of profile extracts different statistical *features* from the transaction attributes, according to the type of model built. A list of the employed attributes is presented in Table 1.

Once the profiles are built, Banksealer processes new transactions and ranks them according to their anomaly score and the predicted risk of fraud. The *anomaly score* quantifies the statistical likelihood of a transaction being a fraud w.r.t. the learned profiles. The *risk of fraud* prioritizes the transactions, combining the anomaly score with the transaction amount. Banksealer provides the analysts with a ranked list of potentially fraudulent transactions, along with their anomaly score.

The *local* profile characterizes each user's individual spending patterns. During training, we aggregate the transactions by user and compute the empirical marginal distribution of the features of each user's transactions (for simplicity, we do not consider correlation between features). This representation is simple and effective, and hence is indeed directly readable by analysts who get a clear idea of the typical behavior by simply looking at the profile. At runtime, we calculate the anomaly score of each new transaction using a modified version of the Histogram Based Outlier Score (HBOS) [14] method. HBOS computes the log-likelihood of a transaction according to the marginal distribution learned. The HBOS score is a weighted sum:

$$HBOS(t) = \sum_{0 < i \leq d} w_i * \log \frac{1}{f(t_i)}; \qquad \sum_{0 < i \leq d} w_i = 1$$

where w_i is the weighting coefficient of the i-th feature, that allows analysts to tune the system. It is worth noting that $f(t_i)$ is the application of the min-max normalization [15, pp. 71–72] to the frequency $hist_i$ of the i-th feature. This normalization was necessary in order to account the variance of each feature.

The *global* profile characterizes "classes" of spending patterns, clustering users together. Each user is represented as a feature vector of six components: total number of transactions, average transaction amount, total amount, average time span between subsequent transactions, number of transactions executed from foreign countries, number of transactions to foreign recipients (bank transfers dataset only). To find classes of users with similar spending patterns, we apply an iterative version of the Density-Based Spatial Clustering of Applications with Noise (DBSCAN), using the Mahalanobis distance [22] between the aforementioned vectors. We assign to each user global profile an anomaly score, which tells the analyst how "uncommon" the spending pattern is with respect to other customers. For this, we compute the unweighted-unweighted-Cluster-Based Local Outlier Factor (CBLOF) [3] score, which considers small clusters as outliers with respect to large clusters. More precisely, the more a user profile

Table 1. Attributes for each type of transaction. Attributes in **bold** are hashed for anonymity needs.

Dataset	Attributes
Bank transfers	Amount, CC_ASN, **IP**, **IBAN**, IBAN_CC, Timestamp
Phone recharges	Amount, CC_ASN, **IP**, Phone operator, **Phone number**, Timestamp
Prepaid cards	Amount, Card type, **Card number**, CC_ASN, **IP**, Timestamp

deviates from the dense cluster of "normal" users, the higher his or her anomaly score will be.

The global profile is also leveraged to mitigate the issue of *undertraining*. Undertrained users are users that performed a low number of transactions, and represent a relevant portion of a typical dataset. For undertrained users, we consider their global profile and select a cluster of similar users.

Finally, the *temporal* profile deals with frauds that exploit the repetition of legitimate-looking transactions over time, by comparing the current spending profile of the user against their history. During training, we extract the mean and standard deviation of the following aggregated features for each user: total amount, total and maximum daily number of transactions. At runtime, according to the sampling frequency, we calculate the cumulative value for each of the aforementioned features for each user, and compare it against the previously computed metrics.

2.1   Research Goal

While the experiments presented in [8] showed that Banksealer is an effective approach in identifying frauds and seems to provide a meaningful support to the banking analysts in fraud investigation, one of its main limitations is the lack of a bi-directional communication channel between the unsupervised system and the banking analyst to collect and exploit analyst's feedback and knowledge to improve detection performance. Furthermore, Banksealer works with an empirical configurations of weights, manually set by analysts.

Therefore, the focus of this work is to overcome this limitation by exploiting banking analysts' feedback to auto-tune Banksealer. With auto-tune we mean to find an optimal features weights configuration used by Banksealer to compute the transaction's anomaly score.

This is basically an instance of the *feature weighting* problem, a variant of the more common *feature selection* one [38]. The difference is that in feature selection we basically assign a binary weight to discard redundant or irrelevant attributes. Feature weighting instead assigns real-valued weights to each feature, based on relevance [10, 33, 37].

Feature weighting algorithms can be classified into two categories, based on whether or not the feature weighting is done independently from the detection learning algorithm. If feature weighting is done independently from the detection task, the technique is said to follow a *filter approach*. Otherwise, it is said to follow the *wrapper approach*. A *filter approach* works exclusively on the data and uses probabilistic dependence measures to determine correlations among features. Being independent of the detection task means that changes to the detection system do not impact assigned weights. Also, it is more efficient than a wrapper approach from the computational point of view, since the detection system does not need to be ran to evaluate the candidate weights configurations.

On the other hand, a *wrapper approach* [19] consists of executing feature weighting, intertwined with the detection task. Wrapper approaches usually

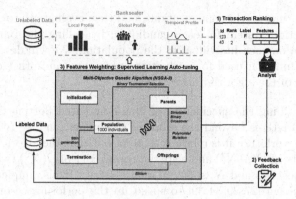

Fig. 1. Logical view of Banksealer integrated with the feature weighting module.

outperform *filter approaches* from a detection point of view, but obviously they are computationally more demanding.

In this work we opted for the more computationally intensive *wrapper approaches*. In particular, we made use of *genetic algorithms*, as they can handle multiple local optima and are designed to support also multiple objective criteria [38]. In fact, as we will detail in the following, our problem requires to trade off between multiple objectives. For a deeper and more exhaustive notion of Genetic Algorithm (GA) we invite the reader to refer to [13, 24, 25] as this is beyond the scope of this paper.

3 Approach Overview

Our approach to solve the feature weighting problem stated in Sect. 2.1 is summarized in Fig. 1. It is composed of three logical steps:

1. **Transaction Ranking.** Banksealer generates in output a ranking of the transactions based on the fraud risk score. As shown in Sect. 2, Banksealer uses the HBOS score to compute the anomaly score of a transaction by combining the weighted contribution of each feature. The formula can be simplified as follow $HBOS(t) = \sum_{0 < i \leq d} w_i \cdot c_i$, where d is the number of features of the dataset, w_i and c_i are respectively the weight and the score contribution of the ith feature.
2. **Feedback Collection.** Analysts, after going through the ordered transaction list, flag as *fraud* transactions that have been verified to be fraudulent, or as *Suspect* the ones that – even if they turned out to be benign – were definitely anomalous enough to warrant investigation. All other transactions are benign. The labeled dataset is the input for the feature weighting process.
3. **Feature Weighting.** After having collected all transaction feedbacks in a labeled dataset, our solution follows the wrapper approach. We opted to use a GA, and specifically Non-dominated Sorting Genetic Algorithm II (NSGA-II) [12]. The basic idea we follow is to generate a population that represents

different feature weight configurations, and then evaluate the accuracy of the fraud detection system for each candidate, by calling the Banksealer testing function on each individual of the population. We will describe in detail the fitness function used for evaluation, operators, and other details of the application of the algorithm in Sect. 4.3.

While describing the specificities of Non-dominated Sorting Genetic Algorithm II (NSGA-II) is beyond the scope of this paper, and we refer the reader to the original work [12], it is relevant to point out that the algorithm exhibits a time complexity of $O(MN^2)$ and a spatial complexity of $O(N^2)$ (where M is the number of objectives and N is the population size). It also implements elitism that can be shown [27,29,34,39] to speed up the performance of GAs significantly. Finally, NSGA-II is parameterless regarding the sharing mechanism used to introduce diversity in the population, which suits our purpose of making the tuning mechanism completely automated.

4   Approach Implementation

In this section we describe in detail the application of the NSGA-II to our problem. We describe the encoding scheme, the handling of constraints, the selection of operators, and the fitness function used.

4.1   Encoding Scheme and Constraints

The first step in designing a GA is the representation of the genes in an individual (i.e., encoding scheme). For our feature weighting problem, this is straightforward. In fact, we decide to implement the weight configuration as a list of real numbers. Each cell of the list is associated to a feature of those used by Banksealer (see Table 1), and the value that each cell contains represents the weight associated to the feature.

Furthermore, at this stage we must consider also possible constraints that may influence our encoding scheme design, but above all the implementation of crossover and mutation operators that we will present later. Our problem has two constraints:

1. $w_i \in [0,1]$, which means that the weight w_i of the ith feature must be a real value between 0 and 1.
2. $\sum_{i=1}^{F} w_i = 1$, which means that the sum of all weights belonging to a configuration must be equal to 1 (F is the total number of features).

Both constraints can be satisfied by normalizing to 1 the sum of all genes values whenever a new individual is created, and by ensuring that each values is positive and real.

4.2 Population

Another very important aspect of GA is the number of individuals in the population. A small population size leads to a faster convergence of the fitness functions. However, the drawback is that this might get the algorithm stuck in local optima.

Therefore, we had to find a trade-off for the population size to get a good solution in a reasonable time. In Fig. 2, we report the evolution of fitness functions according to population size. In all three cases, we get similar results. We can see that for smaller population sizes the algorithm stagnates in local optima for several generations. For population sizes of 500 and 1000, we have almost the same performance, and no relevant improvements are obtained by increasing population to 5000. As a consequence, we chose a population size of 1000 since it represents the optimal trade-off between diversity and performance.

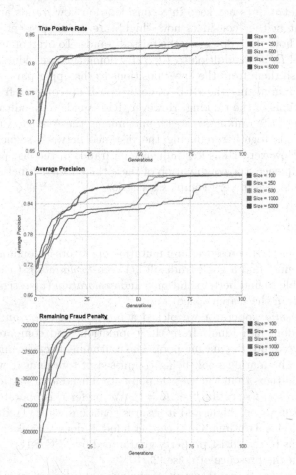

Fig. 2. Population size estimation.

4.3   Fitness Functions

Fitness functions are needed to evaluate feature weights configurations and allow to choose the best ones. We choose three fitness functions: *True Positive Rate*, *Average Precision*, and *Remaining Frauds Penalty*. The first two are defined in Sect. 5.4, and they are both meant to be maximized. The *Remaining Frauds Penalty* (RFP) assigns a penalty (equal to the number of frauds not yet detected) for each normal transaction detected as fraudulent: RFP = $\sum_{k=1}^{R} RF(k) \times N(k)$, where R is the number of total transactions in the ranking and $RF(k)$ is the number of frauds not yet detected at the kth position of the ranking; $N(k) = 1$ if the kth transaction is normal, 0 otherwise. In this case, we want to minimize Remaining Frauds Penalty (RFP).

The reason for choosing TPR is self-explaining: our main goal is to detect as many as possible frauds in the top N positions of the ranking. Its main drawback, however, is that it does not keep into consideration how frauds are arranged in the ranking: it neither considers how "high" are frauds present in the top N positions, nor how "far" down are the missed frauds. To overcome this problem, we use AP and RFP in addition to TPR. By doing this, we reduce the spread of frauds and push them from the lower positions to the upper part of the ranking. In particular, minimizing the RFP value, we push up complex frauds from the very last positions of the ranking. However, RFP tends to penalize more frauds at the bottom of the ranking. Instead, maximizing the Average Precision (AP), we gather frauds together reducing the distance between sequent fraudulent transactions. However, it has less influence on frauds in the last positions.

As a consequence, the combination of these three fitness functions makes the algorithm stable also in very complex scenarios.

4.4   Operators

The choice of selection, crossover and mutation operators is domain specific. Furthermore, we must find a good trade-off between *exploitation* (using knowledge already available to find better solutions) and *exploration* (investigating new and unknown areas in the search space).

For the *selection* operator we picked a *tournament selection*: the operator selects K random individuals from the population, and compares them. The winner is inserted into the mating pool. The tournament repeat until the mating pool is filled. The mating pool, being comprised of tournament winners, has a higher average fitness than the average population fitness. In particular, we used a *binary tournament* selection, i.e. $K = 2$. We prefer this operator because its selection pressure is not high, and it assures genetic diversity, reducing the risk of converging to local optima. On the other hand, it does not take a lot of time to converge thanks to the elitist property guaranteed by NSGA-II. It is also proven to be robust in the presence of noise [23].

For the crossover operator, we chose the *simulated binary crossover* [11]. The operator computes offspring solutions $x_i^{(1,t+1)}$ and $x_i^{(2,t+1)}$ from parent solutions

$x_i^{(1,t)}$ and $x_i^{(2,t)}$ by defining the spread factor β_i as the ratio of the absolute difference in offspring values to that of the parent values:

$$\beta_i = \left| \frac{x_i^{2,t+1} - x_i^{1,t+1}}{x_i^{2,t} - x_i^{1,t}} \right|$$

The spread factor β_i is distributed according to the following probability distribution:

$$P(\beta_i) \begin{cases} 0.5(\eta_c + 1)\beta_i^{\eta} & \beta \leq 1 \\ 0.5(\eta_c + 1)\frac{1}{\beta_i^{\eta+2}} & \text{otherwise} \end{cases}$$

Where η_c is a parameter we set to control the variance of the distribution: A large value of η_c gives a higher probability to create offspring "near" the parents, while a small one allows distant solutions to be selected as offspring. For our problem we set the probability of crossover between two parents to 0.9, and $\eta_c = 5$. This because in our tests greater values of η_c resulted in a slow down of the algorithm convergence, since creating offspring very similar to their parent favors exploitation much more than exploration.

We decide to use the *simulated binary crossover* for three reasons:

- It is designed for offspring with real variables, like our weights.
- It preserves parents schemata in the offspring. By doing this, the crossover operator does not destroy every time the solution creating a new one very different from the parents.
- It has a very interesting self-adaptation property: the location of the offspring solution depends on the difference in parent solutions. If the difference in the parent solution is small, the difference between the offspring and parent solutions is also small and vice-versa.

Finally, we chose the *polynomial mutation* as mutation operator. It attempts to simulate the offspring distribution of binary-encoded bit-flip mutation on real-valued variables. This operator is usually used in pair with simulated binary crossover because it works in a very similar way, favoring mutated offspring nearer to the parents. Adopting the same notation used before for crossover, a new mutated offspring is obtained as $x_i^{t+1} = x_i^t + (x_i^U - x_i^L)\delta_i$, where x_i^U and x_i^L are respectively the upper bound and the lower bound of the variable at position i. Instead, δ_i is defined as:

$$\delta_i = \begin{cases} (2r_i)^{\frac{1}{\eta_m+1}} - 1 & r_i \leq 0.5 \\ 1 - [2(1 - r_i)]^{\frac{1}{\eta_m+1}} & \text{otherwise} \end{cases}$$

where, similarly to crossover, r_i is a random number between 1 and 0, and η_m is the mutation distribution index. We set $\eta_m = 10$, in such a way that resulting offspring are different but rather close to the non-mutated individual. We choose this high mutation rate, because it allows to obtain a high diversity on the Pareto front. Instead, the probability of mutating a single variable is equal to $\frac{1}{F}$ where F is the number of total features used by the algorithm. This results in one mutation per offspring on average (which corresponds roughly to the idea of "shifting the value of one variable").

5   Experimental Evaluation

In this section we describe the experimental evaluation of our learning module integrated with Banksealer.

Table 2. Scenarios of fraudulent activities.

	Type of fraud	IP country	IBAN country
Scenario 1	Information stealing	Foreign	Foreign
Scenario 2	Information stealing	Foreign	Italian
Scenario 3	Information stealing	Italian	Foreign
Scenario 4	Information stealing	Italian	Italian
Scenario 5	Transaction hijacking	-	Foreign
Scenario 6	Transaction hijacking	-	Italian

5.1   Hardware and Computation Times

Our experiments have been executed on a desktop computer with the following specifications: Quad-core 3.40 GHz Intel i7-4770 CPU, 16 GB of RAM, and the Linux kernel 3.7.10 × 86_64. The results of the experiments we made are obtained computing the average on 30 tests to avoid statistical oscillations. In average, a single weighting process of 80 generations on a one-month dataset lasts 1 h and 30 min. We obtained this execution time thanks to the parallel fitness function evaluations that resulted to be about 3 times faster than the non-parallel version.

5.2   Dataset

The dataset in our possession belongs to an important Italian banking institute and is anonymized to protect privacy of customers. The dataset covers the period from April 2013 to August 2013. We split the data in a *training dataset*, used to train Banksealer, consisting of 3 months of data; a *weighting dataset* of one month, containing the analyst feedback and used to learn the optimal configuration of weights; and finally, a *testing dataset*, consisting of the last month of data (and also containing analyst feedback). We show the results on the bank transfer data for brevity, but similar results can be obtained for the other contexts such as prepaid cards and phone recharges.

5.3   Synthetic Fraud Scenarios

The dataset under analysis does not contain frauds. Therefore, as already successfully done in [7,8], we inject fraud scenarios that replicate the typical attacks performed against online banking users. We consider two types of fraudulent attacks, reconstructed using domain expert and analyst advice:

- **Information stealing scenario.** It simulates a banking trojan or a phishing attack in which the customer is deceived into entering its credentials and a one-time-password (OTP). The stolen informations are sent to the fraudster, who uses them to execute a transaction towards an unknown IBAN. In this scenario, we suppose that the fraudster is interested into stealing as much money as he or she can. As a consequence the amount transferred will be very high, from 10.000€ to 50.000€. The fraudster can connect to the bank server from an Italian or a foreign IP address and money can be transfered to an Italian or foreign IBAN. To inject the transactions, we randomly choose a user between those present in the testing dataset and we inject a transaction with a random timestamp.
- **Transaction hijacking scenario.** It simulates the infection of the user's computer by a MitB attack. The malware deceive the user into entering two OTPs and then exploit its capabilities to execute a second transaction using the user's browser. In this scenario, the transaction is still directed toward an unknown IBAN, which can be Italian or foreign, however the connection is executed from the victim's computer. As in the Information stealing scenario, we suppose that the fraudster is interested into stealing as much money as possible and, as a consequence, the amount transferred will be very high, from 10.000€ to 50.000€. To inject the transaction we randomly choose a user between those present in the testing dataset and after that we randomly select one of his or her transaction. The injected transaction will be executed no more than ten minutes after the selected transaction to simulate the MitB session hijacking.

In Table 2 we report the synthetic scenarios and their characteristics.

5.4   Metrics

Given the nature of our system, that "ranks" transactions according to anomaly, we need to slightly redefine the traditional evaluation metrics. We define as "positive" any transaction scored among the top N positions of the ranking, where N is the number of fraudulent transactions in the dataset. In our experiment we inject synthetic fraudulent transactions equivalent to the 1% of the dataset.

A True Positive (TP) is a fraudulent transaction that appears in the first N positions of the ranking. We similarly define True Negative (TN), False Positive (FP), and False Negative (FN). Then we use the traditional definition of *True Positive Rate (TPR)*:

$$TPR = \frac{TP}{TP + FP}$$

We also compute the *Average Precision (AP)*, which takes into account the position of fraudulent transactions in the ranking:

$$AP = \frac{\sum_{k=1}^{R} P(k) \times F(k)}{N}$$

where R is the number of total transactions in the ranking, $P(k) = \frac{\#frauds}{k}$, and $F(k) = 1$ if the kth transaction is fraudulent, 0 otherwise.

Since our dataset is highly unbalanced in favor of normal transactions, we use the Matthews Correlation Coefficient (MCC) and Average Accuracy (AA) metrics, because they are less affected by this problem.

The MCC expresses the relationship between the observed and predicted binary classifications ($MCC = 1$ perfect prediction, $MCC = 0$ no better than random prediction, and $MCC = -1$ total disagreement between prediction and observation):

$$MCC = \frac{TP \times TN - FP \times FN}{\sqrt{(TP+FP)(TP+FN)(TN+FP)(TN+FN)}}$$

The AA is an average of the accuracy obtained for both fraudulent and normal transactions classes:

$$AA = \frac{1}{2}\left[\frac{TP}{TP+FN} + \frac{TN}{TN+FP}\right]$$

5.5   Experiment 1

In this experiment, we want to verify the quality of the *Weighted Banksealer* on single scenarios listed in Table 2, and compare it with *Banksealer* . We use the dataset described in Sect. 5.2. We compute the TPR of both systems as defined in Sect. 5.4. The results of our experiment can be seen in Table 3.

It is evident that feature weighting brings an improvement in almost all scenarios under analysis. Since *Banksealer* already guarantees good fraud detection performance in the information stealing scenario, the most significant gains are in the hijacking scenario, which is the most complex to detect. In particular, when looking at the limited results for Scenario 6, keep in mind that the fraudulent transactions are almost indistinguishable from benign ones in this case.

5.6   Experiment 2

With the objective of verifying the quality of *Weighted Banksealer* on a scenario closer to the real world, we test our approach against a "mixed scenario" obtained injecting in the dataset the same number of frauds, but randomly extracted from all of the scenarios of the first experiment. The results can be seen in Table 4. As we can see, *Weighted Banksealer* gets a *TPR* higher than *Banksealer* with a difference of 23%. But we also improve the ranking, concentrating most frauds in the top positions. This is expressed by the Average Precision, or it can be

Table 3. Experiment 1: TPR, AA and MCC results of Experiment 1. BS = Banksealer, BSW = Weighted Banksealer

Fraud scenario	IP	IBAN	BS			BSW			Improvements		
			TPR (%)	AA (%)	MMC	TPR (%)	AA (%)	MMC	TPR (%)	AA (%)	MMC
1: Information stealing	Foreign	Foreign	97	98	0.97	98	99	0.98	+1	+1	+0.01
2: Information stealing	Foreign	National	91	95	0.91	94	97	0.94	+3	+2	0.03
3: Information stealing	National	Foreign	97	98	0.97	97	98	0.97	0	+0	0
4: Information stealing	National	National	91	95	0.91	92	96	0.92	+1	+1	+0.01
5: Transaction hijacking	-	Foreign	75	87	0.77	95	97	0.95	+20	+10	+0.18
6: Transaction hijacking	-	National	22	68	0.34	57	78	0.63	+35	+10	+0.29

Table 4. Results of Experiment 2.

	TPR	Average precision	Matthews correlation coefficient	Average accuracy
Banksealer	58%	68%	0.67	82%
Weighted Banksealer	81%	88%	0.83	91%
Improvements	**+23%**	**+20%**	**+0.16**	**+9%**

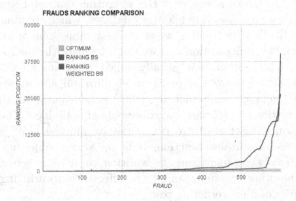

Fig. 3. Ranking comparison for Experiment 2.

seen in Fig. 3, where we plot the cumulative distribution of the detection ordered by ranking. The yellow line models the detection performance of an ideal fraud detection system. It is evident that *Banksealer* diverges earlier than *Weighted Banksealer* from it. For a further comparison of *Banksealer* and *Weighted Banksealer* we report in Fig. 4 also the *Receiver Operating Characteristic* curve (*ROC*), which confirms the better overall performance of *Weighted Banksealer*.

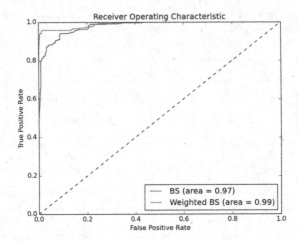

Fig. 4. ROC curve for Experiment 2.

5.7   Overfitting Problem

In the design of the system and during the experimental evaluation, we put great effort in limiting overfitting (a real issue in noisy, unbalanced datasets like ours).

In our problem, overfitting may be caused by an over-weighting of the system caused by the execution of too many generation of the GA. In that case, we could see that the performance over the weighting dataset keeps to increase, while performance decreases on validation dataset, which contains data unseen by the system. To limit overfitting we study how many generations are needed to learn the weights configuration and we stop the algorithm as soon as it starts to learn noise. This approach is usually called *early stopping*. We wait that all three functions converge and reach an equilibrium and as we can see this happens at 80th generation. In fact, after 80th generation in all three functions for several generations no relevant improvements are obtained. The performance on the validation dataset starts to get worse around the 115th generation for the Average Precision fitness function. It is not very visible, but performance on weighting dataset are increasing. In addition, after some generations we can see also that the other functions start to be affected by overfitting. In Fig. 5 we report the results of our overfitting test.

In addition, we put effort to produce synthetic transactions that resemble the real ones to be as realistic as possible and to avoid the overfitting of our approach to "trivial" fraudulent transactions. To evaluate the quality of simulated data with respect to the real one, we empirically compare the distribution of transactions features by applying the kernel density estimation method [30] and box-plot diagrams. In addition, we applied the non- parametric two-sample permutation test for the comparison between central tendency of the features. With respect to other non-parametric tests it does not require verification of any assumption about distribution's shape and variability of the two samples.

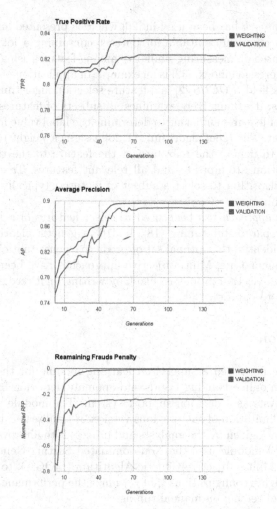

Fig. 5. Overfitting analysis on the different fitness functions.

The null hypothesis specifies that the permutations are all equally likely with a significance level $\alpha = 0.05$. In other words, the distribution of the data under the null hypothesis satisfies exchange-ability. Since we found a $p\text{-}values > \alpha$, we failed to reject the null hypothesis that the samples are drawn from the same distribution.

6   Related Works

Fraud detection, mainly focused on credit card fraud, is a wide research topic, for which we refer the reader to [5,9,31]. In this section we focus on the feature-weighting task.

The *filter approach* has been used in [26] where is presented an unsupervised feature selection algorithm suitable for dataset containing a lot of dimension. The method is based on measuring similarity among features using the maximum information compression index. [32] is an example of application of this algorithm in bioinformatics field. *FOCUS* [2] is a feature selection algorithm for noise-free Boolean domains. It exhaustively examines all subsets of features, selecting the minimal subset of features sufficient to determine the label value for all instances in the training set. The relief algorithm [17,20] assigns a weight to each feature, which is meant to denote the relevance of the feature to the target concept. The relief algorithm attempts to find all relevant features. Tree filters [6] use a decision tree algorithm to select a subset of features, typically for a nearest-neighbor algorithm.

The *wrapper approach* has been used to select features of a Bayesian Classifier [21] or for parameter tuning [18]. In [16] a genetic algorithm has been implemented to identify the optimal set of predictive genes that classify samples by cell line or tumor type. Multi-objective approaches have been implemented in several feature selection problems, like handwriting digit recognition [28] and facial expression recognition [35].

7   Conclusions

In this paper we presented a supervised learning module for the optimization of Banksealer, an online banking frauds and anomaly detection framework used by banking analysts as a decision support system. The module was created to solve one of the limitation of the previous version of Banksealer, the inability to collect the feedback given by the analysts and process it to improve the detection performance. The module uses the Non-dominated Sorting Genetic Algorithm II (NSGA-II), a Multi-Objective-Genetic-Algorithm (MOGA), to automatically learn feature weights configurations that optimize the performance of the overall system, instead of relying on manual tuning.

We field-tested the algorithm on real-world data, showing that it is able to self-tune Banksealer over large, unbalanced datasets, and it improves the detection rates over time. The system is extensible and almost transparent to analysts, who just need to express their feedback on the transaction ranking.

Obviously, the system shows some limits that we wish to address in future works. A first experimental limitation is that, while the dataset of transaction is real, in order to create significant tests we needed to inject synthetically generated frauds to evaluate the quality of the detection task. In the future, we will proceed with further tests on real-world fraud samples.

A more relevant limitation is that, as shown in Sect. 5.6, complex fraudulent transactions can still escape the top of the ranking. While we consider the current results already very successful, we believe that the key to improve them further is to design and test other fitness functions and new features (e.g., sum of the amount) that focuses on solving the complex fraud issue. Redesigning fitness functions can also address different motivations for the analysts: for instance, we are experimenting with a fitness function that aims to maximize the total amount of all fraudulent transactions (as opposed to their number). In a MOGA it is rather easy to add and remove fitness functions, and we are going to exploit this modularity in future works.

Acknowledgment. This work has received funding from the European Union's Horizon 2020 Programme, under grant agreement 700326 "RAMSES", as well as from projects co-funded by the Lombardy region and Secure Network S.r.l.

References

1. Kaspersky Security Bulletin 2016. Technical report, Kaspersky Lab (2017). https://goo.gl/Jzkab2
2. Almuallim, H., Dietterich, T.G.: Learning with many irrelevant features. In: AAAI, vol. 91, pp. 547–552. Citeseer (1991)
3. Amer, M., Goldstein, M.: Nearest-neighbor and clustering based anomaly detection algorithms for RapidMiner. In: Proceedings of the 3rd RapidMiner Community Meeting and Conference (RCOMM 2012), pp. 1–12 (2012)
4. Bolton, R.J., Hand, D.J.: Statistical fraud detection: a review. Stat. Sci. **17** (2002)
5. Bolton, R.J., Hand, D.J., David J.H.: Unsupervised profiling methods for fraud detection. In: Proceedings of Credit Scoring and Credit Control VII, pp. 5–7 (2001)
6. Cardie, C.: Using decision trees to improve case-based learning. In: Proceedings of the Tenth International Conference on Machine Learning, pp. 25–32 (1993)
7. Carminati, M., Caron, R., Maggi, F., Epifani, I., Zanero, S.: BankSealer: an online banking fraud analysis and decision support system. In: ICT Systems Security and Privacy Protection. IFIP Advances in Information and Communication Technology, vol. 428, pp. 380–394. Springer, Heidelberg (2014)
8. Carminati, M., Caron, R., Maggi, F., Epifani, I., Zanero, S.: BankSealer: a decision support system for online banking fraud analysis and investigation. Comput. Secur. **53**, 175–186 (2015) http://dx.doi.org/10.1016/j.cose.2015.04.002
9. Chandola, V., Banerjee, A., Kumar, V.: Anomaly detection: a survey. ACM Comput. Surv. **41**(3), 15:1–15:58 (2009)
10. Cost, S., Salzberg, S.: A weighted nearest neighbor algorithm for learning with symbolic features. Mach. Learn. **10**(1), 57–78 (1993)
11. Deb, K., Agrawal, R.B.: Simulated binary crossover for continuous search space. Complex Syst. **9**(3), 1–15 (1994)
12. Deb, K., Pratap, A., Agarwal, S., Meyarivan, T.: A fast and elitist multiobjective genetic algorithm: NSGA-II. IEEE Trans. Evol. Comput. **6**(2), 182–197 (2002)
13. Eiben, A.E., Smith, J.E.: Introduction to Evolutionary Computing. Springer Science & Business Media, New York (2003)
14. Goldstein, M., Dengel, A.: Histogram-Based Outlier Score (HBOS): a fast unsupervised anomaly detection algorithm. In: KI-2012: Poster and Demo Track, pp. 59–63 (2012)
15. Han, J., Kamber, M.: Data Mining: Concepts and Techniques. The Morgan Kaufmann Series in Data Management Systems Series. Elsevier Science & Tech, Amsterdam (2006)
16. Jirapech-Umpai, T., Aitken, S.: Feature selection and classification for microarray data analysis: evolutionary methods for identifying predictive genes. BMC Bioinform. **6**(1), 148 (2005)

17. Kira, K., Rendell, L.A.: A practical approach to feature selection. In: Proceedings of the Ninth International Workshop on Machine Learning, pp. 249–256 (1992)

18. Kohavi, R., John, G.H.: Automatic parameter selection by minimizing estimated error. In: ICML, pp. 304–312. Citeseer (1995)

19. Kohavi, R., John, G.H.: The wrapper approach. In: Feature Extraction, Construction and Selection, pp. 33–50. Springer (1998)

20. Kononenko, I.: Estimating attributes: analysis and extensions of RELIEF. In: Bergadano, F., Raedt, L. (eds.) ECML 1994. LNCS, vol. 784, pp. 171–182. Springer, Heidelberg (1994). doi:10.1007/3-540-57868-4_57

21. Langley, P., Sage, S.: Induction of selective Bayesian classifiers. In: Proceedings of the Tenth International Conference on Uncertainty in Artificial Intelligence, pp. 399–406. Morgan Kaufmann Publishers Inc. (1994)

22. Mahalanobis, P.C.: On the generalized distance in statistics. In: Proceedings of the National Institute of Science of India, vol. 2, pp. 49–55 (1936)

23. Miller, B.L., Goldberg, D.E.: Genetic algorithms, tournament selection, and the effects of noise. Complex Syst. **9**(3), 193–212 (1995)

24. Mitchell, M.: An Introduction to Genetic Algorithms. MIT Press, Cambridge (1998)

25. Mitchell, T.: Machine Learning. McGraw Hill, New York (1997)

26. Mitra, P., Murthy, C., Pal, S.K.: Unsupervised feature selection using feature similarity. IEEE Trans. Pattern Anal. Mach. Intell. **24**(3), 301–312 (2002)

27. Obayashi, S., Takahashi, S., Takeguchi, Y.: Niching and elitist models for MOGAs. In: Eiben, A.E., Bäck, T., Schoenauer, M., Schwefel, H.-P. (eds.) PPSN 1998. LNCS, vol. 1498, pp. 260–269. Springer, Heidelberg (1998). doi:10.1007/BFb0056869

28. Oliveira, L.S., Sabourin, R., Bortolozzi, F., Suen, C.Y.: Feature selection using multi-objective genetic algorithms for handwritten digit recognition. In: Proceedings of 16th International Conference on Pattern Recognition, vol. 1, pp. 568–571. IEEE (2002)

29. Parks, G.T., Miller, I.: Selective breeding in a multiobjective genetic algorithm. In: Eiben, A.E., Bäck, T., Schoenauer, M., Schwefel, H.-P. (eds.) PPSN 1998. LNCS, vol. 1498, pp. 250–259. Springer, Heidelberg (1998). doi:10.1007/BFb0056868

30. Parzen, E.: On estimation of a probability density function and mode. Ann. Math. Stat. **33**(3), 1065–1076 (1962)

31. Phua, C., Alahakoon, D., Lee, V.: Minority report in fraud detection: classification of skewed data. SIGKDD Explor. Newsl. **6**(1), 50–59 (2004)

32. Phuong, T.M., Lin, Z., Altman, R.B.: Choosing SNPs using feature selection. In: Proceedings of Computational Systems Bioinformatics Conference, 2005, pp. 301–309. IEEE (2005)

33. Punch III, W.F., Goodman, E.D., Pei, M., Chia-Shun, L., Hovland, P.D., Enbody, R.J.: Further research on feature selection and classification using genetic algorithms. In: ICGA, pp. 557–564 (1993)

34. Rudolph, G.: Evolutionary search under partially ordered sets. Dept. Comput. Sci./LS11, Univ. Dortmund, Dortmund, Germany, Technical report CI-67/99 (1999)

35. Soyel, H., Tekguc, U., Demirel, H.: Application of NSGA-II to feature selection for facial expression recognition. Comput. Electr. Eng. **37**(6), 1232–1240 (2011)

36. Wei, W., Li, J., Cao, L., Ou, Y., Chen, J.: Effective detection of sophisticated online banking fraud on extremely imbalanced data. World Wide Web **16**(4), 449–475 (2013). http://dx.doi.org/10.1007/s11280-012-0178-0

37. Wettschereck, D., Aha, D.W., Mohri, T.: A review and empirical evaluation of feature weighting methods for a class of lazy learning algorithms. Artif. Intell. Rev. **11**(1–5), 273–314 (1997)

38. Yang, J., Honavar, V.: Feature subset selection using a genetic algorithm. In: Liu, H., Motoda, H. (eds.) Feature Extraction, Construction and Selection, pp. 117–136. Springer, New York (1998)

39. Zitzler, E., Deb, K., Thiele, L.: Comparison of multiobjective evolutionary algorithms: empirical results. Evol. Comput. **8**(2), 173–195 (2000)

Scalable Attack Path Finding
for Increased Security

Tom Gonda$^{(\boxtimes)}$, Rami Puzis, and Bracha Shapira

Department of Software and Information Systems Engineering,
Ben-Gurion University of the Negev, Beer-Sheva, Israel
tomgond@post.bgu.ac.il
http://bgu.ac.il

Abstract. Software vulnerabilities can be leveraged by attackers to gain
control of a host. Attackers can then use the controlled hosts as stepping stones for compromising other hosts until they create a path to the
critical assets. Consequently, network administrators must examine the
protected network as a whole rather than each vulnerable host independently. To this end, various methods were suggested in order to analyze the multitude of attack paths in a given organizational network, for
example, to identify the optimal attack paths. The down side of many of
those methods is that they do not scale well to medium-large networks
with hundreds or thousands of hosts. We suggest using graph reduction
techniques in order to simplify the task of searching and eliminating optimal attacker paths. Results on an attack graph extracted from a network
of a real organization with more than 300 hosts and 2400 vulnerabilities
show that using the proposed graph reductions can improve the search
time by a factor of 4 while maintaining the quality of the results.

Keywords: Network security · Attack graphs · Planning · Graph
reduction · Attack models

1 Introduction

The software products used in today's corporate networks are vast and diverse
[1]. As a result, software vulnerabilities can be introduced to the network which
an attacker can later leverage in order to gain control of the organization's hosts.
In practice, even organizations that are minded of security can have hosts with
many critical vulnerabilities present in their network [2].

One of the security analyst tasks is to decide which vulnerabilities and which
hosts to patch against attacks. The cost of patching a host, and the effort involved
can some times be extremely high [3]. There is a risk that a patch will break a
production system, on top of the maintenance time it takes to patch the system.

This raises the now-common need to prioritize which vulnerabilities in which
hosts to patch. An important factor in the decision to patch a host or not is
how an attacker can leverage the host as a stepping stone in order to reach

© Springer International Publishing AG 2017
S. Dolev and S. Lodha (Eds.): CSCML 2017, LNCS 10332, pp. 234–249, 2017.
DOI: 10.1007/978-3-319-60080-2_18

critical assets. In order to find the probable path of an attacker, many models have been suggested to represent all attacker's possible paths in a network [4,5].

We chose to use MulVAL (Multi-host, Multi-stage Vulnerability Analysis Language) framework [6] to represent an attacker's possible actions in the network. A brief description of the framework, and the logical attack graphs (Also called LAGs) it produces can be found in Subsect. 3.2.

Using the models that represent the attacker's possible actions, many researchers then applied planning methods to find the optimal attacker's path [7–9]. The downside of many of those methods is that they do not scale well to medium to large networks.

In this paper we aim to reduce the time it takes to find attack paths which an attacker might use, by reducing the size of the attack graph. We intend to do so without effecting the quality of the optimal path. We review the metrics in which we will check that comparison in Sect. 6.

Our contribution is a reduction (described in Sect. 4) that allows finding low-cost attacker paths faster, without compromising the quality of the paths found (experiments in Sect. 7). In results compared to existing approaches on graphs containing more than 200,000 nodes, which represent 309 network hosts with 2398 vulnerabilities the proposed reduction improved the running time in a factor of 4.

2   System Overview

This paper deals with reducing the size of the LAGs, in order to speed the computation time for finding attack paths. Figure 1 shows the overall workflow of our work. First, network scans are being performed to collect data about the network structure and vulnerabilities present in the network as described in Subsect. 3.1. Next, the reductions presented at the related work (Sect. 5) are applied, in order to reduce the input to the graph generation framework (MulVAL). Then, the MulVAL framework is applied to create a logical attack graph. The LAG model is presented in Sect. 3.2. After the LAG was generated, our reduction which is presented in Sect. 4 can be applied in order to reduce the LAG generated in the previous step. At last, the result of the reduced graph is converted to a planning

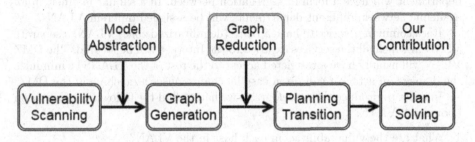

Fig. 1. Work-flow illustration

problem, and solved by a generic planner, as explained in Sect. 3.3. Each experiment described in Sect. 7 will go through all the above steps, although in each experiment only one reduction will be applied, either before or after the graph generation phase.

3   Background

One of the foundations of our work is attack graphs. Attack graphs have been used in multiple variations for over a decade to represent possible attacks on a system [10,11]. Attack graphs usually include information about the preconditions needed to execute actions (exploits, password guessing, network sniffing, etc.) and the possible outcomes of these actions (like agent being installed on target machine, accounts compromised and more). In many cases, attack graphs represent all the attack paths in the target network. Attack path usually represent a series of actions which end with the attacker gaining unauthorized access to an asset. Our main focus is LAGs [12] since they have been the most scalable among the models, and provide open-source implementations.

To produce the attack graph, we had to provide vulnerability scans of the different hosts in the network, and the connections between hosts. We used real-world networks for our work. The way we scanned the networks and produced the topology for the attack graph is outlined in Subsect. 3.1.

We then transformed the attack graph into a planning problem, and used a generic solver to find the optimal attacker path within that attack graph. Scientific background about planning with numeric state variables and the transformation from attack graph to a planning problem can be found in Subsect. 3.3.

3.1   Data Set

In order to create attack graphs as close as possible to the real world, we decided to produce the attack graphs from a large institute with thousands of hosts. For our work we looked at each VLAN in the institute separately. VLAN (Virtual Local Area Network) is a way to unite computers of certain characteristic within an organization. As an example, in corporate network, different departments could be assigned different VLANs so that the sales department and the HR department will have a form of segregation between. In a similar manner, in an academic network, different departments will be assigned different VLANs.

It's a common practice to have a DMZ (demilitarized zone) VLAN, a separate VLAN in which all the services exposed to the Internet will be located. The DMZ VLAN will usually have a restricted access to the rest of the VLANS to minimize the damage an attacker can do in case he compromises a machine in the DMZ.

In order to produce the attack graph we had to find the following information about each VLAN:

1. What are the vulnerabilities in each host in the VLAN?
2. What connections can be made from a VLAN to the rest of the VLANs (some connections can be blocked or enabled through firewall between the VLANs)?

To collect this information we used Nessus vulnerability scanner [13]. We chose Nessus after a comparison with additional vulnerability scanner - OpenVAS [14] since Nessus is more common in attack graph research and has better integration to the MulVAL framework which we used to produce the attack graphs.

We chose 3 different VLANs in the institute which we decided to scan. The scan have been performed in the following manner: First we scanned all the VLANs from the Internet, external to the organization. Then, for two of the VLANs we scanned, we positioned the scanning computer inside the VLAN and scanned the 2 other VLANs, and the VLAN itself from within, An illustration can be seen in Fig. 2.

In order to change the location of the scanning computer between VLANs without the need of physically changing locations we used trunk ports. When using trunk ports, Ethernet frames are tagged with the desired VLAN and then passed to the desired VLAN through ports that are able to handle tagged Ethernet frames. This allowed easier scanning from multiple VLANs without having to physically access the different locations in the organization.

An obvious result we have observed is that scanning a VLAN from different locations produced different results. For example, in some VLANs scanned from the Internet, no hosts were detected. This was probably caused by a firewall filtering connections to this VLAN. In some hosts, we have seen different set of services exposed to different VLANs. For example, when scanning some VLANs from the Internet, only the web service at port 80 was available. When scanning the same host from within the organization we have seen additional services exposed such as web management or network share services.

We used the different scan results achieved when scanning from the different locations to create the topology of the network. For each VLAN, we treated the scan made from within the VLAN as representing the 'true state' of the network. An assumption was made that no device filtered or altered the scan within a VLAN. This is somewhat a possible assumption, since communication within VLAN does not go through any hosts, so the possibility of interference is low. Hence, in the model, the hosts in each VLAN, the services they run and the vulnerability those services have were taken from the scan made from the host inside the VLAN. Before explaining how the connection between hosts were created, a few formal definitions are needed.

Definition 1. *VLAN. A VLAN V is a set of ip addressed such that each ip address is within the VLAN. We denote the different VLANs in the organization as V_1, V_2, V_3. For formality, we will define the internet as a VLAN as well, denoted $V_{internet}$. In our case $V_{internet}$ has only one ip.*

Definition 2. *Connection. A connection c is defined by the tuple:*

$$(ip\_src, ip\_dst, protocol, port)$$

Where ip_src is the source ip of the connection, ip_dst is the destination of the connection. protocol is the network protocol being used (like tcp or udp) and, port is the port used. A connection represent that the source ip can initiate a connecting to the destination ip, in the protocol stated and in the port stated.

Definition 3. *Scan Item. A scan item s is defined by a tuple:*

$$(ip, protocol, port, software, vulnerability)$$

Where ip is an ip of a scanned computer. software is a software installed on the computer and vulnerability is the CVE of the vulnerability found in that software.

Definition 4. *Scan. A scan S_{ij} is a set of scan items. S_{ij} holds:*

$$\forall s \in S_{ij} : p_1(s) \in V_j$$

Where p_n is the n projection of s. A scan represents all the hosts and vulnerabilities found in V_j when scanned from V_i. Since the computer scanning from the internet is irrelevant for us:

$$\forall i \in \{1, 2, 3\} : S_{i,internet} = \emptyset$$

We defined two types of connection: Inner network connection:

$$INNER_i = \{\forall ip \in V_i \quad \forall s \in S_{ii} | (ip, p_1(s), p_2(s), p_3(s))\}$$

Meaning that we model a connection for any two hosts within the VLAN, in all protocols and ports found when scanning the VLAN from within.

Inter network connection:

$$INTER_{ij} = \{\forall ip \in V_i \quad \forall s \in S_{ij} | (ip, p_1(s), p_2(s), p_3(s)\}$$

Meaning that we model a connection between a host in V_i to a host in V_j in some protocol and port only if when scanning V_j from V_i a vulnerability was found in the host at V_j in that protocol and port.

Finally the connections allowed in our model are:

$$K = \{1, 2, 3\}$$

$$\bigcup_{i \in K} INNER_i \cup \bigcup_{i \neq j \in K} INTER_{ij} \cup INTER_{internet,1}$$

We included connection from the internet to only one VLAN in our network because otherwise the graphs produced and solutions found would often be trivial one step solutions.

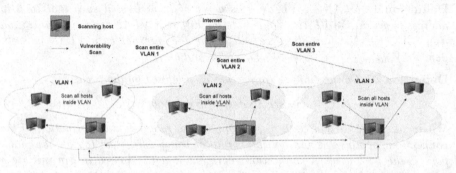

Fig. 2. Scan methodology overview

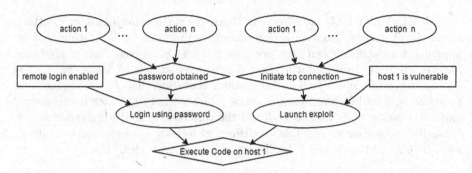

Fig. 3. Example attack graph

3.2 Logical Attack Graph

Logical attack graphs (LAGs) are graphs that represent the possible actions and outcomes of attacker trying to gain a goal asset in a system. The graph contains 3 types of nodes:

Derived fact nodes (can also be referred as privilege nodes) represent a capability an attacker has gained after performing an action (derivation phase). Example of such node can be a node stating that the attacker can execute arbitrary code on a specific machine with certain privileges. These are the diamond shaped nodes seen in Fig. 3.

Derivation nodes (can also be referred as action nodes) usually represent an action the attacker can take in order to gain a new capability in the system. The outcome of performing an action, is an instantiation of a new derived fact. Example of an action node can be seen in Fig. 3 as the oval nodes. One of the possible ways to gain code execution in a host is launching an exploit that allows remote code execution. Another possibility is obtaining a password of a valid user, and logging in with his credentials. A derived fact can be instantiated by **either one** of it's parent nodes, which are action nodes. In order to instantiate an action node (or derivation node) **all** of it's parent nodes need to be instantiated.

Primitive fact nodes are the ground truth nodes of the model, they represent facts about the system. Usually they can represent network connectivity, firewall rules, user accounts on various computer and more. In the example Fig. 3 they are the box shaped nodes.

Definition 5. *Attack Graph. Formally, attack graph is represented as a tuple:*

$$(Np, Ne, Nc, E, L, G)$$

Where N_p, N_e and N_c are three sets of disjoint nodes in the graph, E is a set of directed edges in the graph where

$$E \subseteq (N_e \times N_p) \cup ((N_p \cup N_c) \times (N_e))$$

L is a mapping from a node to its label, and $G \subseteq N_p$ is a set of the attacker goals. N_p, N_e and N_c are the sets of privilege nodes, action nodes and primitive fact nodes, respectively.

The edges in a LAG are directed. There are three types of edges in attack graph: (a, p) an edge from an action node to a predicate node, stating that by applying a an attacker can gain privilege p. (p, a) is an edge from a predicate node to an action node, stating that p is a precondition to action a. (f, a) is an edge from fact node f to an action node a stating that f is a precondition to action a. The labeling function maps a fact node to the fact it represents, and a derivation node (action node) to the rule that is used for the derivation. Formally, the semantics of a LAG is defined as follows: For every action node a, let C be a's child node and P be the set of a's parent nodes, then

$$(\wedge L(P) \Rightarrow L(C))$$

is an instantiation of interaction rule L(a) [12]. In our work we add cost function C to the LAG. $C(a)$ where $a \in N_e$ is the cost the attacker pays to perform an action.

3.3   Planning with Numeric State Variables

Planning is a branch of AI that deals with choosing action sequences in order to achieve a goal. A planning framework is usually given a description of the possible propositions in the world, the possible actions including their preconditions and effects, the initial set of proposition, and the desired propositions in the goal state. It's goal is to find sequence of actions that results in a state that satisfies the goal.

Formally, Numeric planning task is a tuple (V, P, A, I, G) Where P is a set of logical propositions used in the planning task. $V = \{v^1, v^2...v^n\}$ is a set of n numeric variables. A state s is a pair $s = (p(s), v(s))$ where $p(s) \subseteq P$ is the set of true proposition for this state, and $v(s) = (v^1(s), v^2(s)...v^n(s)) \in \mathcal{Q}^n$ is the vector of numeric variables assignments. A is a the set of actions in the problem. An action is a pair $(pre(a), eff(a))$ where pre (precondition) is the precondition needed to be satisfied in order to activate the action. Formally, when planning with numeric states, precondition, con, is also a pair $(p(con), v(con))$ where $p(con) \subseteq P$ is the set of proposition required to be true. $v(con)$ is a set of numeric constraints. In our model, we do not have numeric constraints before activating actions, so we will not go into details about their formal definition. An effect is a triple $eff = (p(eff)^+, p(eff)^-, v(eff))$ where $p(eff)^+ \subseteq P$ is a set of propositions assigned true, as an effect of the action activation, $p(eff)^- \subseteq P$ is a set of propositions assigned false as an effect of the action activation, and $v(eff)$ is a set of effects on the numeric variables in V.

In a numeric planning task, I is a state representing initial state $s = (p(s), v(s))$, and G is a condition representing the goal condition [15]. In our model, attacker's actions comes with a cost (such as risk of detection, or ease of exploitation for vulnerabilities). Our goal is to find a plan with minimal cost for the attacker, assuming an attacker will try to reach his goals with minimal effort or risk.

PDDL is Planning Domain Definition Language, a language build for representing multiple planning problems, specifically it allows modeling numeric tasks.

PDDL also allows specifying optimization criterion, which is an expression the solver will later minimize or maximize. The variable we would like to minimize in our work is the attacker cost for reaching a goal.

Researchers have used planning to represent attacker trying to achieve goals in the network for quite some time [7,9,16]. It seems that the two prominent planners in this domain have been the Metric-FF [15], and SGPlan [17]. In our work we have used Metric-FF planner, since it was able to handle with larger amount of predicates and actions generated when converting attack graphs to a planning problem.

We transformed an attack graph to a planning problem in the following manner: All of the primitive fact nodes have been turned into propositions. Those propositions are initially true in the initial state of the task. All of the derived fact nodes (privilege nodes) where translated into propositions in the model, they are initially false in the model. Each derivation node (action node), became an action $a = (pre(a), eff(a))$ in the planning task. $pre(a) = (p, v)$ where p is a set of the action's precondition, and v is a set numeric constraints. In our model, p contains the proposition of all of the action node's parents in the graph. As an example, in Fig. 3, the action 'Login using password' will have two proposition as preconditions, one that represents the primitive fact 'remote login enabled' and another that represent the derived primitive 'password obtained'.

As stated above, the effect of a is a triple $(p(eff)^+, p(eff)^-, v(eff))$ where $p(eff)^+$ is the predicate representation of the action node's parent. In our example, it will be the predicate of 'Execute Code on host 1'. $p(eff)^- = \emptyset$ since in our current model, the attacker does not lose previously achieved goals by launching new attacks. The numeric effect of an action node is an increase of the numeric variable representing the attacker total effort. Each action node can be assigned with a cost $cost \in \mathcal{N}$. If an action is assigned with a cost, then the numeric effect of the action is: $v(eff) = (total\_effort, + =, cost)$. The goal of the planner is to find a sequence of actions that end in one of the goal predicate true, while minimizing $total\_effort$ variable.

4   PathExpander Algorithm

In this section we will describe our proposed reduction algorithm. In the core of the algorithm we find the shortest path between a source node and a target node, and expand the graph using this shortest path. For this, we assume that our graph has a source node, and that the graph has a single target node. We argue that these are valid assumptions in our model. For source node, all the attack graphs have a fact node representing the attacker initial location. This can be used as the source node for our algorithm. For target node, we choose the goal node in the LAG. In case there are multiple goal nodes in the attack graph, we can easily create a single goal by creating virtual actions applicable only from goal nodes, which lead to a single new goal node. This transition was described in depth in [18].

After we have the shortest path between a goal node and a source node in the LAG, we verify that all of the action node's preconditions are met in that path.

Algorithm 1. PathExpander algorithm

---

1 function PathExpander(G, s, t)
    Input : LAG G, source s, target t
    Output: Reduced LAG G'
2 forall $v \in G$ do
3 | $Color(v) = White$;
4 end
5 $Q \leftarrow WeightedShortestPath(G, s, t)$;
6 while $Q \neq \emptyset$ do
7 | $v \leftarrow Q.pop()$;
8 | if $Type(v) = fact$ then
9 | | $Color(v) \leftarrow Black$;
10 | else if $Type(v) = action$ then
11 | | if $Color(v) = White$ then
12 | | | $Color(v) \leftarrow Grey$;
13 | | | $Q.push(v)$;
14 | | | if $\exists u \in v.parents$ s.t $Color(u)=Grey$ then
15 | | | | Continue;   /* Loop Detected */
16 | | | else
17 | | | | forall $u \in v.parents$ do
18 | | | | | $Q.push(u)$;
19 | | | | end
20 | | | end
21 | | else if $Color(v) = Grey \wedge \forall u \in v.parents : Color(u) = Black$ then
22 | | | $Color(v) \leftarrow Black$
23 | else if $Type(v) = privilege$ then
24 | | if $\exists u \in v.parents$ s.t $Color(u) = Black$ then
25 | | | $Color(v) \leftarrow Black$;
26 | | | continue;
27 | | if $Color(v) = White$ then
28 | | | $Color(v) \leftarrow Grey$;
29 | | $Q.push(v)$;
30 | | $U = FilterGreyNodes(v.parents)$;
31 | | $Q.push(GetMinimumNode(U))$;
32 end
33 return *Subgraph of G induced by black nodes*

A precondition to an action node can be either derived predicate (privilege node) or a primitive fact node. In case it's a primitive fact node, we simply add that node to the reduced graph. In case it's privilege node, we have to decide which action will satisfy that node. We choose to expand the action with minimal cost that can satisfy the privilege node. Careful care should be taken in order to handle possible cycles in the graph. We have solved this complexity by incorporating a mechanism similar to the DFS search algorithm.

The algorithm is specified in Algorithm 1. In line 5, we assign a stack data structure, Q, with the shortest path in the attack graph G, where the source node is at the top of the stack, and the goal node is at the bottom. If we encounter a leaf node (fact node) we can't expand that node further, and mark it as a solved node in line 9. If we encounter an action node for the first time, we add it to the stack, to revisit and make sure all his parent nodes were also satisfied. If one of the action node's parents was visited already (grey) this means we encountered a cycle and should not mark this action node as resolved. If all of the action node's children were resolved (black), we can resolve this action node (line 22). For privilege node, we first check if one of it's parents (action nodes) is satisfied (line 24). If so, then the action node also satisfied the current privilege node (line 25). If the privilege node is not already satisfied by an action node, we expand the privilege node's cheapest parent (which is an action node) that is not already expanded (lines 30 and 31).

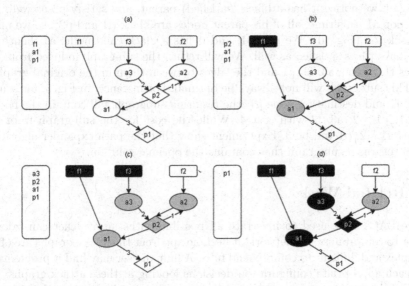

Fig. 4. Example PathExpander execution. Source node is f1 and destination node is p1

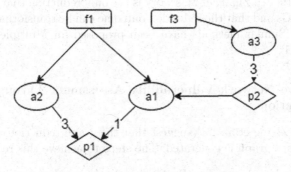

Fig. 5. Example LAG for which PathExpander algorithm returns non-optimal sub-graph

In Fig. 4, the goal node is p1, and the source node is f1. First the shortest path is found. In this case the shortest path includes f1 → a1 → p1 - Fig. 4a. We pop f1, it's a leaf node, so we mark it black, and continue to the next node. We pop a1, it's not yet visited so we mark it gray, and push it back to the stack to revisit. a1 has one parent node, p2 so we push p2 to the stack. p2 is poped, it is a privilege node that has no black children, so we continue. It's color is white, so we mark it gray. We push p2 back since we'll have to revisit and make sure p2 was satisfied. p2 has two parents: a2 and a3. The edge (a2, p2) is cheaper than (a3, p2) so we push a2 to the stack. After we pop a2, we notice it has a gray parent node (p2) meaning a cycle. So we skip a2, after marking him gray (Fig. 4b). Since we pushed p2 before pushing a2, we will pop p2 again. This time, p2 has only one white parent node - a3. So we push a3. This time, all of a 3's children can be satisfied (f3, Fig. 4c). So we satisfy a3, and then when we revisit p2 again we notice it has a black (satisfied) parent, and satisfy p2 as well. We now pop a1, this time, all of his parent nodes are black (f1 and p2), so we mark it black too (Fig. 4d). We pop p1, and it has a black child node: a1, hence we will mark p1 as satisfied as well. We will return the sub-graph induced from the nodes f1, a1, p2, a3, f3, p1 and the edges between them in the original graph.

This sub-graph will not always be optimal. For instance, in Fig. 5, for source node f1 and destination node p1, the resulting subgraph will contain the nodes: f1, a1, p1, p2, a3, f3 with cost 4. While the cost for the sub-graph from the nodes: f1, a2, p1 will be 3. Experiment show that the path expander algorithm often returns a sub-graph that contains the optimal solution.

5   Related Work

Since LAGs were used to illustrate all possible paths an attacker can take in order to compromise the network, it became apparent that these graphs are often complex and difficult to comprehend fully. A human user may find it problematic to reach appropriate configuration decisions looking at these attack graphs.

For this reason, many researchers have set their goal to reduce the size of a LAG, with minimal impact to the conclusion that can be drawn from the reduced attack graph [19–21]. Zhang et al.'s work is the only reduction that could directly be used on LAGs and that the reduction outcome can be transformed into a planning problem. Similar methods have been proposed for Multiple Prerequisites (MP) graphs [11].

5.1   Effective Network Vulnerability Assessment Through Model Abstraction

In their work, Zhang et al. [21] suggest that the graph reduction will take place before the attack graph is generated. The steps to achieve this reduction are:

1. Reachability-based grouping. Hosts with the same network reachability (both to and from) are grouped together.

2. Vulnerability grouping. Vulnerabilities on each host are grouped based on their similarities.
3. Configuration-based breakdown. Hosts within each reachability group are further divided based on their configuration information, specifically the types of vulnerabilities they possess.

Following those steps results in an reduced input to attack graph generators - namely MulVAL, which results in a reduced and easier to understand attack graph. In an experiment described in the article, an attack graph with initially 217 nodes and 281 edges was reduced to 47 nodes and 55 edges. In our experiments we also applied those algorithms with some success. In their work, the authors also examined the effect such reductions have on the quantitative security metrics of the attack graph which represent the likelihood an asset will be compromised [22]. It was shown that using this reductions yields different security metrics for different hosts in the network, compared to the original model. The authors claimed that the new security metrics represent the real world better, since many of the vulnerabilities are dependent of each other. We implemented some of the reductions described here, and tested their effectiveness (This will be described in the results section).

6   Evaluation

Our goal is to evaluate how different reductions affect two main parameters. The first parameter is the time it takes finding minimal attack path.

$$TotalTime = Gen + Reduce + Solve$$

Where Gen is the time it takes MulVAL framework to generate an attack graph, $Reduce$ is the running time of the reduction algorithm and $Solve$ is the time it takes the solver to find a solution. The second parameter we took into account is the cost of the minimal plan found by the solver using the different reductions.

$$TotalCost = \sum_{a \in P} Cost(a)$$

Where P is the plan action sequence found by the solver, and $Cost(a)$ is the cost of an action in the sequence according to our attack graph. Initially, the costs in our experiments were taken from exploitability metric in CVSS (Common Vulnerability Scoring System) [23] to represent the easiest exploitable path in the graph. Meaning the path found is the easiest exploitable path for an attacker.

After some experiments we have noticed that the $TotalCost$ of all the paths found have the same cost which is the number of steps an attacker takes. By investigating the results we have concluded that the vulnerabilities costs using the exploitability metric lack variance. To illustrate: more than 70% out of 2500 vulnerabilities were of cost 1 and 2 (the easiest exploits). In order to produce more varied data, we have randomly assigned the cost for vulnerabilities in our experiments, drawn from a uniform distribution. To test the reductions on datasets with different sizes, we created 4 additional datasets from the original dataset, in which only vulnerabilities above certain CVSS impact metric (representing the damage an attacker can cause by applying a vulnerability) were included. This created 5 different datasets with varying number of hosts and vulnerabilities, and varying cost for vulnerabilities.

7 Results

Figure 6 shows the running time in seconds it took to find the attacker path for each reduction on networks in different sizes. The Y axis, in log scale, shows the overall time it took find an attack path. This includes the time it took to generate the attack graph, the time it took to reduce the attack graph and the time it took the planner to solve the planning problem. "Without" is the baseline, meaning that we do not change the original LAG in any form. "Grouping" Refers to the 1st reduction presented in Subsect. 5.1 in which hosts with similar reachability configuration are grouped. "Aggregate" refers to the 2nd reduction in Subsect. 5.1 in which similar vulnerabilities in a software installed on a host are aggregated together. "Aggregate and Group" means applying the two previous reductions together. "PathExpander" is our algorithm described in Sect. 4. The X axis, in log scale shows the number of nodes in the attack graph. The largest graph which included all the hosts and vulnerabilities had 220,700 nodes and represented 309 hosts containing 2398 software vulnerabilities. The results show that as the size of the network gets bigger, PathExpander algorithms finds an attacker path about 4 times faster than the second best reduction (Aggregate and Group). The trend-line for the PathExpander is $y = 0.3892x + 3.9922$ with $R^2 = 0.9939$ while the trend-line for the Aggregate and Group reduction is $y = 1.5598x - 8.59$ with $R^2 = 0.9622$.

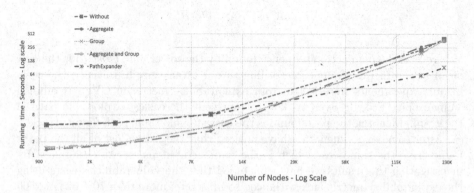

Fig. 6. Total run time in respect to the size of the network and reduction used

Figure 7 shows the cost of the plan found using each reduction in respect to the size of the network. As we can see, using different reductions we found plans with different costs. This is possible due to the fact the planner we have used, Metric-FF is not an optimal planner, and does not guarantee to return the optimal plan. Another important fact we notice is that sometimes by reducing the size of the graph, we find better paths than those found in the non-reduced graph. We have manually checked the planning input files in those cases and made sure that the low-cost plan found on the reduced graph were present in the non-reduced graph, and indeed they were present.

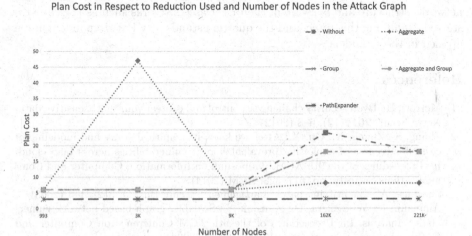

Fig. 7. Total cost in respect to the size of the network and reduction used

8   Conclusion and Discussion

By looking at the results, we can see two interesting trends which are desired for our method. First, in terms of running time, we observe that as the network gets bigger, PathExpander finds solutions much faster than the compared methods. In the largest network which contained 220,700 nodes and represented 309 network hosts with 2398 vulnerabilities, our method found a result more than 3 times faster (93.34 s compared to 356.91 s) than the second best reduction used (Aggregate and Group).

In terms of the quality of the results, our methods consistently found the best attacker path compared to the other methods. We suspect that this is due to the fact that the planner we have used, Metric-FF is not an optimal planner.

Those two results show that using the PathExpander algorithm in order to reduce an attack graph before searching for solution using general planner can both improve the overall running time it takes to find an attacker path, and the quality of the paths found. This can allow security administrators derive better conclusions in regards to which vulnerabilities in which hosts to patch first in order to keep the network secure.

Although in our experiments, drawn from real-life scenarios, the cost of the paths found were always optimal, in the general case this might not always be true. In the future we aim to analyze the conditions in which the results are guaranteed to be optimal. More-over we intend to examine how similar reduction methods can be applied to more complex models that include both costs for actions and probabilities of success.

Another possibility for future work is to relax the assumptions about the attacker. Mainly the fact that this work assumes that the attacker knows the networks structure and is aware of the target assets. Works such as [24,25] have started examining this topic, and the question stands how PathExpander can be applied in those models.

References

1. Morrow, B.: Byod security challenges: control and protect your most sensitive data. Netw. Secur. **2012**(12), 5–8 (2012)
2. Zhang, S., Zhang, X., Ou, X.: After we knew it: empirical study and modeling of cost-effectiveness of exploiting prevalent known vulnerabilities across IaaS cloud. In: Proceedings of the 9th ACM Symposium on Information, Computer and Communications Security, pp. 317–328. ACM (2014)
3. Shostack, A.: Quantifying patch management. Secure Bus. Q. **3**(2), 1–4 (2003)
4. Ammann, P., Wijesekera, D., Kaushik, S.: Scalable, graph-based network vulnerability analysis. In: Proceedings of the 9th ACM Conference on Computer and Communications Security, pp. 217–224. ACM (2002)
5. Sheyner, O.M.: Scenario graphs and attack graphs. Ph.D. thesis, US Air Force Research Laboratory (2004)
6. Ou, X., Govindavajhala, S., Appel, A.W.: MulVAL: a logic-based network security analyzer. In: USENIX Security (2005)
7. Roberts, M., Howe, A., Ray, I., Urbanska, M., Byrne, Z.S., Weidert, J.M.: Personalized vulnerability analysis through automated planning. In: Working Notes of IJCAI 2011, Workshop Security and Artificial Intelligence (SecArt 2011), vol. 4 (2011)
8. Sarraute, C.: New algorithms for attack planning. In: FRHACK Conference, Besançon, France (2009)
9. Ghosh, N., Ghosh, S.: An intelligent technique for generating minimal attack graph. In: First Workshop on Intelligent Security on Security and Artificial Intelligence (SecArt 2009). Citeseer (2009)
10. Poolsappasit, N., Dewri, R., Ray, I.: Dynamic security risk management using Bayesian attack graphs. IEEE Trans. Dependable Secure Comput. **9**(1), 61–74 (2012)
11. Ingols, K., Lippmann, R., Piwowarski, K.: Practical attack graph generation for network defense. In: 22nd Annual Conference on Computer Security Applications Conference, ACSAC 2006, pp. 121–130. IEEE (2006)
12. Ou, X., Boyer, W.F., McQueen, M.A.: A scalable approach to attack graph generation. In: Proceedings of the 13th ACM Conference on Computer and Communications Security, pp. 336–345. ACM (2006)

13. Beale, J., Deraison, R., Meer, H., Temmingh, R., Walt, C.V.D.: Nessus Network Auditing. Syngress Publishing, Rockland (2004)
14. OpenVAS Developers: The Open Vulnerability Assessment System (OpenVAS) (2012)
15. Hoffmann, J.: The Metric-FF planning system: translating "ignoring delete lists" to numeric state variables. J. Artif. Intell. Res. **20**, 291–341 (2003)
16. Obes, J.L., Sarraute, C., Richarte, G.: Attack planning in the real world. arXiv preprint arXiv:1306.4044 (2013)
17. Chen, Y., Wah, B.W., Hsu, C.W.: Temporal planning using subgoal partitioning and resolution in SGPlan. J. Artif. Intell. Res. **26**, 323–369 (2006)
18. Albanese, M., Jajodia, S., Noel, S.: Time-efficient and cost-effective network hardening using attack graphs. In: 2012 42nd Annual IEEE/IFIP International Conference on Dependable Systems and Networks (DSN), pp. 1–12. IEEE (2012)
19. Noel, S., Jajodia, S.: Managing attack graph complexity through visual hierarchical aggregation. In: Proceedings of the 2004 ACM Workshop on Visualization and Data Mining for Computer Security, pp. 109–118. ACM (2004)
20. Homer, J., Varikuti, A., Ou, X., McQueen, M.A.: Improving attack graph visualization through data reduction and attack grouping. In: Goodall, J.R., Conti, G., Ma, K.-L. (eds.) VizSec 2008. LNCS, vol. 5210, pp. 68–79. Springer, Heidelberg (2008). doi:10.1007/978-3-540-85933-8_7
21. Zhang, S., Ou, X., Homer, J.: Effective network vulnerability assessment through model abstraction. In: Holz, T., Bos, H. (eds.) DIMVA 2011. LNCS, vol. 6739, pp. 17–34. Springer, Heidelberg (2011). doi:10.1007/978-3-642-22424-9_2
22. Homer, J., Ou, X., Schmidt, D.: A sound and practical approach to quantifying security risk in enterprise networks. Kansas State University Technical Report, pp. 1–15 (2009)
23. CVSS: A complete guide to the common vulnerability scoring system (2007)
24. Shmaryahu, D.: Constructing plan trees for simulated penetration testing. In: The 26th International Conference on Automated Planning and Scheduling, vol. 121 (2016)
25. Hoffmann, J.: Simulated penetration testing: from "Dijkstra" to "turing test++". In: ICAPS, pp. 364–372 (2015)

Learning Representations for Log Data in Cybersecurity

Ignacio Arnaldo[1]([✉]), Alfredo Cuesta-Infante[2], Ankit Arun[1], Mei Lam[1], Costas Bassias[1], and Kalyan Veeramachaneni[3]

[1] PatternEx Inc, San Jose, CA, USA
iarnaldo@patternex.com
[2] Universidad Rey Juan Carlos, Madrid, Spain
alfredo.cuesta@urjc.es
[3] MIT, Cambridge, MA, USA
kalyan@csail.mit.edu

Abstract. We introduce a framework for exploring and learning representations of log data generated by enterprise-grade security devices with the goal of detecting advanced persistent threats (APTs) spanning over several weeks. The presented framework uses a divide-and-conquer strategy combining behavioral analytics, time series modeling and representation learning algorithms to model large volumes of data. In addition, given that we have access to human-engineered features, we analyze the capability of a series of representation learning algorithms to complement human-engineered features in a variety of classification approaches. We demonstrate the approach with a novel dataset extracted from 3 billion log lines generated at an enterprise network boundaries with reported command and control communications. The presented results validate our approach, achieving an area under the ROC curve of 0.943 and 95 true positives out of the Top 100 ranked instances on the test data set.

Keywords: Representation learning · Deep learning · Feature discovery · Cybersecurity · Command and control detection · Malware detection

1 Introduction

This paper addresses two goals. First, it proposes methods to develop models from *log* and/or *relational* data *via* deep learning. Second, it applies these methods to a cybersecurity application.

Consider advanced persistent threats (APTs). These attacks are characterized by a series of steps: infection/compromise, exploitation, command and control, lateral movement, and data exfiltration [1,19]. In this paper, we focus on the detection of the "command and control," step, i.e. the mechanisms used to maintain a communication channel between a compromised host inside the targeted organization and a remote server controlled by the attacker. Although this phase of the attack can span weeks or months, its detection requires significant sophistication. Savvy attackers minimize their footprints by combining

© Springer International Publishing AG 2017
S. Dolev and S. Lodha (Eds.): CSCML 2017, LNCS 10332, pp. 250–268, 2017.
DOI: 10.1007/978-3-319-60080-2_19

active and stealthy phases, and establish communication channels via unblocked services and protocols, therefore blending in with legitimate traffic.

When *log* data is analyzed over a period of several weeks, these communications exhibit distinctive network profiles [9]. In particular, compromised machines will periodically attempt to communicate with remote server(s), and repeatedly establish lightweight connections through which they receive new instructions. During a minor fraction of these connections, the compromised machine will download a larger amount of data, which corresponds to a software update [19]. The frequency and network profile of these connections will depend on the particular malware family or exploit involved in the attack [15]. Despite these aforementioned observations, most machine learning-based detection techniques only analyze individual connections (see [21] and therein). Given the large volume of data, and the number of connections that must be monitored and analyzed, it is a challenge to identify behavioral patterns over multiple weeks of data.[1]

This example application identifies two pressing needs that could be addressed by deep learning. They are: (1) the development of automated methods to identify patterns, a.k.a *features* by processing data collected over long periods of time and (2) the identification of patterns that can deliver highly accurate detection capability. In recent years, data scientists have made tremendous strides in developing deep learning-based models for problems involving language (which use text as data) and vision (which use images as data). Deep learning has turned out to be so powerful in these two domains because it is able to produce highly accurate models by working with raw text or images directly, without requiring humans to transform this data into features. At its core, almost all deep learning models use multi-layered neural networks, which expect numeric inputs. To generate these inputs, images or text are transformed into numerical representations (often designed by humans).

In order to develop similar solutions for *log* or *relational* data, our first goal is to identify ways to process and generate numerical representations of this type of data. To maximize both quality and efficiency, this step requires a compromise between the amount of human knowledge we incorporate into developing these representations, *vs.* how much we exploit the ability of deep neural networks to automatically generate them. In this paper, we present multiple ways *log* data can be represented, and show how deep learning can be applied to these representations. Our contributions through this paper are:

- **Deep learning for log/relational data:** We present multiple ways to represent log/relational data, and 4 different deep learning models that could be applied to these representations. To the best of our knowledge, we are the first to elucidate steps for generating deep learning models from a *relational* dataset.

---

[1] Depending on an organization's size and level of activity, devices such as next-generation firewalls can generate up to 1TB of log data and involve tens of millions of entities on a daily basis.

- **Applying deep learning to cybersecurity applications:** We apply these methods to deliver models for two real world cybersecurity applications.
- **Comparing to human-driven data processing:** We demonstrate the efficacy of the deep learning models when compared to simple aggregations generated by human-defined standard database operations.

The rest of this paper is organized as follows. Section 2 describes the steps required to process raw logs and obtain data representation suitable for deep learning. Section 3 introduces the collection of deep learning techniques. We build upon these techniques and augment them with human-generated features for better discovery in Sect. 4. The experimental work and results are presented in Sects. 5 and 6. Section 7 presents the related work, and we conclude in Sect. 8.

2 Data Transformations and Representations

In this section, we describe a generic sequence of data transformation steps human data scientists can take to derive *features* from timestamped log data. With these transformations, we identify the data representations that can be fed into a deep learning technology.

Rep 1: Raw logs: In a nutshell, *logs* are files that register a time sequence of events associated with entities in a monitored system[2]. Logs are generated in a variety of formats: `json`, `comma-separated-values`, `key-value` pairs, and `event logs`. (Event types are assigned unique identifiers and described with a varying number of fields).

Rep 2: Structured representation: The first step entails parsing these *logs* to identify entities (e.g., IP addresses, users, customers) relationships between entities (e.g., one-to-many, many-to-many), timestamps, data types and other relevant information about the relational structure of the data. The data is then stored either in the relational data model (`database`) or *as-is*, and the relational model is used to process the data as needed. Either way, in this step, humans either come up with a relational model using their prior knowledge of how data is collected and what it means, or acquire this knowledge by exploring data.

Rep 3: Per entity, event-driven time series: Once the relational structure is identified, a temporal sequence of events associated with a particular instance of an entity is extracted. For example, we may identify all the events associated with a particular IP address. These events are usually irregular in time, and each event is described with a number of data fields. For example, a sequence of network connection events associated with an IP address is described using ports, protocols, the number of bytes sent and received, etc.

---

[2] In enterprises today, logs are generated by network devices, endpoints, and user authentication servers, as well as by a myriad of applications. Each device registers a certain kind of activity, and outputs different information. Note that even devices belonging to the same category (eg. network devices such as firewalls) report different information and use a different format depending on the vendor and version.

Rep 4: Per entity, aggregated, regular time series: The next step involves defining a time interval and performing simple aggregations for each irregular time series.[3] For example, for any given IP address, we can average the number of bytes sent and received per connection over a time interval of an hour. The output of this step is a regular multivariate time series for each entity-instance. The resulting data representation can be viewed as a multi-dimensional array $D \in \mathbb{R}^{n \times t \times p}$ where n is the number of entity-instances, t is the number of time steps, and p is the number of aggregations.[4]

Rep 5: Entity-feature matrix: This last step consists of generating an entity-feature matrix, in which each row corresponds to an instance of the entity and each column is a feature. This can be directly generated from REP 3 through a process known as *"feature engineering"* or REP 4. Given a multivariate time series $D \in \mathbb{R}^{n \times t \times p}$, the simplest way to generate this representation is to "flatten" the data, resulting in a $n \times (t \times p)$ feature matrix. A common alternative is to perform another aggregation step, this time on top of the regular time series. In the latter case, the result is a $n \times p'$ matrix, where p' is the number of second-level aggregations.

Data Representations Amenable for Deep Learning. One important benefit of deep learning models is their potential to alleviate the feature engineering bottleneck. Below we consider the nuances of the application of deep learning models to different representations.

1. Input data must be separated into independent examples, much like images. Thus, it is necessary to identify the relational structure, and to separate data by entity-instances. Automation of this step is possible, but is beyond the scope of this paper.
2. Deep learning techniques can be applied to the third (per entity event-driven time series), fourth (aggregated regular time series), and fifth (entity-feature matrix) representations.
3. Note that while deep learning models can be applied to entity-feature matrices (last representation), we consider that this approach does not leverage their potential for feature discovery, since multiple levels of aggregations are defined by humans.
4. In this paper, we leverage deep learning techniques to learn features on regular aggregated time series (Rep 4.).

3   Learning Representations Using Deep Neural Networks

We describe 4 different methods suitable for learning representations out of REP 4. (1) Feed-forward neural networks, (2) Convolutional networks, (3) Recurrent neural networks with LSTMs, and (4) Autoencoder + random forest

---

[3] For numeric fields, aggregations include *minimum, maximum, average,* and *standard deviation*; for categorical values, common aggregations are *count_distinct* and *mode*.

[4] For example, if we consider a dataset spanning over 10 days with $n = 1000$ entity instances, a time step $t = 1$ day, and $p = 20$ aggregations, the result of this step would be a $1000 \times 10 \times 20$ array.

pipeline. The first three approaches can be categorized as methods for supervised *feature extraction* or *feature learning* and the last method (without random forest) can be categorized as an an unsupervised approach to *feature learning*.

3.1 Feed-Forward Neural Networks

Feed-forward neural networks (FFNNs) are composed of one or more layers of nodes. The input layer consists of $p \times d$ neurons (one for each value in the input data), while the output layer is composed of m nodes, where m is the number of classes in the data. Intermediate layers are composed of an arbitrary number of nodes, with each layer fully connected to the next one. Figure 1 (left) shows a FFNN trained to classify multivariate time-series.

3.2 Convolutional Networks

Convolutional networks (CNNs or ConvNets) are FFNNs with special connection patterns (see Fig. 1), and have been widely applied for image and video recognition. At the core of CNNs are convolutional filters or *kernels*. These filters are trained to identify patterns in reduced regions of the input data (small shapes in the case of images, or patterns in consecutive data points in the case of time series). CNNs are composed of an arbitrary number of such filters, and are therefore capable of identifying a wide variety of low-level patterns in the data. The same set of filters is applied across all the input data, and for each region of the input data where they are applied, the filter generates an output value that indicates how similar the input region is to the filtered pattern. The output of the convolutional layer is generally fed to a pooling layer, which applies a local maximum operation. Intuitively, this operation provides robustness to determine whether a pattern exists in a region of the input data, independent of its exact

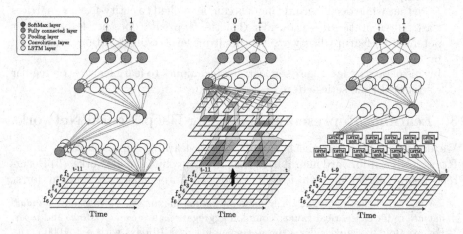

Fig. 1. FFNN-based (left), CNN-based (center), and LSTM-based (right) time series classifiers. For clarity, not all the connections are shown.

location. The outputs of the last convolutional/pooling layers are fed to a fully connected feed-forward neural network. As with standard FFNNs, the final layer is composed of m nodes, where m is the number of classes in the data.

By stacking several layers of convolutional filters and pooling layers, CNNs can identify patterns involving larger regions of the input data. This is a clear example of a "deep" architecture, where lower layers learn to detect building blocks of the input data, while the last layers detect higher-level patterns. It is important to stress that all the parameters (weights) in CNNs are learned during the training process. That is, the networks learns to identify the local patterns that ultimately help them to discriminate between data categories.

In the case of multivariate time-series data, CNNs can exploit locality to learn temporal patterns across one or more variables. Note however, that the relative position of features is generally arbitrary (adjacent features are not necessarily related). This observation motivates the use of convolutional filters of width $= 1$; that is, filters that learn patterns in each feature independently. Another valid possibility explored in this paper considers filters of width $= p$, where p is the total number of input features. In this last case, the network will learn filters or patterns involving all the features.

3.3 Recurrent Neural Networks with LSTMs

Long short-term memory networks (LSTMs) are a special case of recurrent neural networks first introduced in [11]. The main characteristic of these architectures is the use of LSTM cells. LSTM cells maintain a state, and generate an output given a new input and their current state. Several variants of LSTM cells have been proposed, we use the LSTM variant introduced by [17].

Right panel in Fig. 1 shows a high-level representation of a LSTM architecture. The input data is a time-ordered array that is fed sequentially to the network. At each time step, the LSTM cells update their state and produce an output that is related both with the long-term and short-term (i.e. recent) inputs. The final output of the LSTM architecture is generated after propagating all the input sequence through the network.

LSTMs architectures are a solution to the vanishing and exploding gradient; they are said to be superior to recurrent neural networks and Hidden Markov Models to model time series with arbitrarily large time gaps between important events. With respect to FFNNs and CNNs, their main potential advantage is that inputs to LSTM architectures are sequences of arbitrary length, therefore enabling us to train and reuse a single model with time series of different lengths. These two characteristics of LSTMs are particularly relevant for information security analytics, where the goal is to detect attacks that are generally implemented in steps spread over time, and where modeled entities exhibit very different levels of activity, therefore generating time series of varying length.

3.4 Autoencoder + Random Forest Pipeline

Autoencoders are multi-layer feed-forward neural networks. The input and output layers have the same number of nodes, while intermediate layers are composed of a reduced number of nodes. We consider autoencoders that are composed of three hidden layers. The first and third hidden layers count $p/2$ neurons, while the second, central layer is composed of $p/4$ neurons, where p is the dimensionality of the data. The tan-sigmoid transfer function is used as an activation function across the network. The network is trained to learn identity-mapping from inputs to outputs. The mapping from inputs to intermediate layers compresses the data, effectively reducing its dimensionality. Once the network is trained, we can compress the data by feeding the original data to the network, and retrieving the output generated at the central layer of the autoencoder. The output of the central layer is then fed to a random forest classifier.

4 Combining Human Defined and Learnt Features

In this section, we determine whether feature discovery techniques are contributing to improvements in classification accuracy. We do this by separating the aggregated values corresponding to last time step, generated as part of REP 4, from the historic data (previous time steps), and applying the feature discovery methods only to the historic data. All the presented techniques are extensions of the methods described in Sect. 3.

Let $D^i = D^i_{hist} \cup D^i_{last}$ be the multivariate time series associated to entity i, and let d be the number of time steps in the series. Therefore, D^i_{last} is the aggregated time series vector corresponding to the last time step data and D^i_{hist} is the time series composed of the previous $(d-1)$ vectors. In our case, the time unit is 1 day, and we consider $d = 28$ time steps. We introduce a pipeline where:

- Deep learning methods learn a set of "time series features" from D_{hist},
- These learned features are concatenated with D_{last}.
- The combination of learned "time series features" and D_{last} is fed into a random forest classifier.

This way, feature discovery techniques learn a set of "time series features" while the final predictions are generated by interpretable models. By analyzing the grown decision trees, we determine the relative importance of D_{last} and the automatically discovered features. In the following, we describe unsupervised and supervised techniques to discover new features from historic data.

4.1 Extension of Dimensionality Reduction Methods (RNN)

Given a time series dataset $D = D_{hist} \cup D_{last}$, we first apply a dimensionality reduction technique to D_{hist} (historic feature values). The outputs of the dimensionality reduction are combined with the last time step's feature vector and fed into a random forest as depicted in Fig. 2. We use the same dimensionality reduction technique explained in Sect. 3, namely RNNs or autoencoders. In the following, we refer to this extension as 'RNN + RF ext'.

4.2 Extension of Supervised Deep Learning (FFNN, CNN, LSTM)

We consider the models depicted in Fig. 3. The designed models have two separate inputs: D_{hist} and D_{last}. While D_{hist} undergoes a series of nonlinear transformations in the left layers of the network, D_{last} is directly connected to the last layers of the network. With this design, we expect to force the network to learn features from D_{hist} that are complementary to D_{last}. Once trained, these models can be used in two fashions: (1) as standalone models used to·predict on unseen data, (2) as "feature generators" used to extract features for unseen data. In this paper we adopt the second strategy, and feed the extracted features into a random forest classifier. We illustrate this strategy in Fig. 2 (right).

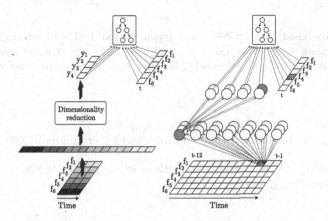

Fig. 2. Dimensionality reduction and random forest pipeline (left), and FFNN-based model used as time-series feature generator (right)

We now detail the steps involved both in model training and deployment using these "feature generators". At training time, we proceed as follows:

1. Train the models in Fig. 3 via backpropagation using the dataset D.
2. Once the model is trained, propagate D through the network and retrieve the outputs generated at the last layer of the left section. Note that the output D_{ts} will be a matrix of shape $n \times q$, where n is the number of examples and q is the number of learned features.
3. Concatenate D_{last} and D_{ts} to obtain D_{conc}, a new dataset with shape $n \times (p + q)$. Note that p is the number of human-engineered features.
4. Train a decision tree classifier with D_{conc}.

To predict on unseen data D', we proceed as follows:

1. Propagate D' through the network and retrieve the outputs generated at the last layer of the left section of the network. As in the training scenario, the output D'_{ts} will be a matrix of shape $n \times q$

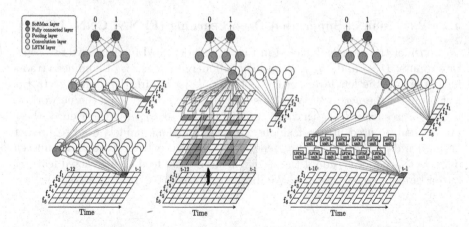

Fig. 3. FFNN-based (left), CNN-based (center), and LSTM-based (right) models designed to learn time series features. These models present a novel structure that enables to complement a set of existing features with new features learned from historic data.

2. Concatenate D'_{last} and D'_{ts} to obtain D'_{conc}
3. Feed D'_{conc} to the trained random forest and generate predictions.

In the following, we refer to these method extensions as 'FFNN + RF ext', 'CNN + RF ext', and 'LSTM + RF ext'.

5   Experimental Work

This section describes the two datasets studied in this paper, as well as the parameterization of the models introduced in previous sections.

5.1   Real-World Command and Control Detection Dataset

We consider two months' worth of logs generated by an enterprise next generation firewall and target the detection of command and control. These log files register approximately 50 million log lines and 150K active entities daily, summing to a total of 3 billion log lines and 12 million analyzed entities.

Extracting daily features: In this step, we extract 32 features, describing the activity of each entity within a 24-hour time window. The features capture information about the number of connections, the bytes sent and received per connection, the packets sent and received per connection, the duration of the connections, and the intervals between connections, as well as information about relevant ports, applications and protocols, and alerts raised by the firewall.

Labeling the dataset: To label the dataset, we use a combination of outlier analysis and validation through VirusTotal's [4] threat intelligence. We perform outlier analysis (see methods in [23]) on the feature data on a daily basis, and investigate the top outliers using VirusTotal. VirusTotal provides the latest files detected by at least one antivirus program that communicates with a given IP address when executed in a sandboxed environment. In addition, it provides the fraction of considered antivirus software that reported the file. We consider an IP address to be malicious if at least 5 different antivirus programs or scanners report that malicious files are communicating with that address.

It is important to note that this intelligence-based labeling process is noisy, not only because antivirus programs themselves might present false positives, but because malicious files might communicate with IP addresses for purposes other than command and control. In fact, creating and labeling a real-world dataset is challenging in itself, both because the labeling must be performed by human analysts, a scarce resource with very limited bandwidth, and because the context required to determine whether a particular entity is involved in an attack is often missing. This severely limits the number of labeled attack examples available for offline modeling and experimentation.

Building a control dataset: We preserve all the attack examples, and sub-sample 1% of the remaining entities, which are considered benign. The result is a dataset composed of $89K$ examples. The data pertaining to the first month ($53K$ entities) is used to train the models, while data from the second month ($36K$ entities) is used for testing. It is worth noting that, although we analyze a subsampled dataset, malicious activities remain a minor fraction (0.56%) of the examples. This results in an extreme class imbalance, which increases the difficulty of the detection problem.

From daily features to multi-week time series: For each sampled entity, we build a multivariate time series in which the time step is a day, the length of the series is $d = 28$ days, and the activity at each time step is described with 32 features. Therefore, our dataset can be viewed as a $89K \times 28 \times 32$ array (num. examples $\times$ time steps $\times$ features).

5.2   ISCX Botnet Dataset

Introduced in 2014, the ISCX Botnet dataset [5] is a comprehensive dataset released in packet capture (pcap) format which contains activity traces of 16 types of botnets, as well as legitimate traffic. To build the dataset, the authors combined traces extracted from the ISOT botnet detection dataset [27], the ISCX 2012 IDS dataset [18], and traffic generated by the Malware Capture Facility Project [3]. The botnet traffic is either captured by honeypots, or through executing the bot binaries in safe environments. Table 1 summarizes the characteristics of the data. It is important to note that the dataset is divided into a training (4.9 GB) and testing set (2.0 GB), where the training split includes traffic generated by 7 botnet types, while the testing set contains traffic generated by 16 botnet types. This way, the authors propose a challenging dataset to evaluate whether models that have been trained to detect a reduced set of botnets can accurately detect unseen botnets. In their best-performing effort, the authors report a detection (true positive) rate of 75% and a false positive rate of 2.3%.

Table 1. Characteristics of the ISCX 2014 Botnet dataset

Split	#Flows	#Src IPs	#Dst IPs	#Src/Dst IPs	#Flow TS	#Mal. TS	#Ben. TS
Training	356160	7355	40502	57321	65737	38485	27252
Testing	309206	6392	17338	28657	36532	13480	23052

From pcap to flow features: We use FlowMeter [8], a network traffic flow generator, to separate the packet capture data into individual flows. FlowMeter aggregates flows on the basis of the 5-tuple set (Source IP, Source Port, Destination IP, Destination Port, Protocol) and a timeout parameter. Each flow is described with the following 23 features: *Duration, Bytes per second, Packets per second, Min/Max/Avg/Std packet inter-arrival times, Min/Max/Avg/Std inter-arrival times of sent packets, Min/Max/Avg/Std inter-arrival times of received packets, Min/Max/Avg/Std active time*, and *Min/Max/Avg/Std idle time*.

Labeling the dataset: The dataset includes a list of malicious IPs and their corresponding botnet types. In some cases, individual IPs are reported, but in others, the authors report source and destination IPs as a pair. Therefore, we label as malicious all flows that include one of the individually listed IPs (either as source or destination), and all flows where both the source and destination IPs match a reported pair. All remaining flows are considered benign. Although the authors report the botnet type associated with the malicious IPs, we approach the problem as a binary classification problem (malicious vs benign).

Flow features to time series of flows: We first aggregate all the flows that involve the same pair of source and destination IPs (independently of ports and protocols). Thus, for each pair of source/destination IPs, we obtain a $t \times p$ matrix, where t represents the number of flows, and $p = 23$ represents the number of features. For symmetry with the real-world dataset, we split time series into segments of (at most) 28 flows. Note that this last step is only applied when the pair of source/destination IPs presents more than 28 flows. This way, the preprocessing of the training split results in a $65737 \times 28 \times 23$ array (num. examples $\times$ flows $\times$ features), while the testing split results in a $36532 \times 28 \times 23$ array.

5.3   Model Implementation, Training, and Validation

We compare the models proposed in this paper against random forests [6] and against a pipeline composed of a dimensionality reduction step performed with PCA followed by a random forest classifier. Note that we consider data composed of n examples, p features, and d days. In order to apply these approaches, the data is flattened to obtain a feature matrix with n examples and $p \times d$ features. The resulting entity-feature matrix is suitable for training random forests. In the case of the PCA + Random Forest pipeline, the data is projected to the principal component space using the top j principal components, and the projected data is fed to a random forest classifier. Its extended counterpart is referred to as 'PCA + RF ext' and is analogous to the RNN-based method explained in Sect. 4.1.

Table 2. Description and number of features generated by the compared models

Method	#discov. features	#layers	Training algorithm
PCA + RF	16	-	-
RNN + RF	8	3 (16-8-16)	Adam
FFNN	16	3 (16-16-16)	Stoch. grad. descent
CNN	16	2 (conv + pool) + 1 fully conn	Adam
LSTM	16	1 layer with 100 LSTM cells	Adam
RNN + RF ext	8	3 (16-8-16)	Adam
PCA + RF ext	16	-	-
FFNN + RF ext	16	3 (16-16-16)	Stoch. grad. descent
CNN + RF ext	16	2 (conv + pool) + 1 fully conn	Adam
LSTM + RF + int	16	1 layer with 100 LSTM cells	Adam

Table 2 shows the details of the implementation and training of the models compared in this paper. To enable a fair comparison with methods such as random forests or PCA, we did not parameter-tune any of the neural network-based methods (FNN, CNN, LSTM, Autoencoders (RNN)). While this can lead to a poor model parametrization, we are interested in these methods' "out-of-the-box" performance, since it is a better performance proxy for how well they will detect malicious behaviors other than command and control.

6   Results

In this section, we compare the detection performance of the learning methods introduced in this paper on the real-world command and control dataset and on the ISCX 2014 botnet dataset. We also analyze the importance of automatically-discovered features.

6.1   Real-World Command and Control Dataset

Table 3 shows the AUROC and true positives in the top 100 compared methods when evaluated on unseen data.

Effect of longer time span data: Our first observation is that the AUROC achieved using 1 day of data reaches 0.923 for *RF* and 0.928 for *PCA + RF*. However, if we use more days for training, the performance of these two methods degrades. This degradation is noteworthy since we do not know beforehand the length of the time necessary for the successful detection malicious behaviors.

Did augmentation help? On average, the AUROC and number of TP in the Top 100 for the extended methods that try to complement human generated features (i.e. methods labeled with *ext*) is higher than the ones that don't. Note that, by design, the methods CNN, LSTM, PCA + RF ext, RNN + RF ext, FFNN + RF ext, CNN + RF ext, and LSTM + RF ext require more than one day of data; therefore, for those methods, we do not present performance metrics for the one-day case. We also notice that the performance of these methods does not degrade as we increase the time span.

The best and the worst: The best AUROC is achieved using *PCA + RF ext* with 28 days of data, and using *CNN + RF ext* with 7 days of data. These models present AUROCs of 0.943 and 0.936 respectively when evaluated on unseen data. However, this is only marginally better than the 0.923 baseline AUROC obtained with a random forest classifier using one day of data. In particular, our results show that the use of RNN + RF (autoencoders) achieves the worst detection performance since it is unable to either detect attacks or discover new features.

Table 3. AUROC and true positives in the top 100 of the compared methods when evaluated on unseen data. Data sets are represented by their time span (1, 7, 14 and 28 days).

Method	AUROC				True positives in top 100			
	1 day	7 days	14 days	28 days	1 days	7 days	14 days	28 days
RF	0.923	0.895	0.881	0.883	95	84	89	82
PCA + RF	0.928	0.830	0.816	0.867	86	66	68	74
RNN + FR	0.814	0.747	0.686	0.701	37	35	4	19
FFNN	0.906	0.840	0.829	0.869	7	0	0	0
CNN	-	0.901	0.718	0.873	-	0	1	4
LSTM	-	0.898	0.877	0.869	-	8	26	31
PCA + RF ext	-	0.920	0.927	0.943	-	89	92	87
RNN + RF ext	-	0.747	0.678	0.756	-	9	30	3
FFNN + RF ext	-	0.929	0.888	0.912	-	92	93	92
CNN + RF ext	-	0.936	0.876	0.837	-	95	89	74
LSTM + RF ext	-	0.904	0.914	0.923	-	88	89	89

Key findings: The goal of our exploration was to examine, how these methods perform "out-of-box". For the real world data set, based on our results, we cannot conclusively say whether the new learning methods helped. We also posit that:

- perhaps the information present in the last day's features is enough to accurately detect command and control communications.
- the performance of those methods using FFNN, CNN, LSTM, and RNN (autoencoders) can be improved via parameter tuning.

6.2   ISCX 2014 Botnet Dataset

Method Comparison: Given that the training and testing splits contain traces of different botnets, we perform two series of experiments. First, we compute the 10-fold cross-validation AUROC on the training set for the models being compared (left section of Table 4). This setup allows us to compare the models' capacity to identify known botnets on unseen data. Second, we compute the testing set AUROC to analyze the models' capacity to identify unseen botnets (right section of Table 4). The resulting detection rates are in accordance with the results reported in [5], where the authors present very high accuracies in cases where traces of the same botnets are included in the training and testing splits, while the detection rates on unseen botnets drop significantly. For instance, the AUROC of the random forest trained on individual flows drops from 0.991 to 0.768. As stated in [5], this performance drop shows that the trained models do not generalize well when it comes to accurately detecting unseen botnets.

Detecting previously seen botnets: With the exception of the CNN method, all methods achieve high cross-validation detection metrics. In particular, when 7 consecutive flows are modeled, the AUROCs range from 0.904 to 0.997. The best cross-validation results are achieved by the random forest and PCA + Random Forest methods, which achieve AUROCs of 0.997 and 0.992 respectively. In both cases, considering multiple flows yields incremental benefits (from 0.991 to 0.997, and from 0.990 to 0.992). However, in general, considering multiple flows does not systematically improve detection. While it yields improvements for RF, PCA + RF, FFNN, FFNN + RF ext, CNN + RF ext, and LSTM + RF ext, it results in performance decreases for RNN + RF, CNN, LSTM, PCA + RF ext, and RNN + RF ext.

Did augmentation help? The AUROCs of methods that complement human-engineered features (i.e. methods labeled with *ext*) are higher than those for the complementary subset for FFNN, CNN, and LSTM models, and lower for PCA and RNN.

Detecting previously unseen botnets: When it comes to detecting unseen botnets, the best AUROCs are achieved by the models PCA + RF ext (0.811) and LSTM + RF ext (0.808), which in both cases model segments of 28 consecutive flows. This represents an improvement of 5.60% and 5.21% with respect to the baseline random forest trained with individual flows (0.768). In this case, the AUROC of

Table 4. Cross-validation AUROC on the training split of the ISCX training split, and AUROC on the testing split of the compared methods. Data sets are represented by the number of considered flows (1, 7, 14 and 28 flows).

Method	CV AUROC Training Set				AUROC Training/Testing Set			
	1 flow	7 flows	14 flows	28 flows	1 flow	7 flows	14 flows	28 flows
RF	0.991	0.997	0.997	0.997	0.768	0.753	0.766	0.748
PCA + RF	0.990	0.992	0.992	0.992	0.769	0.753	0.743	0.774
RNN + FR	0.975	0.971	0.965	0.955	0.746	0.747	0.741	0.641
FFNN	0.905	0.947	0.947	0.948	0.724	0.713	0.751	0.744
CNN	-	0.737	0.632	0.620	-	0.644	0.633	0.449
LSTM	-	0.907	0.903	0.899	-	0.624	0.744	0.542
PCA + RF ext	-	0.997	0.830	0.832	-	0.788	0.802	0.811
RNN + RF ext	-	0.995	0.809	0.805	-	0.715	0.701	0.728
FFNN + RF ext	-	0.978	0.949	0.986	-	0.731	0.739	0.788
CNN + RF ext	-	0.929	0.936	0.849	-	0.748	0.747	0.759
LSTM + RF ext	-	0.904	0.932	0.901	-	0.672	0.779	0.808

all the extended methods that complement human-engineered features (i.e. methods labeled with *ext*) is higher than the complementary subset.

Feature Analysis: We analyze the features discovered with the models *PCA + RF ext* and *LSTM + RF ext* using 28 days of data (see Sects. 4.1 and 4.2). These models are chosen for analysis because they present the highest AUROCs (0.811 and 0.808 when evaluated on unseen data).

Given that the training and testing sets contain traces generated by different botnets, we merge the two splits, obtaining a single dataset composed of 102246 examples (65734 train + 36512 test). Also, since many of the modeled pairs of source and destination IPs present a reduced number of flows, we analyze feature importance over varying lengths of the time series. This way, we consider four different views of the data:

1. All pairs of source and destination IPs. This results in a dataset composed of 102246 examples, out of which 51946 are malicious.
2. Pairs of source and destination IPs presenting at least 7 flows. This results in 22093 examples, out of which 9093 are malicious
3. Pairs of source and destination IPs presenting at least 14 flows. This results in 18849 examples, out of which 8336 are malicious
4. Pairs of source and destination IPs presenting at least 28 flows. This results in 16414 examples, out of which 7529 are malicious.

Table 5 reports the sum of the importance of all human-engineered features, as well as the sum of the importance of all automatically discovered features. The results show that the discovered features are used by the classifier in all four scenarios. The aggregated importance of learned features is 13.4% and 14.4% for PCA-based and LSTM-based features when all examples are considered.

Table 5. Aggregated importance of human-engineered and discovered features.

Method	Aggregated feature importance							
	All examples		7 flows or more		14 flows or more		28 flows	
	Human	Auto	Human	Auto	Human	Auto	Human	Auto
PCA + RF ext	86.6%	13.4%	54.1%	45.9%	52.7%	47.3%	51.2%	48.8%
LSTM + RF ext	85.7%	14.4%	54.6%	45.4%	50.6%	49.4%	52.3%	47.7%

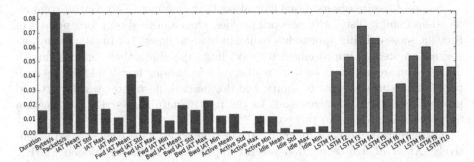

Fig. 4. Feature importance as determined by a random forest classifier of human-engineered features (blue) and automatically discovered features (red) with a LSTM-based model. The aggregated importance of human-engineered features is 50.6%, while that of discovered features is 49.4%. Only pairs of source/dest IPs with 14 or more flows are considered for the analysis presented in this figure. (Color figure online)

These low values are explained by the fact that most pairs of source/destination IPs present a single flow, and so the classifier relies on the individual flow features. However, the aggregated importance of discovered features increases as we consider examples composed of 7, 14, and 28 of flows. In particular, the importance of human and LSTM-based features is close to parity (50.6% vs. 49.4%) when examples composed of 14 or more flows are considered. This case is highlighted in Fig. 4, in which we show the importance of the 23 original flow features (blue) and the 10 features (red) as determined by a random forest classifier learned with the LSTM model. The most important human-engineered features are *Bytes per second* (1st overall), *Packets per second* (3rd overall), and *Avg inter-arrival time* (5th overall). All LSTM features are used and considered important. Features LSTM-3, LSTM-4, and LSTM-8 are ranked 2nd, 4th, and 6th in overall importance.

7   Related Work

There is a large research community focused on addressing InfoSec use cases with machine learning [12]. The command and control detection problem, and botnet detection in particular, has been widely studied (see [10,21] and therein). Two key aspects differentiate this paper from existing work. First, most research initiatives consider publicly available datasets that are either synthetic or generated in controlled environments. Working with public datasets allows researchers to replicate reported methodologies and to compare results fairly. However, these datasets generally suffer from a lack of generality, realism, and representativeness [5], and results obtained using them do not necessarily transfer to real-world deployments. In this paper, we work with a dataset obtained over two months from a real-world system. (Obtaining representative, real-world datasets is a challenge in itself, and has been discussed in previous sections.)

Second, despite observations indicating that command and control communications exhibit distinctive network profiles when analyzed over long periods of time [9], most existing approaches model individual flows [21]. In [5], the authors suggest a potential improvement for modeling capabilities that includes "multidimensional snapshots of network traffic, e.g., combining flow level features with pair level features (a pair of source and destination, no matter which port and protocol used)." This corresponds to the multivariate time-series classification approaches introduced in this paper.

Classification methods for multivariate time series have been studied extensively. Xi et al. [26] compared several approaches, including hidden Markov models [13], dynamic time warping decision trees [16], and a fully connected feed-forward neural network [14]. Wang et al. [24] explore different methods for projecting time series data into 2D images. The authors then explore the use of convolutional networks to tackle several regression problems from the UCR repository [7].

While deep learning-based solutions have been used for problems involving computer vision and natural language processing, only a few examples exist in the domain of information security. Staudemeyer et al. [20] explore the use of recurrent neural networks with LSTMs to tackle the intrusion detection problem on the 1999 KDD Cup dataset [2]. A recent work by Tuor et al. [22] explores deep learning methods for anomaly-based intrusion detection. There are also reported approaches that leverage LSTM-based models to differentiate the algorithmically-generated domains used for command and control from legitimate ones [25]. This paper is, to the best of our knowledge, the first paper to introduce a generic framework for discovering features from any set of timestamped log data. Moreover, this is the first attempt to automatically discover features that complement existing human-engineered features.

8   Conclusions

In this paper, we have presented multiple ways to represent log/relational data, and 4 different deep learning models that can be applied to these representations. We apply these methods to deliver models for command and control detection on a large set of log files generated at enterprise network boundaries, in which attacks have been reported. We show that we can detect command and control over web traffic, achieving an area under the ROC curve of 0.943 and 95 true positives out of the Top 100 ranked instances on the test data set. We also demonstrate that the features learned by deep learning models can augment simple aggregations generated by human-defined standard database operations.

References

1. Adversarial tactics, techniques and common knowledge. https://attack.mitre.org
2. KDD Cup 99. http://kdd.ics.uci.edu/databases/kddcup99/kddcup99.html
3. Malware capture facility project. http://mcfp.weebly.com/
4. VirusTotal. https://www.virustotal.com
5. Beigi, E.B., Jazi, H.H., Stakhanova, N., Ghorbani, A.A.: Towards effective feature selection in machine learning-based botnet detection approaches. In: 2014 IEEE Conference on Communications and Network Security, pp. 247–255 (2014)
6. Breiman, L.: Random forests. Mach. Learn. **45**(1), 5–32 (2001)
7. Chen, Y., Keogh, E., Hu, B., Begum, N., Bagnall, A., Mueen, A., Batista, G.: The UCR time series classification archive (2015)
8. Draper-Gil, G., Lashkari, A.H., Mamun, M.S.I., Ghorbani, A.A.: Characterization of encrypted and VPN traffic using time-related features. In: Proceedings of the 2nd International Conference on Information Systems Security and Privacy, ICISSP, vol. 1, pp. 407–414 (2016)
9. García, S., Uhlíř, V., Rehak, M.: Identifying and modeling botnet C&C behaviors. In: Proceedings of the 1st International Workshop on Agents and CyberSecurity, ACySE 2014, NY, USA, pp. 1:1–1:8. ACM, New York (2014)
10. Garcia, S., Zunino, A., Campo, M.: Survey on network-based botnet detection methods. Secur. Commun. Netw. **7**(5), 878–903 (2014)
11. Hochreiter, S., Schmidhuber, J.: Long short-term memory. Neural Comput. **9**(8), 1735–1780 (1997)
12. Jiang, H., Nagra, J., Ahammad, P.: Sok: applying machine learning in security-a survey. arXiv preprint arXiv:1611.03186 (2016)
13. Kim, S., Smyth, P., Luther, S.: Modeling waveform shapes with random effects segmental hidden Markov models. In: Proceedings of the 20th Conference on Uncertainty in Artificial Intelligence, UAI 2004, pp. 309–316. AUAI Press, Arlington (2004)
14. Nanopoulos, A., Alcock, R., Manolopoulos, Y.: Information processing and technology. In: Feature-based Classification of Time-series Data, pp. 49–61. Nova Science Publishers Inc, Commack (2001)
15. Plohmann, D., Yakdan, K., Klatt, M., Bader, J., Gerhards-Padilla, E.: A comprehensive measurement study of domain generating malware. In: 25th USENIX Security Symposium (USENIX Security 2016), pp. 263–278. USENIX Association, Austin (2016)
16. Rodríguez, J.J., Alonso, C.J.: Interval and dynamic time warping-based decision trees. In: Proceedings of the 2004 ACM Symposium on Applied Computing, SAC 2004, NY, USA, pp. 548–552. ACM, New York (2004)
17. Sak, H., Senior, A.W., Beaufays, F.: Long short-term memory based recurrent neural network architectures for large vocabulary speech recognition. CoRR abs/1402.1128 (2014)
18. Shiravi, A., Shiravi, H., Tavallaee, M., Ghorbani, A.A.: Toward developing a systematic approach to generate benchmark datasets for intrusion detection. Comput. Secur. **31**(3), 357–374 (2012)

19. Sood, A., Enbody, R.: Targeted Cyber Attacks: Multi-staged Attacks Driven by Exploits and Malware, 1st edn. Syngress Publishing, Burlington (2014)
20. Staudemeyer, R.C., Omlin, C.W.: Evaluating performance of long short-term memory recurrent neural networks on intrusion detection data. In: Proceedings of the South African Institute for Computer Scientists and Information Technologists Conference, SAICSIT 2013, NY, USA, pp. 218–224. ACM, New York (2013)
21. Stevanovic, M., Pedersen, J.M.: On the use of machine learning for identifying botnet network traffic. J. Cyber. Secur. Mobility **4**(3), 1–32 (2016)
22. Tuor, A., Kaplan, S., Hutchinson, B., Nichols, N., Robinson, S.: Deep learning for unsupervised insider threat detection in structured cybersecurity data streams (2017)
23. Veeramachaneni, K., Arnaldo, I., Korrapati, V., Bassias, C., Li, K.: AI^2: training a big data machine to defend. In: 2016 IEEE 2nd International Conference on Big Data Security on Cloud (BigDataSecurity), IEEE International Conference on High Performance and Smart Computing (HPSC), and IEEE International Conference on Intelligent Data and Security (IDS), pp. 49–54 (2016)
24. Wang, Z., Oates, T.: Imaging time-series to improve classification and imputation. In: Proceedings of the 24th International Conference on Artificial Intelligence, IJCAI 2015, pp. 3939–3945. AAAI Press (2015)
25. Woodbridge, J., Anderson, H.S., Ahuja, A., Grant, D.: Predicting domain generation algorithms with long short-term memory networks. arXiv preprint arXiv:1611.00791 (2016)
26. Xi, X., Keogh, E., Shelton, C., Wei, L., Ratanamahatana, C.A.: Fast time series classification using numerosity reduction. In: Proceedings of the 23rd International Conference on Machine Learning, ICML 2006, NY, USA, pp. 1033–1040. ACM, New York (2006)
27. Zhao, D., Traore, I., Sayed, B., Lu, W., Saad, S., Ghorbani, A., Garant, D.: Botnet detection based on traffic behavior analysis and flow intervals. Comput. Secur. **39**, 2–16 (2013)

Attack Graph Obfuscation

Hadar Polad$^{(\boxtimes)}$, Rami Puzis, and Bracha Shapira

Ben Gurion University of the Negev, Beer Sheva, Israel
poladh@post.bgu.ac.il

Abstract. Before executing an attack, adversaries usually explore the victim's network in an attempt to infer the network topology and identify vulnerabilities in the victim's servers and personal computers. In this research, we examine the effects of adding fake vulnerabilities to a real enterprise network to verify the hypothesis that the addition of such vulnerabilities will serve to divert the attacker and cause the adversary to perform additional activities while attempting to achieve its objectives. We use the attack graph to model the problem of an attacker making its way towards the target in a given network. Our results show that adding fake vulnerabilities forces the adversary to invest a significant amount of effort, in terms of time, exploitability cost, and the number of attack footprints within the network during the attack.

1 Introduction

Protecting a network is always a difficult task, because attackers constantly explore new ways to penetrate security systems by exploiting their vulnerabilities. These vulnerabilities often go unpatched, due to a lack of resources, negligence, or a variety of other reasons.

Although network professionals have offered various versions of the attack process over the years, today the general anatomy of the attack process is thought to be comprised of five steps [19]:

1. Reconnaissance
2. Scanning
3. Gaining access
4. Maintaining access
5. Covering tracks

Some networking professionals estimate that an adversary routinely spends up to 95% of its time planning an attack, while only spending the remaining 5% on the execution of the attack [15]. During the reconnaissance step, the attacker attempts to gather as much information about the designated network as possible. While doing so, the adversary generates traffic on the network, making it vulnerable [14].

In this research, we try to sabotage the reconnaissance and scanning steps of the attack process by obfuscating the information acquired by the adversary. While making the attacker repeat steps 1 and 2 repeatedly, after failing to achieve step 3.

© Springer International Publishing AG 2017
S. Dolev and S. Lodha (Eds.): CSCML 2017, LNCS 10332, pp. 269–287, 2017.
DOI: 10.1007/978-3-319-60080-2_20

It is well known, that attackers rely upon the ability to accurately identify the operating system and services running on the network in order to plan and execute successful attacks [19].

The desire to mislead a possible attacker underlies this research to explore the possibilities of obfuscating the information acquired by an adversary. This has been achieved by adding fake vulnerabilities that distract the attacker and contribute to the erroneous construction of an attack path. Misleading the attacker with false information can set the attacker on a path that will deplete its resources, increase the likelihood of detection due to the increased activity, and keep the attacker away from essential targets. Our hypothesize claims that adding fake vulnerabilities will cause the attackers to perform additional activities while attempting to achieve their goals.

In this study we assume that the attacker will choose the path with the lowest total cost in the resulting attack graph. In addition, the attack graph is constructed from the information known to the adversary. In addition, we assume that the attacker knows the structure of the given network, and the vulnerabilities in each host.

In this research, we make the following contributions:

- We add fake vulnerabilities to hosts in the network in order to make it harder for the adversary to reach its goal in the target network. This will be reflected in the time of the attack planning, the overall cost of the attack and the attack path length.
- We gathered a collection of guidelines for fake vulnerabilities placement.

  This study was conducted with the knowledge that when a layer of deception is added to a host in the network, it inhibits the network's routine activity. Therefore, the client, i.e., the enterprise aiming to protect its network, should decide the desired level of security it wishes to apply and what resources it can and will provide for the task of protecting its network.

  Another consideration pertains to the fake information provided to attackers is if the fake information given to the attacker is naive or poorly chosen, the attacker may immediately become suspicious and assume that the responses obtained are deceptive [24].

  In our research, we try to add the deceptive information carefully and sensibly, in such a way that it cannot be easily detected by an attacker. While applying deceptive information to a specific host, it should be consistent and it should be selected as conclusions from the information provided.
- We provide good modeling for the effect of adding fake vulnerabilities, on an attacker. We are using attack graphs as modeling approach to the problem of an adversary attacking the network. While creating good parameters for calculating the adversary's efforts affected by the fake vulnerabilities.
- In contrast to the many articles that use small synthetic networks, this study examines the impact of our algorithm on a real enterprise network, and attempts to solve the above problem in a real organization.

The paper is organized as follows. We provide background on our way of modeling in Sect. 2. We review related work at Sect. 3. Then we explain the problem, describe our approach and experimental evaluations in Sect. 4. In Sect. 5 we draw our conclusions.

2   Background

2.1   Modeling the Problem

In order to model the problem of an attacker penetrating a network, we must find a model that describes all the possible paths an attacker could use to achieve its goal in a specific network.

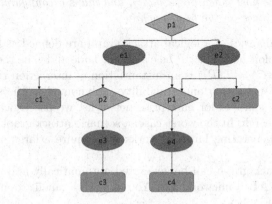

Fig. 1. Example of logical attack graph·

The earliest work on attack graphs was done manually by Red Teams [27]. Even with advancements in the technology, it is understood that the attack graph generation is not, by any means, a trivial task. The main challenges are scalability and visualization of the resulting graph, such that it is useful and comprehensible to the human user. When trying to generate attack graphs for large scale networks, which can include thousands of vulnerabilities and hosts, the task of building a graph that represents the network and the relationships between all of the components in the graph is difficult. Attack graph generation techniques require [16]:

1. List of existing vulnerabilities in the network hosts.
2. The network topology.

These two requirements serve as input to the attack graph generation model.

Due to the difficulties mentioned above, the task of gathering the necessary data about the network structure and the vulnerabilities present in the hosts should be automated as well. In order to achieve that, one should use a vulnerability scanner, such as: Nessus vulnerability scanner [5], openVAS [4] etc.

In previous studies two major attack graph models are presented: **scenario graphs** [27, 28] and **logical graphs** [20].

Logical attack graphs are defined as follows [20]:

Definition 1.

$$(Np, Ne, Nc, E, L, G)$$

Is a logical attack graph, where Np, Ne and Nc are three sets of disjoint nodes in the graph.

$$E \subseteq (Np \times Ne) \cup (Ne \times (Np \cup Nc))$$

L is a mapping from a node to its label, and $g \in Np$ is the attackers goal. Np, Ne and Nc are the sets of privilege nodes, exploit nodes and fact (leaf) nodes, respectively. The labeling function maps a privilege node to the privilege that can be run on the host machine (exploit), and maps a configuration node to the configuration in place on the host machine.

Formally, the semantics of a logical attack graph are defined as follows:

For every exploit node e, let P be e's parent node and C be the set of e's child nodes, then $(\wedge L(C) \Rightarrow L(P))$ is an instantiation of interaction rule L(R) [8,20].

Due to its visualization and scalability (polynomial in the size of the network) which are suitable for enterprise networks, we use logical attack graph model in our research. In the worst case, a scenario attack graph's size could be exponential, so generating large networks can require a large amount of CPU resources.

A logical attack graph can be generated automatically and relatively easily using the **MulVAL** framework [21]. You can see a small example of a logical attack graph in Fig. 1.

Monotonic assumption: Amman et al. [7] were the first to present the term, monotonic assumption, which states that an attacker does not decrease his ability by executing attacks and thus does not need to lose privileges he already gained. Under this assumption attacker privileges always increase during the analysis.

2.2   Vulnerabilities Representation

Known and common vulnerabilities have been compiled and listed in a system operated by the MITRE Corporation and the U.S. National Vulnerability Database [3]. Each vulnerability is tagged with a unique CVE (Common Vulnerabilities and Exposures) identifier and has a CVSS score, which is an open industry standard for assessing the severity of computer system security vulnerabilities.

CVSS scores are based on two subscores:

- Impact Subscore - reflects the direct consequence of a successful exploit and represents the consequence to the impacted component.
- Exploitability Subscore - reflects the ease and technical means by which the vulnerability can be exploited.

3 Related Work

The approach presented in this research utilizes a defense strategy based on providing false information to attackers, and a defensive tool that modifies the network representation. In this section we review studies related to this approach, specifically, we will describe studies that apply deception and attack graph games.

3.1 Deception

Throughout history the use of deception has evolved in societies around the world, and today, its use has expanded and it has become an integral part of our technical systems.

In cyber security, deception often starts in the initial reconnaissance phase and can be used to create a controlled path for attackers to follow.

An understanding of this area was needed in order to effectively conduct our research. In our work, the main role of deception is to make the attacker believe that the information obtained is real. When this is successfully accomplished, the defender can gain a significant advantage over the adversary.

Several empirical experiments have been conducted in order to demonstrate the effect of deception in Cyber security. Cohen et al. [10,11] showed how deception can control the path of an attack using red teams in experiments attacking a computer network.

Repik [22] work makes a strong argument in favor of using deception as a tool of defense. He discusses why planned actions taken to mislead hackers have merit as a strategy and should be pursued further.

Honeypots were first used in computer security in the late 1990s. The use of deception in honeypots involves temptation which is used to mislead the attacker, so that the attacker erroneously believes that the parts of the system it is interacting with are entirely legitimate. In fact, parts of the system are actually isolated and monitored, and this gives the defender the ability to detect, deflect, or, in some manner, counteract attempts at unauthorized use of information systems. In literature, honeypots are used in several ways [6]:

- As a detection tool - Honeypots result in a low false positive rate, because they are not intended to be used as part of the user's routine tasks in a system; thus any interaction with the honeypots is illegitimate.
- As prevention tools by slowing down attackers or discouraging them from continuing their attacks. For example, the LaBrea "sticky" honeypots [2] which answer connection attempts in a way that causes the machine on the other end to get "stuck", sometimes for a very long time.

The recognition that deception could be an integral part of the network defense field, resulted in the need for a tool for deception, and Fred Cohen designed the *Deception ToolKit (DTK)* [9], a tool which makes it appear to attackers as though the system running the DTK has a large number of widely known vulnerabilities.

When the attacker issues a command or request, the DTK generates a predefined response, in order to encourage the attacker to continue its exploration of the host, or results in a shutdown of the service.

Since honeypots threaten the secrecy of attack methods, cyber attackers try to avoid honeypots. Rowe et al. [24] suggest the idea of fake honeypots, in which a system might pretend to be a honeypot in order to scare away attackers, reducing the number of attacks and their severity.

If the deception is obvious to the adversary, even unintentionally, the attacker can avoid, bypass, and even overcome the deceptive traps. We aim to avoid this situation and make the deception hard to identify by an adversary.

Honeypots are also used for gathering information about attacks. The Honeynet Project [1] is an international security research organization, which invests its resources in the investigation of the latest attacks and the development of open source security tools to improve Internet security.

3.2 Attack Graph Games

Recently there has been significant interest in game theory approaches to security. Durkota et al. introduced the term, "attack graph game" [12], and presented a new leader-follower game-theory model of the interaction between a network administrator and an attacker who follows a multistage plan to attack the network. In order to determine the best strategy for the defender, they used the Stackelberg game (a two phase game), in which the defender (the leader in this game) takes actions in order to strengthen the network by adding honeypots. Then, the attacker selects an optimal attack plan based on knowledge about the defender's strategy. The Stackelberg equilibrium is found by selecting the pure action of the defender that minimizes the expected loss under the assumption that the attacker will respond with an optimal attack.

The researchers presented the problem by using a type of logical attack graph that they refer to as dependency attack graphs which are generated by MulVAL.

The results in this paper were based on an experiment conducted on a small business network (20 hosts) which scales well (less than 10 s) with 14 honeypots, but hasn't been tried on a large network.

Korzhyk et al. [17] compared the Stackelberg framework to the more traditional Nash framework. A recent survey [18] explored the connections between security and game theory more generally.

After a thorough review of studies that are directly related to our approach, we can say that to the best of our knowledge, there is no approach that adds false vulnerabilities and considers that the act of adding these "lies" has consequences to the routine operation of the network. Therefore, we were unable to find a method that carefully selects which fake vulnerabilities to add and indicates where to add them in the network's PCs (without adding dedicated decoys), with the aims of making it **harder** for the attacker, forcing the attacker to use a significant amount of resources, and increasing the attack execution time dramatically.

In attack graph games, there is a choice of where to put decoys in the network by minimizing the expected loss of the defender. In contrast, our main goal is to make it more strenuous for the attacker, causing attrition and a waste of the adversary's valuable resources until the adversary is detected or waste its resources.

In deception the work conducted has largely been based on adding decoys, however no work has been performed to evaluate the effect of the deception cost. In addition, a large amount of work has been done in luring the attacker to a particular path, attacker intimidation, and identifying the attacker, as well as gathering information on the activities of the attacker. Based on our careful review of the literature, we were unable to identify previous research that meets all of our requirements and addresses the network defense problem in this particular way.

4 Attack Graph Obfuscation-Based Defense

4.1 The Problem

Achieving security cannot be accomplished with a single, silver-bullet solution. In order to ensure effective and comprehensive security, one should combine numerous defense mechanisms. There are four protection mechanisms commonly used in computer systems [6]: denial and isolation, obfuscation, deception and adversary characterization.

In the area of deception, there are few approaches that can provide the following:

1. The ability to model the problem in a way that considers all of the paths that lead the adversary toward its goal in the target network. Attack graphs has this capability, which can serve as an advantage with experienced and inexperienced adversaries.
2. Invisibility to the adversary. A poorly designed deception layer may not adequately hide the defense technique from the adversary, enabling the attacker to identify the false information. If the deception can be detected by the attacker (even partially), it will most likely know how to avoid it; therefore, the deception layer should be consistent and carefully designed.
3. The ability to make enough changes in a functioning network to mislead an adversary, while maintaining a stable system that functions. This can be addressed by carefully selecting the network components to be defined as deceptive. Note that a decision that all network parts should be deceptive results in an enormous waste of resources on the part of the organization trying to defend itself from cyber attacks.

   An organization should determine the extent of security which will satisfy its needs and the amount of resources it can dedicate toward implementing this.

4.2   Attacker Model

We believe that the adversary's goal is to obtain privileges in its designated target in the penetrated network, while minimizing its use of resources and making as little noise in the network as possible. Too much noise can lead to detection by an IDS. Limiting the consumption of resources is an objective shared by all - defenders and attackers alike.

This led us to conclude that the best approach for modeling the attacker is:

- Adding cost to the action nodes by their matching exploitability subscore (see Sect. 2.2), which is derived from the CVSS score.
- The attacker aims to choose the path to its goal in the targeted network with the lowest cost.

The attack graph will constructed considering the above, while using the information the attacker can get. In this research we assume the attacker have all the information about the network - topology and vulnerabilities in each host.

4.3   Overview of Our Solution

Our main goal is to present a new, novel defense method that will change an adversary's attack graph of a targeted enterprise network. In doing so, deceive the attacker that believes the information provided is genuine.

The major question that we attempt to answer in our research is:

Given an attack graph G of a specific network and a maximal amount of effort that the defender can put toward its defense, in order to change G to G', what value of G' will require the most effort on the part of the attacker, as the attacker works to achieve the attack's goal?

While considering the following:

Maximal amount of effort that the defender can put toward its defense, refers to the number of hosts and/or servers characterized as deceptive.

There is a significant effect on the routine functioning of the network when false information about a host is provided. We assume that when a host is characterized as deceptive, the amount of wrong information provided is negligible.

Additionally, we need to be careful when choosing the vulnerabilities to add. A conflict between two (or more) of the vulnerabilities selected, may lead attackers to identify the deception mechanism and adapt itself to the new situation.

Our goal is to create additional realistic, but fake, activities that the attacker must tend to in order to achieve its goals; parameters to measure this effect are defined below in Sect. 4.4.

We used Fred Cohen's *Deception ToolKit (DTK)* [9] in order to preform a deceptive information about a host (runs Linux OS). We used [23] guidelines in order to do so. The DTK makes it appear to attackers as if the system running the DTK has a large number of widely known vulnerabilities. When the attacker issues a command or request the DTK's results in a pre-defined response, in order to encourage the attacker to continue his exploration of the host or result

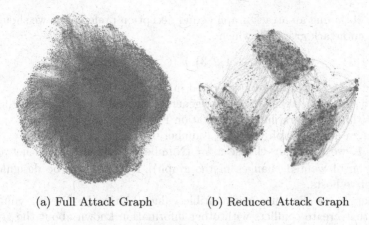

(a) Full Attack Graph   (b) Reduced Attack Graph

Fig. 2. Attack graphs

a shutdown of the service. The Deception tool kit made up of C programs, Perl scripts and additions/modifications to host configuration files.

In order to provide further clarification regarding our research question, we provide an algorithm, presented in the following section.

Evaluation Algorithm

We first create the graph G, then modify it to G' (see Fig. 3) and then we get the attack plan and evaluate the parameters:

– **Create attack graph G:**

1. Scan the network - We gather the information on the network hosts using the Nessus [5] vulnerability scanner. The Nessus scanner provides the host's OS, open ports, existing vulnerabilities, and other information.
2. Create the topology - We obtain information about the connectivity between the hosts in the organization by changing the translated Nessus file.
3. Assign nodes' cost - In our research we find it necessary to construct the attack graph from the adversary's point of view. Based on the assumption that the attacker's main concern is reaching its goal, we assume that the consequence of a specific exploit is not one of its primary goals when attacking the target network. We assume that its main concerns are the cost and difficulties encountered while performing an exploit. This led us to conclude that the best approach for modeling the attacker is adding cost to the action nodes by their matching exploitability subscore (see Sect. 2.2), which derived from the CVSS score.
4. Create attack graph G - After gathering all of the information needed in order to construct an attack graph, we produce a logical attack graph using the MulVAL framework [21].

– **Create attack graph G':**

Then, we obfuscate the attack graph by following this algorithm: (see Algorithm 1 and graphic representation in Fig. 3):

After obtaining an attack graph G and deception preferences, we should generate an attack graph G' which:

$$\forall i \in \{1, 2, 3\}, p_i(G') >> p_i(G) \tag{1}$$

where p_i is the parameters described in Sect. 4.4.

In order to achieve the above, we generate the following algorithm (see also Algorithm 1 and graphic representation Fig. 3):

Given an attack graph G and N (number of changes),

step 1, we randomly choose x [= (Number of IPs in original network) × (Portion of wanted changes in the graph)] IPs that will be designated as deceptive hosts.

In **step 2,** we eliminate incompatible vulnerabilities, by filtering vulnerabilities that create conflicts with other information known about the given IP address. This is done by filtering vulnerabilities that do not match the operating system that exists on the targeted computer. The information gathered about the IP address was collected earlier by the Nessus scanner [5].

In **step 3,** we randomly choose how many vulnerabilities to add to each IP address. This is done for the following reasons:

- To be as unpredictable as possible, in order to be invisible and avoid detection by the attacker.
- Our assumption is that once a host is characterized as deceptive, the amount of wrong information it provides is negligible.

In **step 4,** we choose x vulnerabilities for each IP address chosen as deceptive. These vulnerabilities are randomly selected, with preference to low-cost vulnerabilities. This is done based on the assumption that adding the same vulnerabilities is a strategic mistake, which can lead to discovery of the defender's deception.

In **step 5,** we create new nodes in the attack graph. For every chosen IP address, we add the new chosen vulnerabilities to the .nessus file. Then, we regenerate the attack graph with Mulval, using the same connectivity as before. As a result, new paths to the target are created, some of them consists of only fake vulnerabilities, and some have a combination of fake and real vulnerabilities.

– **Get attack plan:**

In order to obtain the attack plan, we use Jorg Huffman's POMDP [25,26] planner and convert it to an MDP. The output of this MDP converter is a type of PDDL (Planning Domain Definition Language) file.

We solve the PDDL file obtained in the previous step using the fast-downward planner [13] which we executed with the landmark-cut heuristic search.

4.4 Formulating Parameters for Success

We evaluate our method by four main parameters:

1. p_1 = Number of recalculation - these are increased each time the adversary's attack plan consists of fake vulnerabilities and replanning is needed.

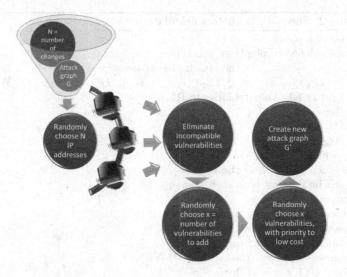

Fig. 3. Create G′ process

2. p_2 = Total time - total planning time (the time it took for the attacker to plan its path to reach the goal).
3. p_3 = Plan cost - the total cost the attacker should pay through its attack.
4. p_4 = Plan length - the number of steps the attacker made in the attack graph through its attack.

The procedure of calculating the parameters for success is presented in Algorithm 2.

Algorithm 1. Create G'

Input: G (AttackGraph), numofChanges

Output: D=ip1:[vul1,vul2...], ip2:[vul1,vul2...] ...

1: G' = G
2: $pickedIps$ ← list of random chosen IPs from $AttackGraph$
3: **for** each IP $ip \in pickedIps$ **do**
4:     $currentVulList$ ← list of all vulnerabilities exist in ip
5:     $OSVersion$ ← getOSVersion(RV_u)
6:     $vulList$ ← list of all vulnerabilities associate with $OSVersion$
7:     $validOSList$ ← $vulList \setminus currentVulList$
8:     $numOfVulToAdd$ ← $Random(0, validOSList ← Size())$
9:     $chosenVul$ ←
        $chooseRandomList(validOSList, numOfVulToAdd)$
10:     $AddNodes(G', IP, chosenVul)$
11: **end for**
12: **return** G'

Algorithm 2. Success parameters calculation

Input: p (attack plan)
Output: reCalculation, planCost, totalTime
1: Let $p_i = (u_1, u_2, ...u_m)$ be the attack plan at iteration i
2: Let $D = (IP_1 : (v_1, v_2, ...v_{n_1}), IP_2 : (v_1, v_2, ...v_{n_2})...IP_m : (v_1, v_2, ...v_{n_m})) \leftarrow$ all the assignment of fake vulnerabilities to IPs.
3: flag = True
4: reCalculation = 0
5: **while** flag **do**
6:    flag = False
7:    **if** $\exists i$, such that $u_i \in D$ **then**
8:       Let i be the first node such that $u_i \in D$
9:       flag = True
10:      time = time + $p_{counter}.time$
11:      planCost = planCost + $\sum_{j=1}^{i} cost(u_j)$
12:      planLength = planLength + i
13:      removeNodeFromAttackGraph(IP,CVE)
14:      **for** $k \in (1, 2...i - 1)$ **do**
15:         cost(k)=0
16:      **end for**
17:      $p_{(reCalculation + 1)}$ =getNewAttackPlan()
18:      break
19:   **end if**
20:   reCalculation = reCalculation + 1
21: **end while**

Correctness: Due to the monotonic assumption [7] (mentioned in Sect. 2), we assume that if an attacker gains some privileges in the network, it does not lose them. For this reason, each of the parameters is calculated as if the attacker doesn't lose the privileges gained during the attack.

We will divide the parameters calculation into two cases:

1. If the attacker tries to exploit a fake vulnerability, it invests the associated cost and then replans a path to the goal, taking into account the achievements made thus far. In this case, the parameters are calculated as follows:
   - Total time is calculated as the sum of the time it took to plan the path to reach the current target and the total time the attacker spent until this point (in all of its previous plans).
   - Expected cost is calculated as the sum of:
     - The costs of all of the nodes in the attack plan that preceded the node which contains the fake vulnerability (including the node itself).
     - The total cost the attacker spent until this point (in all of its previous plans).
   - Path length is calculated as the number of all of the nodes in the attack plan that preceded the node which contains the fake vulnerability, and the path length of the attacker spent until now (in all of its previous plans).

– Number of recalculations is calculated as the number of attack plans incorporates fake vulnerabilities.
2. If the attacker successfully reaches the goal, it means that the attack plan was devoid of fake vulnerabilities. In this case, the parameters are calculated as follows:
   – Total time is calculated as the sum of the time it took to plan the path to the target and the total time the attacker spent until this point (in all of its previous plans).
   – Expected cost is calculated as the sum of the costs of all of the nodes in the attack plan and the total cost the attacker spent until this point (in all of its previous plans).
   – Path length is calculated as the number of all of the nodes in the attack plan and the total path length the attacker spent until this point (in all of its previous plans).
   – Number of recalculations is calculated as the number of recalculations so far plus 1.

We compute the *planCost* and *planLength*, taking into account the fact that the adversary stops carrying out its attack plan when it is interrupted by a fake vulnerability. Therefore, the cost and length of the attack plan is the sum of all of the exploitable vulnerabilities, including the cost of trying to exploit the fake vulnerability.

In order to follow the monotonic assumption, after computing the current parameters, we nullify the cost of all of the nodes that don't include fake vulnerabilities, which the attacker could have exploited during the attack plan, to the first node that contains a fake vulnerability. From this vertex on, the attacker cannot execute the attack as planned or gain additional privileges. In the next attack plan, these nodes will not be considered in the attack path length and cost.

4.5   Data Collection

Enterprise Network Dataset. In order to measure the effectiveness of our method we needed a real data that represents a real network. We collected data using a Nessus [5] scan of an enterprise organization. This network consists of 150 hosts and 394 vulnerabilities. We used an automatic tool for constructing the attack graph from the real network. We chose a state of the art attack graph generation tool called MulVAL [21]. The created attack graph consists of 157,387 nodes and 250,753 edges. The attack graph generated from the described network is presented in Fig. 2(a).

In order to create a more manageable dataset, that is interesting and also reflects the real network, we reduced the network dataset to 79 IPs, which contain a total of 140 vulnerabilities and created the following topology:

In order to create an attack graph, two inputs are needed: list of existing vulnerabilities in the network hosts, which gathered by Nessus scan [5] and the topology of the network. The network topology needs to be created according to the real network construction or some other interesting topology. We simulated topology similar to the topology of the organization. The topology contained three main networks:

- DMZ - the network accessible from the Internet.
- Internal network.
- Secure network - the network in which all the important assets were located.

We randomly created connections between hosts in the DMZ network to hosts in the internal network, and between hosts in the internal network and hosts in the secure network.

This process provided us the opportunity to obtain a challenging and interesting network that we can apply our algorithm on.

Figure 2(b) contains the attack graph generated from the reduced network with the new topology.

Vulnerabilities and Exploitability Subscore Dataset. In order to apply our method, we need a dataset of all known vulnerabilities' CVE IDs and their exploitability subscores; we used the NVD CVE dataset [3]. From this list of vulnerabilities (containing more than 90,000 vulnerabilities) we chose the vulnerabilities to add to our network, based on the vulnerabilities' exploitability subscores. The exploitability subscore is used to determine the cost of the nodes in the generated attack graph.

We also needed a list that matches an operating system to vulnerabilities can be found in this OS. This was done by parsing the vulnerability description provided by the NVD dataset [3].

4.6   Experiments

Implementation. In order to produce reliable experiments which can be generalized, we followed each step as described above.

Platform. We used a system with a quad cores Intel(R) Xeon(R) CPU E5-2620 v2 @ 2.10 GHz, 2 GB of memory and running Ubuntu 12.04.5 with a Linux kernel 3.13.0-95. The resources were fully dedicated to our experiments.

In our experiments we followed the following steps:

1. **Create attack graph G:** see Fig. 2(b).
2. **Create attack graph G′:** For our experiments We created three types of networks that G′ represents:
   - Network with 10% of the total number of computers defined as deceptive (eight computers).
   - Network with 30% of the total number of computers defined as deceptive (24 computers).
   - Network with 50% of the total number of computers defined as deceptive (40 computers).

On average, each deceptive IP address was assigned 3.3 fake vulnerabilities; each vulnerability added randomly was selected with a preference to vulnerabilities with a low exploitability cost.

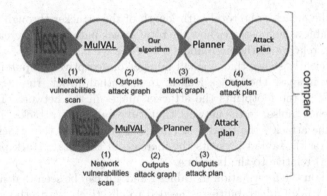

Fig. 4. Evaluation process

Evaluation. Algorithm 2 was used in order to calculate the parameters we defined in Sect. 4.4.

We planned five trials for every portion of computers selected, which produced a total of 16 attack graphs, one of which is the original attack graph (G) which represents the given network, and selected as a baseline. Throughout the experiment each graph had the same topology, with the same connectivity and attack goals.

Results. The results are presented in Fig. 5 and further explanations below.

Table 1 provides the attack graphs' data, which includes the attack graphs' average number of nodes, edges, and vulnerabilities relative to the percentage of computers defined as deceptive.

Table 1. Attack graph data

Deceptive IPs	Nodes	Edges	Vulnerabilities
0%	10147	16591	140
10%	10815	18421	164.2
30%	12392.8	22591	220.6
50%	13936.8	26734.4	257.2

- During the planning phase - the total time of the attacker's planning phase. Table 2 shows how the planning time increased (more than double) relative to the base line.

  Based on the information from Tables 1 and 2 it can be concluded that even with making 10% of the computers deceptive, good results are observed. With the addition of just only 24 vulnerabilities (on average), the total time of the planning phase more than doubled.

- During the attack - the total path length of the attacker through the attack graph. Table 3 demonstrates how the total attack path length increased, again doubling, relative to the base line.

  This clearly shows that a small number of deceptive computers can dramatically increase the attack length. In an actual attack, the attack length translates to the footprints the attacker makes in the network. These footprints create "noise" in the network that can result in the attacker's detection.
- During the attack - the total cost (the amount of effort the attacker invests in the attack). Table 4 presents the increase in the total attack path length (doubling) relative to the base line.

  Based on the information in Tables 1 and 4 it can be seen that just a small amount of fake vulnerabilities is needed to effectively waste the adversary's resources, and more specifically that by adding vulnerabilities to less than 30% of the computers in the network can cost the adversary almost 1.5 times more.

Table 2. Planing time relative to base line

% of deceptive computers	Total time
10%	2.5
30%	3
50%	4.7

Table 3. Total attack path length relative to base line

% of deceptive computers	Total attack length
10%	1.7
30%	1.8
50%	2

Table 4. Total attack path cost relative to base line

% of deceptive computers	Total attack cost
10%	1.46
30%	1.53
50%	1.6

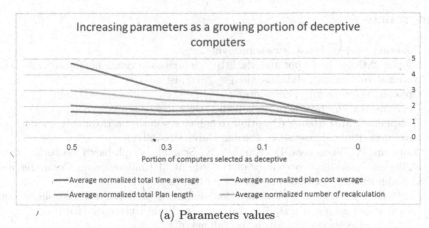

(a) Parameters values

Number of vulnerabilties added to the
original network

(b) Average number of added vulnerabilities

Fig. 5. Results

5   Conclusions

Our experiments show that the direction taken by our approach delivers good results. We demonstrate that by adding just a small number of fake vulnerabilities can significantly affect the amount of resources that the attacker will invest in finding the path to its goal.

Thus, the use of a relatively simple algorithm can serve as the basis for a strong defensive mechanism. As expected, when the rate of deceptive computers increases, the effort expended by the attacker increases as the effect on the adversary time and resources is significantly high.

In future work, we will use a heuristic function in order to find which hosts will be selected as deceptive.

References

1. Honeynet project. https://www.honeynet.org/
2. Labrea: "Sticky" honeypot and ids. http://labrea.sourceforge.net/labrea-info.html
3. National vulnerability database. https://nvd.nist.gov/
4. Openvas. http://www.openvas.org/
5. Deraison, R.: The Nessus project. http://www.nessus.org
6. Almeshekah, M.H.: Using deception to enhance security: a taxonomy, model, and novel uses. Ph.D. dissertation, Purdue University (2015)
7. Ammann, P., Wijesekera, D., Kaushik, S.: Scalable, graph-based network vulnerability analysis. In: Proceedings of the 9th ACM Conference on Computer and Communications Security, pp. 217–224. ACM (2002)
8. Chatterjee, S.: Dragon: a framework for computing preferred defense policies from logical attack graphs. Ph.D. dissertation, Iowa State University (2014)
9. Cohen, F.: Deception tool kit. http://all.net/dtk/
10. Cohen, F., Koike, D.: Leading attackers through attack graphs with deceptions. Comput. Secur. **22**(5), 402–411 (2003)
11. Cohen, F., Koike, D.: Misleading attackers with deception. In: Proceedings from the Fifth Annual IEEE SMC on Information Assurance Workshop, pp. 30–37. IEEE (2004)
12. Durkota, K., Lisỳ, V., Bošanskỳ, B., Kiekintveld, C.: Optimal network security hardening using attack graph games. In: Proceedings of IJCAI, pp. 7–14 (2015)
13. Helmert, M.: The fast downward planning system. J. Artif. Intell. Res. (JAIR) **26**, 191–246 (2006)
14. Huber, K.E.: Host-based systemic network obfuscation system for windows. Technical report, DTIC Document (2011)
15. Kewley, D., Fink, R., Lowry, J., Dean, M.: Dynamic approaches to thwart adversary intelligence gathering. In: Proceedings of DARPA Information Survivability Conference and Exposition II, DISCEX 2001, vol. 1, pp. 176–185. IEEE (2001)
16. Khaitan, S., Raheja, S.: Finding optimal attack path using attack graphs: a survey. Int. J. Soft Comput. Eng. **1**(3), 2231–2307 (2011)
17. Korzhyk, D., Yin, Z., Kiekintveld, C., Conitzer, V., Tambe, M.: Stackelberg vs. nash in security games: an extended investigation of interchangeability, equivalence, and uniqueness. J. Artif. Intell. Res. (JAIR) **41**, 297–327 (2011)
18. Manshaei, M.H., Zhu, Q., Alpcan, T., Bacşar, T., Hubaux, J.-P.: Game theory meets network security and privacy. ACM Comput. Surv. (CSUR) **45**(3), 25 (2013)
19. Murphy, S., McDonald, T., Mills, R.: An application of deception in cyberspace: Operating system obfuscation1. In: International Conference on Information Warfare and Security, p. 241. Academic Conferences International Limited (2010)
20. Ou, X., Boyer, W.F., McQueen, M.A.: A scalable approach to attack graph generation. In: Proceedings of the 13th ACM Conference on Computer and Communications Security, pp. 336–345. ACM (2006)
21. Ou, X., Govindavajhala, S., Appel, A.W.: MulVAL: a logic-based network security analyzer. In: USENIX Security (2005)
22. Repik, K.A.: Defeating adversary network intelligence efforts with active cyber defense techniques. No. AFIT/ICW/ENG/08-11. Air Force Inst of Tech Wright-Patterson AFB OH School of Engineering and Management (2008)
23. SANS Institute Reading Room: Installing, Configuring, and Testing The Deception Tool Kit on Mac OS X (2006). https://www.sans.org/reading-room/whitepapers/detection/installing-configuring-testing-deception-tool-kit-mac-os-1056

24. Rowe, N.C., Custy, E.J., Duong, B.T.: Defending cyberspace with fake honeypots. J. Comput. **2**(2), 25–36 (2007)
25. Sarraute, C., Buffet, O., Hoffmann, J.: Penetration testing==POMDP solving? arXiv preprint arXiv:1306.4714 (2013)
26. Sarraute, C., Buffet, O., Hoffmann, J.: POMDPs make better hackers: accounting for uncertainty in penetration testing. arXiv preprint arXiv:1307.8182 (2013)
27. Sheyner, O., Haines, J., Jha, S., Lippmann, R., Wing, J.M.: Automated generation and analysis of attack graphs. In: Proceedings of 2002 IEEE Symposium on Security and Privacy, pp. 273–284. IEEE (2002)
28. Sheyner, O.M.: Scenario graphs and attack graphs. Ph.D. dissertation, US Air Force Research Laboratory (2004)

Malware Triage Based on Static Features and Public APT Reports

Giuseppe Laurenza[1(✉)], Leonardo Aniello[1], Riccardo Lazzeretti[1],
and Roberto Baldoni[1,2]

[1] Department of Computer and System Sciences "Antonio Ruberti",
Research Center of Cyber Intelligence and Information Security (CIS),
Sapienza Università di Roma, Rome, Italy
{laurenza,aniello,lazzeretti,baldoni}@dis.uniroma1.it
[2] CINI Cybersecurity National Laboratory, Rome, Italy

Abstract. Understanding the behavior of malware requires a semi-automatic approach including complex software tools and human analysts in the loop. However, the huge number of malicious samples developed daily calls for some *prioritization mechanism* to carefully select the samples that really deserve to be further examined by analysts. This avoids computational resources be overloaded and human analysts saturated. In this paper we introduce a *malware triage* stage where samples are quickly and automatically examined to promptly decide whether they should be immediately dispatched to human analysts or to other specific automatic analysis queues, rather than following the common and slow analysis pipeline. Such triage stage is encapsulated into an architecture for semi-automatic malware analysis presented in a previous work. In this paper we propose an approach for sample prioritization, and its realization within such architecture. Our analysis in the paper focuses on malware developed by Advanced Persistent Threats (APTs). We build our knowledge base, used in the triage, on known APTs obtained from publicly available reports. To make the triage as fast as possible, only static malware features are considered, which can be extracted with negligible delay, without the necessity of executing the malware samples, and we use them to train a random forest classifier. The classifier has been tuned to maximize its precision, so that analysts and other components of the architecture are mostly likely to receive only malware correctly identified as being similar to known APT, and do not waste important resources on false positives. A preliminary analysis shows high precision and accuracy, as desired.

Keywords: Malware analysis · Advanced Persistent Threats · Static analysis · Malware triage

1 Introduction

Cyber threats keep evolving relentlessly in response to the corresponding progress of security defenses, resulting in an impressive number of new malware that are being discovered daily, in the order of millions [6]. To cope with

S. Dolev and S. Lodha (Eds.): CSCML 2017, LNCS 10332, pp. 288–305, 2017.
DOI: 10.1007/978-3-319-60080-2_21

this enormous volume of samples there is the necessity of a malware knowledge base, to be kept updated over time, and to be used as the main source of information to realize novel and powerful countermeasures to existing and new malware. Some malware are part of sophisticated and target-oriented cyber attacks, which often leverage customized malware to remotely control the victims and use them for accessing valuable information inside an enterprise or institutional network target. According to NIST Glossary of Key Information Security Terms[1], such *"adversary that possesses sophisticated levels of expertise and significant resources which allow it to create opportunities to achieve its objectives by using multiple attack vectors (e.g., cyber, physical and deception)"* is known as Advance Persistent Threats (APT). APTs typically target Critical Infrastructures (CIs) as well as important organizations, stealthily intruding them and persisting there over time spans of months, with the goal of exfiltrating information, undermining or impeding critical aspects of a mission, program, or organization; or positioning itself to carry out these objectives in the future. Therefore, among the large amount of collected malware, those belonging to some APT should be considered as the most dangerous. In addition, the sooner an APT malware is identified, the smaller is the loss it can cause. Within this scenario, it is important to define an efficient workflow for APT malware analysis, aimed first at quickly identifying malware that could belong to APTs and increase their priority in successive analysis (i.e., *APT malware triage*), and then determine whether these suspicious samples are really related to APTs (i.e., *APT malware detection*). This early identification can be embedded in the malware analysis architecture recently presented in [15], which provides semi-automatic malware analysis, and supports a flow of analysis, continuous over time, from the collection of new samples to the feeding and consequent growth of the malware knowledge base. Such an architecture includes totally automated stages, in order to keep up with today's pace of new malware, and also manual stages, where human analysts have to reverse engineer and study in details the samples that have not been completely understood through automatic means. Although the architecture is framed in a scenario tailored for CIs, its employment can be naturally extended to any situation where a malware knowledge base is desired. Within the architecture, a rank is produced for each sample as the intermediate output of some automatic analyses, based on current malware knowledge base, representing to what extent such sample resembles something that is already known and included in that knowledge base. Such rank determines whether the sample should be further analyzed by a human analyst. This can be seen as a specific instance of sample prioritization, where samples follow different paths within a complex analysis workflow depending on priority scores they get assigned during the first stages. To this end, in this paper we introduce a *malware triage* stage, where samples are timely analyzed to understand as soon as possible whether they likely belong to some known APT campaign and should be dispatched, with highest priority, to human analysts or other components of the architecture for further analysis. Such prioritization is mainly

---

[1] http://nvlpubs.nist.gov/nistpubs/ir/2013/NIST.IR.7298r2.pdf.

aimed at giving precedence to what we deem to be more important to analyze. In fact understanding whether we are being threaten by an APT is much more urgent than dissecting an unknown variant of some adware. It is to note that the objective of this triage stage is not APT malware detection, which is instead pursued at a later stage by human analysts and specialized architecture components, rather the final goal of the triage is spotting with the highest precision samples that seem to be related to known APTs. The addition of this triage stage does not call for any relevant change in the architecture, rather it can be easily realized using already envisaged components. A prompt priority assignment is thus fundamental when designing a malware triage stage, which translates to employing analysis techniques very efficient in terms of time performance. There is hence the necessity of a triage stage that, in order, (i) has low computational complexity to timely analyze a great number of samples, (ii) has high precision, i.e. does not prioritize non-APT malware to not overload human analysts and/or complex components with urgent but not necessary analyses, and (iii) has high accuracy, i.e. a high number of samples are correctly identified. For the first purpose, we adopt an approach based only on static analysis methods. Although it is known that static features are not as effective as dynamic features for malware analysis [7], we choose anyway to only rely on static features because for the triage stage we deem prioritization speed more important than accuracy, and we leave more accurate analyses to successive stages along the analysis pipeline. We leverage on publicly available reports on APTs, which include MD5s of related malware. We collect these malware from public sources, such as VirusTotal[2], then the content of sample binaries are examined to extract *static features* to produce required feature vectors. No sample execution is hence needed, and no expensive virtualized or emulated environment needs to be setup and activated. These static features are then used in machine learning tools, where APT name represents the class label, to identify whether the sample is similar to some known APT malware. A part of the samples is used in the training phase and the other samples are used in the validation of the classifier, together with non-APT samples. For classification we train a random forest classifier, because it guarantees high efficiency and low complexity. The classifier has been tuned to provide the precision and accuracy constraints. Results are encouraging, as they suggest this approach can be easily realized within the architecture proposed in [15], and effective in identifying samples similar to malware realized by known APTs with a precision of 100% and accuracy up to 96%. The contributions of this paper are (i) the definition of a novel policy for malware triage, based on the similarity to malware developed by known APTs, (ii) the design of the malware triage stage within the architecture proposed in [15], (iii) a prototype implementation of such architecture, and (iv) an experimental evaluation regarding the performance of the proposed malware triage, using a number of public reports about APTs as dataset. The rest of the paper is structured as follows. Section 2 reports on the state of the art in the field of malware analysis architectures and malware triage. Section 3 describes the reference architecture. The malware triage approach is

---

[2] https://www.virustotal.com/.

detailed in Sect. 4, while prototype implementation and experimental evaluations about triage accuracy are reported in Sect. 5. Conclusions are drawn and future works presented in Sect. 6.

2   Related Work

The analysis of *Advanced Persistent Threats* is an important topic of research within the cyber-security area: many researchers focus on the *avoidance* and/or *detection* of this type of attacks. In [16,24] methodologies are shown to detect the presence of these advanced intruders through *anomaly detection* systems that use network traffic features. In [23] a framework is proposed to leverage dynamic analysis to find evidences of APT presence. Other researchers concentrate their effort in the hardening of organizations [5,11,26,27]. They propose procedures to raise security levels through the implementation of various precautions based on the analysis of previous attacks.

On the contrary, our work is not oriented to develop a monitor to detect suspicious activity or to improve the robustness of organization's defenses. Rather, we aim to develop a triage approach to support expert analysts in their work, trough a prioritization of *interesting* threats. Several works in literature try to face the problem of malware triage by using the same basic principle: finding similarities among malware to identify variants of samples already analyzed in the past, so that they are not analyzed in details and thus do not waste resources such as human analysts.

One famous work in this field is Bitshred [10]: it is a framework for fast large-scale malware triage using features hashing. The main idea is to reduce the size of the feature vector to speed up the machine learning analysis.

VILO [14] is another tool for malware classification: it is based on nearest neighbor algorithm with weighted opcode mnemonic permutation features and it aims to be a *rapid learner of malware families*. It is well suited for malware analysis because it makes minimal assumptions about class structures and thus it can adapt itself to the continuous changes of this world.

An interesting triage approach is the one use by SigMal [13]: using signal processing to measure the similarity among malware. This approach permits to define more noise-resistant *signatures* to quickly classify malware.

All these works propose a triage approach based on the idea of performing deep analysis only on malware that are not similar to known classes (like new malware families), instead our approach prioritizes malware that are related to already known malicious samples in order to find novel samples possibly developed by APTs. We base our system on static features extracted by static analysis. While it is quite unreliable for malware detection [19], in our application static analysis represent a lightweight and efficient tool for classification of detected malware among APTs campaigns. Structural properties [28] would add important knowledge to the classifier, however we discarded them because of their high complexity

3   Architecture

Malware triage is a pre-processing phase, aiming at proritize APT malware analysis in the architecture presented in [15]. In this section we summarize a description of the given malware analysis framework, showing how sample analysis flow is arranged through a staged view of the architecture. For a detailed description of the building blocks composing the architecture, and interactions of the framework within multiple organizations and Critical Infrastructures (CIs), we remind to the original paper [15].

Sample analysis is organized in a series of stages, from sample retrieval to information sharing, shown in Fig. 1.

In the *Loading Stage*, malware samples are gathered from known datasets, malware crawlers, honeypots and other distinct sources. Also APT reports are collected and related malware are retrieved. In the *Input Stage*, samples collected are stored together with a set of metadata characterizing the samples themselves, including the APT they belong to, if any, and their source.

Samples collected are then analyzed in the *Analysis Stage*. *Analysis Tools* are used to examine sample content and analysis in order to extract significative features representing the samples. Machine Learning *Classifiers* are in charge of assigning samples to predefined learned classes on the base of features values. *Clustering tools* group samples according to their similarity, with the goal of isolating specific groups of malware and link unknown samples to them. *Correlators* try to match samples with information about cyber threats retrieved from external sources.

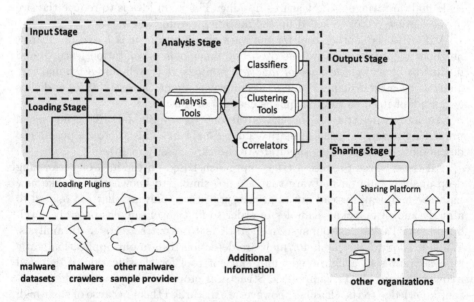

Fig. 1. Staged view of the architecture of the malware analysis framework.

Results obtained by these tools are pushed to the *Output Stage* and eventually made available to other organizations in the *Sharing Stage*.

We underline that both input and output stage share the same storage space, hence the output of the analysis also enrich the information associated to samples. Samples can pass through the analysis stage several times, in the case new tools are added to the architecture, some tool is updated, or samples deserve special attention by analysts. As shown in Sect. 4, the triage approach aims to promptly analyze malware samples to associate them a rank indicating whether samples can be related to some known APT and hence they deserve further investigation.

4   Triage Approach

We propose an approach for malware triage based on the identification of samples similar to malware known to be developed by APTs. From now on, we say that these samples are *related to known APTs*. The basic idea is to generate a dataset by collecting public APT reports (such as [17, 20, 25]) and retrieving the binaries of the malware referenced in these reports. Each malware is assigned to an APT based on what is written in the reports. Static features are extracted from these binaries and used to train a classifier to be used for the triage.

Figure 2 highlights the components of the architecture (see Sect. 3 and [15]) that are involved in the malware triage process. The flow starts with the *APT loading plugin*, which continuously crawls different public sources in order to obtain

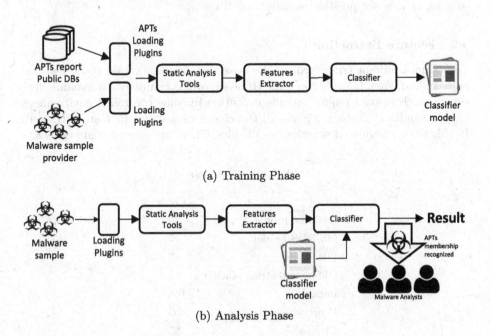

(a) Training Phase

(b) Analysis Phase

Fig. 2. Data classification flow.

reports about APTs and feeds the system with the information contained in them. For the training phase, the *Features Extraction* component periodically analyses all new static analysis reports of malware related to known APTs to produce the feature vectors required to train the *Classifier*. For the analysis phase, novel samples have to first pass through the malware triage stage. This includes a static analysis phase (performed in the *Sample Analysis Section* of the architecture) aimed at producing the feature vectors (done by *Features Extraction* component) to feed to the trained *Classifier*. If a sample is classified as related to known APT, then it is directed to an alternative path within the analysis chain, e.g., it is submitted to some human analyst for manual examination. We first present the phase of malware retrieval based on public APT reports (Sect. 4.1), then describe what static features are considered (Sect. 4.2), and finally detail how the classifier is trained and then used for the actual triage (Sect. 4.3).

4.1 APT Malware Loading

This component crawls external, publicly-available sources to collect reports related to malicious campaigns, activities and software that have been associated to APTs. These reports are produced by security firms and contain different *Indicator Of Compromises* [IOCs] related to specific APTs, including domains, IPs and MD5s of malware. The loading plugin parses them and add these information into the Knowledge Base. When a malware is added through its MD5, the architecture searches for the corresponding binary file in order to store and analyze it. Unfortunately, many of these malware are not available on public sources, so it is not possible to collect all of them.

4.2 Feature Extraction

To obtain a prompt triage process, we base the classification on static features only. Indeed, they take shorter time to be extracted compared to dynamic features, which instead require sample execution in some controlled environment (e.g., a sandbox). Table 1 reports all the classes of features that are extracted. In this work, considered samples are PE files. There are seven feature classes.

Table 1. Features classes

Class	Count
Optional header	30
MS-DOS header	17
File header	7
Obfuscated string statistics	3
Imports	158
Function lengths	50
Directories	65

Optional Header Features. These features are extracted from the optional header of the PE. It contains information about the logical layout of the PE file, such as the address of the entry point, the alignment of sections and the sizes of part of the file in memory.

MS-DOS Header Features. Features related to the execution of the file, including the number of bytes in the last page of the file, the number of pages or the starting address of the *Relocation Table*.

File Header Features. These features are extracted from the file header, which is a structure containing information about the physical layout and the properties of the PE file in general, like number of sections, timestamp and the CPU platform which the PE is intended for.

Obfuscated String Features. FireEye Labs Obfuscated String Solver (FLOSS) is a tool to automatically de-obfuscate strings from files through static analysis. The result of this tools is used to compute some statistics, like how many entry-points or relocations are present in the file, that compose the features of this class.

Imports Features. Functions can be imported from other executables or from DLLs. We are interested in the import of a specific set of known DLLs and APIs, and use their occurrences as feature. We also use three counters representing the total number of imported APIs, the total number of imported DLLs, and the total number of exported functions.

Function Lengths Features. FLOSS also provides measurements of function lengths. This class contains different counters to store that information. Due to the huge number of different functions, we use *bucketing* to reduce the number of possible features.

Directories Features. PE header includes an array containing all the *DATA DIRECTORY* structures, so, similarly to what we do for imports, we check the occurrence in the file of some particular directory names. We use their size as features, similarly to function lengths features.

4.3 Classification

We firstly tried to setup the classifier using a class for each known APT, representing the malware collected on the base of APT reports, and an additional class to represent all the samples that have not been created by any known APT. If a sample were assigned to the latter class, then it would be considered not related to any known APT. Otherwise, the classifier would establish the most likely APT which developed malware similar to that sample. The problem of this approach lies in training the classifier on such additional class. Indeed, the overwhelming majority of samples belongs to that class, including most of malicious samples, as a really tiny percentage of malware have been actually created by some APT. This translates to an excessive heterogeneity among samples of that class, and an extreme imbalance among classes in the training set, which makes this approach infeasible. Hence, we give up such additional class and only use classes representing known APTs for training. Given $C = \{c_i\}$ be the set of classes, with $N = |C|$ being the number of classes, equal to the number of actually used APTs, we train the classifier on N classes (one for each APT in the dataset). In the analysis flow we label the input samples in $N + 1$ classes, where the additional class is composed by all the outlier samples, that can be non-APT malware samples as well as samples of unknown APTs. Being the classifier trained on N samples, the idea is that all the samples too distant from all the centroids belong to such $(N + 1)$-th class. We use random forest [4] as learning method for the classification, as it turned out to be really effective in several works related to malware analysis [9,12,22], also because of its ensemble approach. Moreover random forest permits to classify samples by using different types of features (numbers, binary and non-numeric labels). A random forest operates by generating a set of decision trees at training time, and using them as classifier by picking the most frequent chosen class among them. Let $T = \{t_j\}$ be the set of decision trees of the random forest. $N_T = |T|$ is the number of trees. In order to determine whether a sample is related to a known APTs or classified as non-APT, we rely on the confidence score of the classification: if this score exceeds a threshold, then the sample is considered as related to the relative APT, otherwise it is not. As one of the main goals is minimizing how many irrelevant samples are delivered to human analysts or keep scarce resources busy, we are interested in using only those APTs where the classifier can perform with higher precision. We train the classifier with malware of all known APTs, and use a K-fold cross validation to obtain accuracy results. We then remove those APTs where both precision and recall are below a given threshold, and use only remaining APTs to train the actual model. We also have to remove the APTs where available malware are less than K. In the experimental evaluation, we show triage accuracy results for two distinct thresholds.

Classification Confidence Computation. The class assigned by a decision tree depends on the leaf node where the decision path terminates. Each leaf node corresponds to a subset of its training samples, which can belong to distinct classes, and the output class of the leaf node is the most frequent one among them. For a decision tree t_j, let $l_j = \{l_{j,k}\}$ be the set of its leaf nodes, with $N_j = |l_j|$. Let $N_{j,k}$ be the number of training samples of leaf node $l_{j,k}$. We define $class_{i,j,k}$ as the number of training samples of $l_{j,k}$ that belong to class c_i.

Intuitively, the diversity of classes among the training samples of a leaf node reflects how much the decision tree is confident about its output, when this output is determined by that leaf node. Thus, as confidence score for the single decision tree, we use the percentage of training samples that belong to the same class output by the leaf node. We then assign a confidence score to the classification of the whole random forest by averaging the confidence scores of all its decision trees. In a similar way, we can assign to a classified sample a confidence score for each class, to represent to what extent that sample relates to each class.

We assign a confidence vector $confidence_{j,k}$ to each leaf node $l_{j,k}$, where the i-th entry represents the confidence for class c_i, defined as follows

$$confidence_{j,k}[i] = \frac{class_{i,j,k}}{\sum_{m=1}^{N} class_{m,j,k}} \quad i = 1 \ldots N \tag{1}$$

For each sample to analyze, we setup the classifier to output a confidence value for each class, which represents the likelihood that the sample resembles malware created by the APT corresponding to that class. Given a sample s to classify with the random forest, we introduce the function $decision_j(s)$ which determines the leaf node of t_j where the decision path for s ends. Let $l_{j,k}$ be such leaf node, then $decision_j(s) = k$. We define the confidence vector $confidence(s)$ assigned to a sample s classified with the random forest as follows, where the i-th entry represents the confidence for class c_i

$$confidence(s)[i] = \frac{1}{N_T} \sum_{j=1}^{N_T} confidence_{j,decision_j(s)}[i] \quad i = 1 \ldots N \tag{2}$$

Confidence Threshold Computation. Malware developed by a same APT can be very different among each other. For example, they may relate to different phases of an attack (e.g., the payload for intruding target system, and the remote access trojan to enable external control), or they may have been used for attacks to distinct victims. Furthermore, we empirically observe that collected malware are distributed really unevenly among known APTs. This implies that confidence scores obtained for distinct classes cannot be fairly compared. Thus, rather than using a unique confidence threshold to discriminate whether a sample can be considered as related to a known APT, we compute a different threshold for each APT.

We first compute the confidence vector for each sample of the training set TS by using *leave-one-out* cross validation: for each training sample $s \in TS$, we use all the other training samples to train the random forest and then classify s to identify the leaf nodes to use to compute $confidence(s)$. Let $class(s)$ be a function that returns i if the class of training sample s is c_i. Let $TS_i = \{s \in TS : class(s) = i\}$ be the subset of the training set containing all and only the samples of class c_i. We then calculate the threshold vector as follows

$$threshold[i] = \frac{\sum_{s \in TS_i} confidence(s)[i]}{|TS_i|} - \Delta \quad i = 1 \ldots N \tag{3}$$

For each class, rather than directly using the average of its confidence scores as threshold, we decrease it by a *tolerance band* Δ in order to avoid having too many false negatives. During the actual triage in production, a test sample s is classified by the random forest and assigned a confidence vector $confidence(s)$, which is compared to the threshold vector to check whether the following condition holds

$$\exists i \quad confidence(s)[i] > threshold[i] \quad i = 1 \ldots N \tag{4}$$

In positive case, s is considered related to known APTs and dispatched accordingly, together with its confidence vector which may guide the subsequent analyses, as it suggests to what extent s resembles malware developed by each of the APTs used for training the random forest.

5   Experimental Evaluation

In this section we present details about the prototype and the preliminary results achieved. As explained in previous sections, we design our system in order to require the minimum amount of time to produce an evaluation for the triage: we use static analysis that is the faster type of analysis, due to the fact that it does not require sample execution. Moreover, we use a classifier based on Random Forest, which requires a shorter period of time for the classification with respect to other algorithms.

5.1   Prototype Implementation

We implement a prototype of the architecture presented in [15], by developing custom tools and adapting and extending open-source products. The description of prototype implementation is organized according to the same layered view presented in Sect. 3.

Visual Analytics Layer. For this layer we extend *CRITs* [18], an open-source malware and threat repository developed by MITRE. It offers two important characteristics: a well organized knowledge base and an extendible service platform, both accessible trough a web interface. To integrate CRITs into our architecture, we have to develop a set of services to enable the communication with the other components, and to modify some interfaces to show additional information.

Analysis Layer. For the analysis layer we adapt different open-source analysis tools, both for static and dynamic analysis. For example, we extend PEFrame [3] with functions from Flare-Floss [8] in order to have more information at disposal. The modified version of PEFrame is also the source for the Features Extractor Component described in Sect. 4 (developed in R language [21]), which in turn feeds the random forest classifier (implemented in R as well). The details of feature extractor and classifier are reported in Sect. 4.

Storage Layer. For the storage layer we use a MongoDB [2] cluster.

Loading Plugins. We also develop various plugins to gather required data from public sources. We adapt some open-source crawlers and develop some other by ourselves. The APT loader plugin is based on the IOC Parser [1], modified to collect APT reports from some public sources, extract data if interest and insert them into the storage.

5.2 Triage Evaluation

To validate the effectiveness of our approach we perform some preliminary experiments, using datasets prepared by including samples retrieved on the base of the MD5s found in the APT reports crawled by loading plugins. Unfortunately, many referenced malware are not available in public sources, thus some APTs have not enough samples to be properly represented. Furthermore, distinct APTs have very different number of samples, which leads to a highly unbalanced datasets, thus we choose to include only the most distinguishing APTs, basing our decision on the average precision and recall that the default random forest classifier would obtain.

Dataset. We collect 403 different reports about APTs, containing overall references to 9453 MD5. From public sources we manage to collect only 5377 binaries. The resulting dataset contains 5685 sample belonging to 199 different APTs. We discard all the APTs with less than 20 samples to avoid classes not sufficiently represented, which leads to a dataset with 4783 samples and 47 APTs. We also collect from public sources 2000 additional malware that are not known to be developed by any APT.

Training Phase. We build two datasets by using distinct thresholds for precision and recall (see Sect. 4.3): dataset D1 with threshold 0,95 and dataset D2 with threshold 0,90. Table 2 shows details about these datasets. For each dataset, we trained three diverse classifiers by using distinct confidence thresholds Δ: 5%, 10% and 15%.

Table 2. Dataset composition

	APTs	Samples	Mean class size
D1	7	1308	187
D2	15	2521	168

Test Phase. For the test we choose to use a K-fold cross validation with k equals to 10, a common value in literature for this kind of tests. For each execution, we generate the model with $k - 1$ folds and test it with both the remaining fold and all the collected malware not developed by APTs (2000 samples). We consider the triage as a binary classification, and measure its quality by using *Accuracy, Precision, Recall* and *F1*. If a sample is classified as related to known APTs we say it is *positive*, otherwise *negative*. As explained in Sect. 4.3, the most

Table 3. Binary confusion matrix D1 [APTs/non-APTs]

	Δ	5%		10%		15%	
	Triage	Pos	Neg	Pos	Neg	Pos	Neg
Ground	Pos	1209	99	1232	76	1250	58
Truth	Neg	0	20000	0	20000	0	20000

Table 4. Binary confusion matrix D2 [APTs/non-APTs]

	Δ	5%		10%		15%	
	Triage	Pos	Neg	Pos	Neg	Pos	Neg
Ground	Pos	2125	396	2197	324	2251	270
Truth	Neg	2	19998	18	19982	53	19947

Table 5. Confusion matrix D1 [$\Delta = 5\%$]

	A	B	C	D	E	F	G	NA
A	23	0	0	0	0	0	0	2
B	0	36	0	0	0	0	0	8
C	0	0	275	0	0	0	0	28
D	0	0	0	414	0	0	0	22
E	0	0	0	0	16	0	0	6
F	0	0	0	0	0	313	0	28
G	0	0	0	0	0	0	132	5
NA	0	0	0	0	0	0	0	20000

important measure for the triage is the Precision (i.e., minimize false positives), due to the fact that human analysts are a limited resources and we have to reduce as much as possible their workload by striving to deliver them only samples that are highly confidently related to known APTs. Tables 3 and 4 show the results of the classification test for both datasets, which highlight that obtained false positives are indeed very low: we are able to reduce false positives to zero for D1 and to less than 70 for D2. It is to note that these numbers are computed over 20000 actual tests, in fact each of the 2000 negative samples is tested for each of the 10 folds.

Table 6. Confusion matrix D1 [$\Delta = 10\%$]

	A	B	C	D	E	F	G	NA
A	23	0	0	0	0	0	0	2
B	0	39	0	0	0	0	0	5
C	0	0	277	0	0	0	0	26
D	0	0	0	419	0	0	0	17
E	0	0	0	0	17	0	0	5
F	0	0	0	0	0	325	0	16
G	0	0	0	0	0	0	132	5
NA	0	0	0	0	0	0	0	20000

Table 7. Confusion matrix D1 [$\Delta = 15\%$]

	A	B	C	D	E	F	G	NA
A	24	0	0	0	0	0	0	1
B	0	40	0	0	0	0	0	4
C	0	0	283	0	0	0	0	20
D	0	0	0	424	0	0	0	12
E	0	0	0	0	19	0	0	3
F	0	0	0	0	0	328	0	13
G	0	0	0	0	0	0	132	5
NA	0	0	0	0	0	0	0	20000

Tables 5, 6, 7, 8, 9 and 10 display the confusion matrices obtained by considering the classifier as an $N + 1$ classifier, for both the datasets and the same three Δ considered before. Results are coherent with the previous ones: with D1 we always achieve zero false positives, while with D2 we incorrectly label less than 0,002% as related to known APTs.

Table 11 reports quality metrics for all the tests, and shows that our approach is really promising as it scores high levels of accuracy and precisions.

Table 8. Confusion matrix D2 $[\Delta = 5\%]$

	A	B	C	D	E	F	G	H	I	J	K	L	M	N	O	NA
A	74	0	0	0	0	0	0	0	0	0	0	0	0	0	0	27
B	0	20	0	0	0	0	0	0	0	0	0	0	0	0	0	5
C	0	0	33	0	0	0	0	0	0	0	0	0	0	0	0	11
D	0	0	0	66	0	0	0	0	0	0	0	0	0	0	0	38
E	0	0	0	0	122	0	0	0	0	0	0	0	0	0	0	28
F	0	0	0	0	0	476	0	0	0	0	0	0	0	0	0	79
G	0	0	0	0	0	0	39	0	0	0	0	0	0	0	0	26
H	0	0	0	0	0	0	0	22	0	0	0	0	0	0	0	9
I	0	0	0	0	0	0	0	0	272	0	0	0	0	0	0	31
J	0	0	0	0	0	1	0	0	0	407	0	0	0	0	0	28
K	0	0	0	0	0	0	0	0	0	0	9	0	0	0	0	13
L	0	0	0	0	0	0	0	0	0	0	0	299	0	0	0	42
M	0	0	0	0	0	0	0	0	0	0	0	0	120	0	0	17
N	0	0	0	0	0	0	0	0	0	0	0	0	0	27	0	8
O	0	0	0	0	0	0	0	0	0	0	0	0	0	0	138	34
NA	0	0	0	0	0	2	0	0	0	0	0	0	0	0	0	19998

Table 9. Confusion matrix D2 $[\Delta = 10\%]$

	A	B	C	D	E	F	G	H	I	J	K	L	M	N	O	NA
A	76	0	0	0	0	0	0	0	0	0	0	0	0	0	0	25
B	0	22	0	0	0	0	0	0	0	0	0	0	0	0	0	3
C	0	0	34	0	0	0	0	0	0	0	0	0	0	0	0	10
D	0	0	0	69	0	0	0	0	0	0	0	0	0	0	0	35
E	0	0	0	0	125	0	0	0	0	0	0	0	0	0	0	25
F	0	0	0	0	0	494	0	0	0	0	0	0	0	0	0	61
G	0	0	0	0	0	0	47	0	0	0	0	0	0	0	0	18
H	0	0	0	0	0	0	0	22	0	0	0	0	0	0	0	9
I	0	0	0	0	0	0	0	0	275	0	0	0	0	0	0	28
J	0	0	0	0	0	1	0	0	0	414	0	0	0	0	0	21
K	0	0	0	0	0	0	0	0	0	0	12	0	0	0	0	10
L	0	0	0	0	0	0	0	0	0	0	0	309	0	0	0	32
M	0	0	0	0	0	0	0	0	0	0	0	0	128	0	0	9
N	0	0	0	0	0	0	0	0	0	0	0	0	0	29	0	6
O	0	0	0	0	0	0	0	0	0	0	0	0	0	0	140	32
NA	0	0	0	0	0	18	0	0	0	0	0	0	0	0	0	19982

Table 10. Confusion matrix D2 $[\Delta = 15\%]$

	A	B	C	D	E	F	G	H	I	J	K	L	M	N	O	NA
A	78	0	0	0	0	0	0	0	0	0	0	0	0	0	0	23
B	0	22	0	0	0	0	0	0	0	0	0	0	0	0	0	3
C	0	0	38	0	0	0	0	0	0	0	0	0	0	0	0	6
D	0	0	0	73	0	0	0	0	0	0	0	0	0	0	0	31
E	0	0	0	0	128	0	0	0	0	0	0	0	0	0	0	22
F	0	0	0	0	0	501	0	0	0	0	0	0	0	0	0	54
G	0	0	0	0	0	0	50	0	0	0	0	0	0	0	0	15
H	0	0	0	0	0	0	0	23	0	0	0	0	0	0	0	8
I	0	0	0	0	0	0	0	0	277	0	0	0	0	0	0	26
J	0	0	0	0	0	1	0	0	0	419	0	0	0	0	0	16
K	0	0	0	0	0	0	0	0	0	0	15	0	0	0	0	7
L	0	0	0	0	0	0	0	0	0	0	0	324	0	0	0	17
M	0	0	0	0	0	0	0	0	0	0	0	0	131	0	0	6
N	0	0	0	0	0	0	0	0	0	0	0	0	0	29	0	6
O	0	0	0	0	0	0	0	0	0	0	0	0	0	0	142	30
NA	0	0	0	0	0	53	0	0	0	0	0	0	0	0	0	19947

Table 11. Quality measures

	Dataset	D1				D2			
	Measures	Accuracy	Precision	Recall	F1	Accuracy	Precision	Recall	F1
Δ	5%	0.995	1.000	0.886	0.938	0.982	1.000	0.764	0.859
	10%	0.996	1.000	0.910	0.952	0.985	0.998	0.807	0.888
	15%	0.997	1.000	0.938	0.968	0.986	0.994	0.850	0.913

6 Conclusion and Future Works

Among the huge amount of malware produced daily, those developed by Advanced Persistent Threats (APTs) are highly relevant, as they are part of massive and dangerous campaigns that can exfiltrate information and undermine or impede critical operations of a target. This paper introduces an automatic malware triage process to drastically reduce the number of malware to be examined by human analysts. The triage process is based on a classifier which evaluates to what extent an incoming malicious sample could have been developed by a known APT, hence relieving analysts from the burden of analyzing these malware. The classifier is trained with static features obtained by static analysis of available malware known to be developed by APTs, as attested by public reports. Although static features alone are not sufficient to completely exclude relations with APTs, they allow to perform a quick triage and recognize malware that deserve higher attention, with minimal risk of wasting analysts time. In fact the experimental evaluation has shown encouraging results: malware realized by known APTs have been identified with a precision of 100% and an accuracy up to 96%.

At the time of this writing, we are testing our approach in the *real world*, i.e., we are analyzing large malware datasets. As future work, we want to study more effective functions for the evaluation of the threshold (see Sect. 4), in order to improve the overall accuracy of the system. Moreover, we plan to include an additional prioritization step for the samples that result nearer to the chosen threshold: as this situation indicates a higher degree of uncertainty about these sample, they can be sent to a second classifier trained with dynamic features of malware known to be developed by APTs.

Acknowledgments. This present work has been partially supported by a grant of the Italian Presidency of Ministry Council, and by CINI Cybersecurity National Laboratory within the project FilieraSicura: Securing the Supply Chain of Domestic Critical Infrastructures from Cyber Attacks (www.filierasicura.it) funded by CISCO Systems Inc. and Leonardo SpA.

References

1. IOC parser. https://github.com/armbues/ioc_parser/. Accessed 17 Mar 2017
2. MongoDB. https://www.mongodb.com/. Accessed 13 Mar 2017
3. PEFrame. https://github.com/guelfoweb/peframe/. Accessed 17 Mar 2017
4. Breiman, L.: Random forests. Mach. Learn. **45**(1), 5–32 (2001)
5. Chen, P., Desmet, L., Huygens, C.: A study on advanced persistent threats. In: Decker, B., Zúquete, A. (eds.) CMS 2014. LNCS, vol. 8735, pp. 63–72. Springer, Heidelberg (2014). doi:10.1007/978-3-662-44885-4_5
6. CNN: Nearly 1 million new malware threats released every day (2014). http://money.cnn.com/2015/04/14/technology/security/cyber-attack-hacks-security/
7. Damodaran, A., Di Troia, F., Visaggio, C.A., Austin, T.H., Stamp, M.: A comparison of static, dynamic, and hybrid analysis for malware detection. J. Comput. Virol. Hacking Tech. **13**, 1–12 (2015)
8. Fireeye: FireEye labs obfuscated string solver. https://github.com/fireeye/flare-floss/. Accessed 17 Mar 2017
9. Islam, R., Tian, R., Batten, L.M., Versteeg, S.: Classification of malware based on integrated static and dynamic features. J. Netw. Comput. Appl. **36**(2), 646–656 (2013)
10. Jang, J., Brumley, D., Venkataraman, S.: BitShred: Fast, scalable malware triage. Technical report CMU-Cylab-10-022, Cylab, Carnegie Mellon University, Pittsburgh, PA (2010)
11. Jeun, I., Lee, Y., Won, D.: A practical study on advanced persistent threats. In: Kim, T., Stoica, A., Fang, W., Vasilakos, T., Villalba, J.G., Arnett, K.P., Khan, M.K., Kang, B.-H. (eds.) SecTech 2012. CCIS, vol. 339, pp. 144–152. Springer, Heidelberg (2012). doi:10.1007/978-3-642-35264-5_21
12. Khodamoradi, P., Fazlali, M., Mardukhi, F., Nosrati, M.: Heuristic metamorphic malware detection based on statistics of assembly instructions using classification algorithms. In: 2015 18th CSI International Symposium on Computer Architecture and Digital Systems (CADS), pp. 1–6. IEEE (2015)

13. Kirat, D., Nataraj, L., Vigna, G., Manjunath, B.: SigMal: a static signal processing based malware triage. In: Proceedings of the 29th Annual Computer Security Applications Conference. pp. 89–98. ACM (2013)

14. Lakhotia, A., Walenstein, A., Miles, C., Singh, A.: VILO: a rapid learning nearest-neighbor classifier for malware triage. J. Comput. Virol. Hacking Tech. **9**(3), 109–123 (2013)

15. Laurenza, G., Ucci, D., Aniello, L., Baldoni, R.: An architecture for semi-automatic collaborative malware analysis for CIs. In: 2016 46th Annual IEEE/IFIP International Conference on Dependable Systems and Networks Workshop, pp. 137–142. IEEE (2016)

16. Marchetti, M., Pierazzi, F., Colajanni, M., Guido, A.: Analysis of high volumes of network traffic for advanced persistent threat detection. Comput. Netw. **109**, 127–141 (2016)

17. Trend Micro: IXESHE: an APT campaign. Trend Micro Incorporated Research Paper (2012)

18. MITRE: CRITS: collaborative research into threats. https://crits.github.io/. Accessed 17 Mar 2017

19. Moser, A., Kruegel, C., Kirda, E.: Limits of static analysis for malware detection. In: Twenty-Third Annual Computer Security Applications Conference, ACSAC 2007, pp. 421–430. IEEE (2007)

20. O'Gorman, G., McDonald, G.: The elderwood project. Symantec Whitepaper (2012)

21. R Development Core Team: R: A language and environment for statistical computing. R Foundation for Statistical Computing, Vienna, Austria (2008). ISBN 3-900051-07-0, http://www.R-project.org

22. Santos, I., Devesa, J., Brezo, F., Nieves, J., Bringas, P.G.: OPEM: a static-dynamic approach for machine-learning-based malware detection. In: Herrero, Á., et al. (eds.) International Joint Conference CISIS'12-ICEUTE'12-SOCO'12 Special Sessions. AISC, pp. 271–280. Springer, Heidelberg (2013). doi:10.1007/978-3-642-33018-6_28

23. Su, Y., Lib, M., Tang, C., Shen, R.: A framework of APT detection based on dynamic analysis (2016)

24. Tankard, C.: Advanced persistent threats and how to monitor and deter them. Netw. Secur. **2011**(8), 16–19 (2011)

25. Villeneuve, N., Bennett, J.T., Moran, N., Haq, T., Scott, M., Geers, K.: Operation "KE3CHANG" Targeted Attacks Against Ministries of Foreign Affairs (2013)

26. Virvilis, N., Gritzalis, D., Apostolopoulos, T.: Trusted computing vs. advanced persistent threats: can a defender win this game? In: 2013 IEEE 10th International Conference on Ubiquitous Intelligence and Computing and 10th International Conference on Autonomic and Trusted Computing (UIC/ATC), pp. 396–403. IEEE (2013)

27. Vukalović, J., Delija, D.: Advanced persistent threats-detection and defense. In: 2015 38th International Convention on Information and Communication Technology, Electronics and Microelectronics (MIPRO), pp. 1324–1330. IEEE (2015)

28. Wicherski, G.: peHash: a novel approach to fast malware clustering. LEET **9**, 8 (2009)

Author Index

Printed in the United States
By Bookmasters